CZECHOSLOVAK ACADEMY OF SCIENCES

CZECHOSLOVAK ACADEMY OF SCIENCES

TRANSACTIONS

of the

TENTH PRAGUE CONFERENCE

on

INFORMATION THEORY, STATISTICAL DECISION FUNCTIONS, RANDOM PROCESSES

held at

Prague, from July 7 to 11, 1986

VOLUME A

D. REIDEL PUBLISHING COMPANY

A MEMBER OF THE KLUWER ACADEMIC PUBLISHERS GROUP

DORDRECHT / BOSTON / LANCASTER / TOKYO

Library of Congress Cataloging-in-Publication Data

CIP

Prague Conference on Information Theory, Statistical Decision Functions, Random Processes (10th : 1986 : Prague, Czechoslovakia)
Transactions of the Tenth Prague Conference on Information Theory, Statistical Decision Functions, Random Processes, held at Prague, from July 7 to 11, 1986.

Sponsored by Czechoslovak Academy of Sciences.
1. Probabilities--Congresses. 2. Statistical decision--Congresses. 3. Information theory--Congresses. I. Československá akademie věd. II. Title.
QA273.A1P73 1986 519.2 87-16658
ISBN 978-94-010-8216-7 ISBN 978-94-009-3859-5 (eBook)
DOI 10.107/978-94-009-3859-5

TRANSACTIONS include contributions of authors reprinted directly in a photographic way. For this reason the authors are fully responsible for the correctness of their text.

Published by D. Reidel Publishing Company
P.O. Box 17, 3300 AA Dordrecht, Holland, in co-edition with
Academia, Publishing House of the Czechoslovak Academy of Sciences,
Prague, Czechoslovakia.

Sold and distributed in the U.S.A. and Canada by Kluwer Academic Publishers,
101 Philip Drive, Norwell, MA 02061, U.S.A.

Sold and distributed in Albania, Bulgaria, China, Czechoslovakia, Cuba, German Democratic Republic, Hungary, Mongolia, Northern Korea, Poland, Rumania, U.S.S.R., Vietnam, and Yugoslavia by
Academia, Publishing House of the Czechoslovak Academy of Sciences,
Prague, Czechoslovakia.

Sold and distributed in all remaining countries by
Kluwer Academic Publishers Group,
P.O. Box 322, 3300 AH Dordrecht, Holland.

Soft cover reprint of the hard cover 1st edition 1988

CONTENTS

PREFACE

The Conference was organized by the Institute of Information Theory and Automation of the Czechoslovak Academy of Sciences from July 7 - 11, 1986, in Prague.

The round number of the conference was only one of the jubilees connected with its organization. Namely, thirty years of the Prague Conferences (the first one was organized in autumn 1956 in Liblice near Prague), and two anniversaries of Professor Antonín Špaček, the inspirator and first organizer of the Prague Conferences - 75 years of his birth and 25 years of his untimely death. (More about Professor Špaček can be found in the Transactions of the Sixth Prague Conference).

The Tenth Prague Conference kept the traditional style and orientation typical for the previous Prague Conferences. Almost two hundred of participants from 23 countries (Algerie, Austria, Bulgaria, Canada, Czechoslovakia, Federal Republic of Germany, Finland, France, German Democratic Republic, Great Britain, Hungary, Iran, Italy, Japan, Netherlands, Poland, Roumania, Soviet Union, Sweden, Switzerland, United States, Vietnam and West Berlin) took part in its sessions and discussions. There were 14 invited lectures and 92 short contributions included in four parallel sections of the Conference programme; further, 12 contributions were presented as posters. The invited lectures and submitted contributions covered the three traditional subjects of the Prague Conferences introduced in their title, as well as lots of further applications of the probability theory and mathematical statistics. Most of the presented lectures and contributions are published in the present Conference Transactions.

Not only the past history of the Prague Conferences but also their future conception was frequently discussed in connection with their jubilee. The long tradition of the Conferences which is rather rare in case of mathematical meetings obliges the organizers to keep the authority and good scientific level of the Prague Conferences. The character of widely oriented meetings covering practically all the stochastic mathematics and its applications was typical for the previous Prague Conferences and formed their specific atmosphere. The Conferences gave a good opportunity of personal meeting for specialists in different branches of the probability theory and theoretical cybernetics. On the other hand, the development of all those branches causes difficulties in communication between the respective specialists and rather decreasing attractivity of many strictly specialized contributions for some participants.

Thirty years mean much time in the development of science, and also the world of 1986 differs from the one of 1956. The intention of the organizers of the First Prague Conference was to arrange a wide international meeting of mathematicians interested in the probability theory, mathematical statistics and their applications. In fiftieth, the branches covered by the Conference title were relatively new and narrow, the number of probabilists and applied probabilists was rather low, and the possibilities of international scientific contacts were solitary. It is not commonly known, for example, that the First Prague Conference, with about fifty participants from seven countries was the first personal meeting of probabilists from the Soviet Union and United States and one of the first east-west scientific contacts after the world war II.

Many things have changed since that time. Only the historians of mathematics can responsibly measure the quantity of new fundamental methods and results derived in the three main branches of the Prague Conferences between the First and the Tenth one. It is not easy even to enumerate all the new branches and sub-branches of mathematics applying the stochastic or probabilistic concepts or dealing with different forms of uncertainty. Universities pour new generations of mathematicians and engineers schooled in the probabilistic and statistical methods. Scientific meetings on different level of their specialization from the narrow directed workshops and seminars up to the European Meetings of Statisticians are announced month by month.

It was not easy to preserve the reputation and the existence of the Prague Conferences during the whole period of such remarkable changes. To continue their organization means above all to determine the place of the Prague Conferences in the rich scale of the scientific meetings of nowadays. The enormous development of the probabilistic methods and their applications made the original topics of the Prague Conference too wide to be covered by a single conference. The optimal way of the organization of the future Prague Conferences could be to organize rather more specialized monothematic conferences (e. g. on information theory, on complexity theory, on stochastic processes, on games and decisions, etc.) alternatively, with acceptable periodicity and under the common head and numeration of the Prague Conferences. The organizers of the Prague Conferences believe that the more compact subjects of such conferences in connection with their traditional reputation can guarantee the optimistic prospects of the Prague Conferences for the future years.

Organizing Committee

INVITED PAPERS

ASYMPTOTIC NORMALITY AND LARGE DEVIATIONS

Hermann Dinges
Frankfurt am Main

Key words: Asymptotic normality, large deviations, saddlepoint approximations, Wiener germs

1. INTRODUCTION

Many sequences of random variables which turn up in statistics are asymptotically normal

$$L(X_n) \sim N(x^*, \frac{1}{n} \cdot \mathbb{C}).$$

In most practical cases it is worthwhile to look for more precise information about the asymptotic behaviour of the distributions. For the range of small deviations one can frequently use Edgeworth–expansions. For a description of the distributions in the range of large deviations one is led to a function K(x) which is usually called the entropy function. It turns out that in many cases one can go even further and find an asymptotic expansion similar to that one which has first been established by the so–called saddlepoint approximation (Daniels (1954)).

In order to develop a theory of such asymptotic expansions we have introduced the concept of a Wiener germ. We are interested in the behaviour of $L(X_n)$ in an arbitrarily small but fixed neighbourhood U of x^*. Such a fixed set carries almost all mass; in most applications we have $1 - O(c^n)$ with some $c < 1$; for a Wiener germ of order m it is sufficient to stipulate $Pr(X_n \in U) = 1 - O(n^{-m})$.

The existence of moments is not assumed in the theory of Wiener germs although of course one can identify numbers which in the classical applications correspond to the familiar moments.

Before I come to the definition of a Wiener germ let me formulate a rather striking new result for a very classical one-dimensional situation.

THEOREM 0

Let $Y_1, Y_2, \ldots$ be i.i.d. random variables with an integrable characteristic function and

$$\psi(\vartheta) := \ln(E \exp(\vartheta Y)) < \infty$$

for all ϑ in a neighbourhood V of the origin. Then there exist functions

$$A(x),\ A_1(x),\ A_2(x),\ \ldots$$

in some neighbourhood U of $x^* = EY$ such that for the tails of the averages $\overline{X}_n = \frac{1}{n}(Y_1 + \ldots + Y_n)$ we have uniformly in every compact subset of U

$$Pr(\overline{X}_n \leq x) = \Phi(\sqrt{n}[A(x) + \frac{1}{n} A_1(x) + \ldots + (\frac{1}{n})^m A_m(x) + o(n^{-m})]) .$$

The $A_j(x)$ can be obtained from $\psi(\vartheta)$. $A(x)$ and $A_1(x)$ are as follows

$$\frac{1}{2} A^2(x) = K(x) = \sup\{\vartheta x - \psi(\vartheta) : \vartheta \in V\} \text{ for } x \in U$$

$$A_1(x) = -\frac{1}{A(x)} \ln \frac{\sqrt{K''(x)}}{A'(x)}$$

In this presentation we restrict our attention to distributions with density; the corresponding theory for arithmetic distributions is slightly more delicate.

COROLLARY

Uniformly in $x^* \leq x \leq x''$ we have

$$\frac{\Pr(\overline{X}_n \geq x)}{\Phi(-\sqrt{n}\ A(x))} = O(1)$$

$$\frac{\Pr(\overline{X}_n \geq x)}{\Phi(-\sqrt{n}\ A(x) - \frac{1}{\sqrt{n}} A_1(x))} = 1 + O(\frac{1}{n})$$

$$\Pr(\overline{X}_n \geq x) = \Phi(-\sqrt{n}\ A(x) - \frac{1}{\sqrt{n}} A_1(x) - n^{-3/2} A_2(x))\ [1 + O(n^{-2})]$$

This last estimate of the tail probabilities goes beyond older results on large deviations (compare Petrov's book).

It may be interesting to see how our result can be related to Cramér's classical result from 1938. Cramér finds a function $\lambda(y)$ which is suitable to describe the tail probabilities

$$\Pr(\frac{1}{n}(Y_1 + \ldots + Y_n) \geq x)$$

for $x = x_n + x^* + \frac{1}{\sqrt{n}} \sigma \cdot z_n$, where

$$z_n \geq 0 \text{ and } z_n = o(\sqrt{n})\ .$$

Cramér shows

$$\Pr(\overline{X}_n \geq x) = \Phi(-z) \exp(\frac{z^3}{\sqrt{n}} \lambda(\frac{z}{\sqrt{n}}))\ [1 + O(\frac{1}{\sqrt{n}} + \frac{z}{\sqrt{n}})]$$

It is well-known how Cramér's $\lambda(y)$ is related to the "entropy function" $K(x)$ which is the Legendre transform of the cumulant generating function $\psi(\vartheta)$, as we saw above. One has

$$K(x) = \frac{1}{2}(\frac{x-x^*}{\sigma})^2 - (\frac{x-x^*}{\sigma})^3 \lambda(\frac{x-x^*}{\sigma})$$

We now notice

$$\Phi(-\sqrt{n}\ A(x)) = \Phi(-\sqrt{n}[2K(x)]^{1/2}) = \Phi(-\sqrt{n}\ \frac{x-x^*}{\sigma}[1-2\ \frac{x-x^*}{\sigma}\ \lambda(\frac{x-x^*}{\sigma})]^{1/2})$$

$$= \Phi(-z[1-2\ \frac{z}{\sqrt{n}}\ \lambda(\frac{z}{\sqrt{n}})]^{1/2}) \sim \Phi(-z + \frac{1}{\sqrt{n}}\ z^2\ \lambda(\frac{z}{\sqrt{n}}))$$

Notice that for small $|\Delta|$ and z bounded from below

$$\Phi(-z+\Delta) \sim \Phi(-z)\ \exp(q(z)\ \Delta)$$

with $q(z) = \frac{\varphi(z)}{\Phi(-z)}$ $\quad (\sim z$ for $z\to\infty)$.

This gives

$$\Phi(-\sqrt{n}\ A(x)) \sim \Phi(-z)\ \exp(q(z)\ \frac{1}{\sqrt{n}}\ z^2\ \lambda(\frac{z}{\sqrt{n}}))$$

and Cramér's result follows easily.

We hope that our theorem will be interpreted as an advice to the people in the "rough" theory of large deviations. These researchers should not try to improve their estimates of

$$-\frac{1}{n}\ \ln\ \Pr(\frac{1}{n}(Y_1+\ \dots\ +Y_n) \in B)\ ;$$

rather they ought to observe

$$-\ \ln\ p \sim \frac{1}{2}\ [\Phi^{-1}(p)]^2 \quad \text{for } p\to 0$$

and study

$$\frac{1}{\sqrt{n}}\ \Phi^{-1}(\Pr(\frac{1}{n}(Y_1+\ \dots\ +Y_n) \in B))\ .$$

Even in much more general situations there is hope to show the existence of an asymptotic expansion

$$\sim a_0 + \frac{1}{n}\ a_1 + (\frac{1}{n})^2 a_2 + \dots$$

which might even turn out to be uniform in an interesting class of sets which includes sets which come arbitrarily close to the "center".

2. BASICS OF A MATHEMATICAL THEORY OF WIENER GERMS

DEFINITION 1 (Wiener germ of order m; $m \in \{1,2,\ldots\}$)

Let U be a neighbourhood of x^* in $\mathbb{R}^d$. A **Wiener germ** of order m on U with center x^* is a family of densities

$$f_\varepsilon(x)dx_1 \ldots dx_d$$

of the following form:

$$f_\varepsilon(x) = (2\pi\varepsilon)^{-d/2} \exp(-\frac{1}{\varepsilon} K(x))\, D(x) \exp(\varepsilon S(\varepsilon,x))$$

where

a) K(x), the "entropy function", is (m+1)-times continuously differentiable

$$K(x^*) = 0, \quad K(x) \geq 0, \quad K''(x) \text{ positive definite.}$$

b) D(x), the "modulating density", is positive and m-times continuously differentiable.

c) $\varepsilon S(\varepsilon,x) = \varepsilon S_1(x)+\varepsilon^2 S_2(x)+ \ldots +\varepsilon^{m-1}S_{m-1}(x)+\varepsilon^m R(\varepsilon,x)$

with $S_j(x)$ (m−j)-times continuously differentiable,

$R(\varepsilon,x)$ uniformly bounded in U.

d) $$\int_{U^*} f_\varepsilon(x)dx = 1 - O(\varepsilon^m)$$

for every subset U^* of U, which is a neighbourhood of x^*.

EXAMPLE 1

Let $\{W_\varepsilon : \varepsilon\to 0\}$ be a standard Wiener process in $\mathbb{R}^d$. Let $T_0(w), \ldots, T_m(w)$ be sufficiently smooth mappings in a neighbourhood of U. Assume furthermore that for $0\leq\varepsilon\leq\varepsilon^*$

$$T(\varepsilon,w) = T_0(w) + \varepsilon T_1(w) + \ldots + \varepsilon^m T_m(w)$$

is a diffeomorphism. Consider

$$X_\varepsilon = T(\varepsilon, W_\varepsilon) .$$

Then $L(X_\varepsilon)$ has a density $f_\varepsilon(x)dx$ in some neighbourhood U of $x^* = T_0(0)$. The family $\{f_\varepsilon(x)dx : \varepsilon\to o\}$ then forms a Wiener germ. It turns out that the entropy function is

$$K(x) = \frac{1}{2} A_0^T(x) A_0(x)$$

where $A_0(x)$ is the inverse of $T_0(w)$.

The modulating density is determined by $T_0(w)$ and $T_1(w)$. More generally the $S_j(x)$ are determined by $T_0(w), \ldots ,T_{j+1}(w)$.

DEFINITION 2

Let $\{X_n : n\to\infty\}$ be a sequence of random variables. We say that these X_n (considered under the hypothesis H) follow (along ε_n) the Wiener germ on U given by

$$(K(x), D(x)dx, \{S(\varepsilon,x) : \varepsilon\to 0\})$$

if $\varepsilon_n \to 0$ and $f_{\varepsilon_n}(x)dx$ is the density of $L_H(X_n)$ restricted to U.

The following example is a reformulation of a result of H. Daniels (1954) originally obtained by the saddlepoint method.

EXAMPLE 2

Let f(y)dy be a probablity density in $\mathbb{R}^d$ such that the Fouriertransform is integrable and

$$\int e^{\vartheta y} f(y)dy < \infty$$

for all ϑ in a neighbourhood V of the origin.

Let $Y_1, Y_2, \ldots$ be i.i.d. with density

$$f_\vartheta(y)dy = \exp(\vartheta y)f(y)dy \exp(-\psi(\vartheta))$$

under the hypothesis H_ϑ, $\vartheta \in V$. Put

$$\overline{X}_n = \frac{1}{n}(Y_1 + \ldots + Y_n).$$

Then $\{\overline{X}_n : n \to \infty\}$ considered under the hypothesis H_ϑ (for $\vartheta \in V$) follows along $\varepsilon_n = \frac{1}{n}$ a Wiener germ.

The entropy functions satisfy

$$K(\vartheta,x) = K(x) - \vartheta x + \psi(\vartheta) .$$

The modulating density is

$$D(x)dx = |K''(x)|^{1/2}dx .$$

The correcting functions $S_j(x)$ are independent of ϑ.

THEOREM 1

Let $\{L(X_\varepsilon) : \varepsilon \to 0\}$ be a Wiener germ of order m on U. Let $T(\varepsilon,x)$ for $0 \leq \varepsilon \leq \varepsilon^*$ and $x \in U$ be such that $T(\varepsilon,\cdot)$ is a diffeomorphism and uniformly on U

$$T(\varepsilon,x) = T_0(x) + \varepsilon T_1(x) + \ldots + \varepsilon^m T_m(x) + O(\varepsilon^{m+1})$$

with $T_j(x)$ $(m+1-j)$-times continuously differentiable. Assume furthermore that the Jacobi-matrices satisfy

$$J(\varepsilon,x) = J_0(x) + \varepsilon J_1(x) + \ldots + \varepsilon^{m-1} J_{m-1}(x) + O(\varepsilon^m) .$$

The distributions of $Y_\varepsilon = T(\varepsilon, X_\varepsilon)$ restricted to a sufficiently small neighbourhood of $y^* = T_0(x^*)$ then form a Wiener germ of order m.

THEOREM 2

Let $\{L(X_\varepsilon) : \varepsilon \to 0\}$ be a Wiener germ on $U \subseteq \mathbb{R}^d$. $d = d_1 + d_2$. Let T_1, T_2 be mappings such that $T = (T_1, T_2)$ is a diffeomorphism. Let $K(y,z)$ denote the entropy function of $T(X_\varepsilon)$. Put $y^* = T_1(\text{center})$.

a) Then $\{L(T_1(X_\varepsilon)) : \varepsilon \to 0\}$ is a Wiener germ with center y^*.

b) For every y near y^*

$$\{L(T_2(X_\varepsilon) \mid \{T_1(X_\varepsilon) = y\}) : \varepsilon \to 0\}$$

is a Wiener germ with center $s(y)$, where $s(y)$ is that point in $\mathbb{R}^d$ where $K(y, \cdot)$ assumes its minimum.

c) The entropy function of $\{T_1(X_\varepsilon) : \varepsilon \to o\}$ is

$$L(y) := K(y, s(y)) \qquad y \text{ near } y^*.$$

d) For fixed y near y^* the entropy function of $\{T_2(X_\varepsilon) \mid \{Y_\varepsilon = y\}$ is

$$K(y, \cdot) - K(y, s(y)) .$$

THEOREM 3

Let $\{L(X_\varepsilon) : \varepsilon \to 0\}$ be a one-dimensional Wiener germ of order m. Then there exist mappings $T_0(w), \ldots, T_m(w)$ such that

$$L(X_\varepsilon) = L(T(\varepsilon, \sqrt{\varepsilon}\, Z))$$

where Z is standard normally distributed and

$$T(\varepsilon, w) + T_0(w) + \varepsilon T_1(w) + \ldots + \varepsilon^m T_m(w) + o(\varepsilon^{m+1/2})$$

(compare theorem 0 above).

The proofs of these theorems are being worked out in a doctoral thesis at the University of Frankfurt.

REFERENCES

1) Alfers, D. & Dinges, H. (1984) : "A Normal Approximation for Beta and Gamma Tail Probabilities". Z. Wahrscheinlichkeitstheorie u. verw. Geb. 65 399–420.

2) Daniels, H.E. (1954) "Saddlepoint approximations in statistics". Ann. Math. Stat. 25, 631–50.

3) Daniels, H.E. (1983) "Saddlepoint approximations for estimating equations". Biometrika 70, 80–96.

4) Dinges, H. (1985) "Wiener Germs Applied to the Tails of M-Estimators". Proceedings of the IVth Vilnius Conference on Prob. Theory and Math. Stat., June 1985.

5) Dinges, H. (1986). Theory of Wiener Germs II. "Tail Probabilities". Technical Report, SFB 123, Heidelberg-Frankfurt.

6) Höglund, T. (1979) "A unified formulation of the central limit theorem for small and large deviations from the mean". Z. Wahrscheinlichkeitstheorie u. verw. Geb. 49, 105-117.

7) Lugannani & Rice (1980) "Saddlepoint approximation for the distribution of the sum of independent random variables". Adv. Appl. Prob. 12, 479-90.

8) Petrov, V.V. "Sums of independent random variables". Berlin-Heidelberg-New York. Springer 1975 (in particular page 248)

9) Varadhan, S.R.S. (1984) "Large Deviations and Applications". SIAM Publications 46, Philadelphia, Pa.

Fachbereich Mathematik
J.W. Goethe-Universität Frankfurt
Robert-Mayer-Str. 10
D - 6000 Frankfurt am Main
West Germany

SUPERIMPOSED CODES IN R^n

Thomas Ericson

Linköping

Key words: superimposed codes, multiple-access communication, asymptotic bounds.

ABSTRACT

A superimposed code in R^n is a finite set C of unit norm vectors x in Euclidean n-space, R^n, with the proporty that any two sums of at most m vectors from C are separated by at least unity distance. The code C has dimension n, order m, and size $T \triangleq |C|$. The concept has application in various multiple-access communication problems. We derive upper and lower bounds for the maximum size T(n,m) of a superimposed code given the dimension n and order m, and in particular we investigate the asymptotic properties of these bounds.

INTRODUCTION

Let C be a finite set of unit norm vectors x in R^n (R^n denotes Euclidean n-space; the usual inner product $(x,y) = \sum_{i=1}^{n} x_i y_i$ and norm $\|x\| = (x,x)^{1/2}$ are assumed). For $A \subseteq C$, denote by x(A) the sum of all vectors $x \in A$:

$$x(A) \triangleq \sum_{x \in A} x$$

and denote by $A(m)$ the ensemble of all subsets $A \subseteq C$ of cardinality at most equal to m:

$$A(m) \triangleq \{A \subseteq C: |A| \leq m\} .$$

Finally, let $C^{(m)}$ denote the set of vectors $x(A)$ corresponding to subsets $A \in A(m)$:

$$C^{(m)} \triangleq \{x(A): A \in A(m)\} .$$

The code C is said to be a <u>superimposed code of order m</u> if the vectors in $C^{(m)}$ are of at least unit distance from each other:

$$d(C^{(m)}) \triangleq \min_{A \neq B} \| x(A)-x(B) \| = 1; \quad A,B \in A(m).$$

(Notice that in case $A = \{x\}$ is a singleton and $B = \emptyset$ is the empty set we have $\|x(A)-x(B)\| = \|x\| = 1$, so $d(C^{(m)})$ is never larger than unity).

Superimposed codes arise in certain multiple-access problems. We refer in particular to Mazo [1] and chapter 5 in the textbook by Simon-Omura-Scholtz-Lewitt [2]. Our terminology is inspired by the work of Kautz-Singleton [3] and Dyachkov-Rykov [4], which deals with superimposed codes for the binary collision channel. The concept of superimposed codes is also closely related to the concept of B_s-sequences of vectors (Lindström [5], Dyachkov-Rykov [6]).

We characterize a superimposed code C by its <u>dimension</u> n, <u>order</u> m, and <u>size</u> $T \triangleq |C|$. The set of all superimposed codes with parameters (n,m,T) will be denoted $B(n,m,T)$. The problem we address can be formulated as follows: given n and m, how large can T be? Formally our desire is to characterize the function

$$T(n,m) \triangleq \max \{T: B(n,m,T) \neq \emptyset\} .$$

In particular we study the asymptotic increase of $T(n,m)$ as $n \to \infty$ with m fixed. More precisely we are interested in the quantity

$$E(m) \triangleq \limsup_{n\to\infty} \frac{1}{n} \log T(n,m)$$

which we will refer to as the maximal rate (all logarithms are taken to the base 2). We summarize (without proofs) in this presentation a few of the results we have regarding bounds on $E(m)$. A full account will be given in a forthcoming paper.

SPHERE PACKING AND RANDOM CODING

Let C be a code in $B(n,m,T)$ and consider the associated code $C^{(m)}$. As each $x(A) \in C^{(m)}$ clearly must be located inside a sphere of radius m, and as different vectors $x(A)$ are separated by at least unity distance we have the following sphere packing bound:

$$|C^{(m)}| = \sum_{i=1}^{m} \binom{T}{i} \leq \left(\frac{m+1/2}{1/2} \right)^n = (1+2m)^n .$$

From this bound we readily derive the following upper bound on the rate $E(m)$:

$$E(m) \leq E_{SP}(m) \triangleq \frac{1}{m} \log (1+2m)$$

A lower bound can be obtained by the following random coding argument. Choose a code C at random, assuming binary code vectors, with components chosen independently with the values $n^{-1/2}$ and $-n^{-1/2}$ equally probable. It can be shown that the probability that C belongs to $B(n,m,T)$ is positive for large enough n as long as

$$\frac{1}{n} \log T < \min_{1 \leq v \leq m} \frac{1}{2v} \cdot G\left(\frac{3}{4}, c_v \right)$$

where G is a function defined as

$$G(p,q) \triangleq p \log \frac{p}{q} + (1-p)\log \frac{1-p}{1-q} \qquad 0 \leq p,\ q \leq 1$$

and where c_v is defined by

$$c_v \triangleq 2^{-v} \binom{2v}{v} .$$

The function G(p,q) is positive for $0 < p,\ q < 1$ (see Gallager [7] p. 146) and c_v is bounded as follows.

$$\sqrt{\frac{1}{4v}} < c_v < \sqrt{\frac{1}{\pi v}}$$

(Gallager [7] p. 530).

We conclude that E(m) is positive for all m = 1,2,... . For large m the following aymptotic bound applies:

$$E(m) > \frac{3}{16}\frac{1}{m} \log m \ ;$$

which apart from the factor $\frac{3}{16}$ is similar to the asymptotic form of the sphere packing bound:

$$E_{SP}(m) = \frac{1}{m} \log m \ (1+o(1)); \quad m \to \infty .$$

ALGEBRAIC CONSTRUCTIONS

Explicit codes are often characterized by a parameter called absolute correlation. This quantity is defined as follows:

$$c(C) \triangleq \max_{x \neq y} \ |(x,y)| \ ; \quad x,\ y \in C .$$

Absolute correlation is related to the parameter m by the following result.

Theorem 1: Let $C \subseteq R^n$ have absolute correlation $c(C) = c$.

Then C belongs to the set $B(n,m,T)$, where

$$T \triangleq |C|; \quad m \triangleq \min\{T, \left\lceil \frac{1}{2c} \right\rceil\}$$

($\lceil x \rceil$ denotes the smallest integer larger than or equal to x).

This result can be used in combination with the Gilbert-Varshamov bound for binary codes (MacWilliams - Sloane [8] p.33), using the conventional mapping $0 \leftrightarrow n^{-1/2}$; $1 \leftrightarrow -n^{-1/2}$. The result is

Theorem 2: There exists a superimposed code C of length n, size $T = 2^k$, and order m, provided that

$$1 + \binom{n-1}{1} + \ldots + \binom{n-1}{d-2} < 2^{n-k-1}$$

where $d \triangleq \left\lceil n \cdot \frac{2m-1}{4m} \right\rceil$.

Using this result we obtain the following lower bound on the rate:

$$E(m) \geq E_{GV}(m) \triangleq 1 - h\left(\frac{1}{2} - \frac{1}{4m}\right)$$

where h(x) is the binary entropy function

$$h(x) \triangleq - x \log x - (1-x) \log (1-x) \ .$$

The asymptotic form of this bound is

$$E_{GV}(m) = \frac{\log e}{8\ m^2} \log m\ (1+o(1) = 0.180 \cdot \frac{1}{m^2} \log m\ (1+o(1))$$

We notice that this algebraic construction is much weaker than the random coding bound, which might indicate that absolute correlation may be too weak a parameter for characterization of superimposed codes.

A COMPARISON WITH SOME KNOWN CODES

It is of interest to evaluate the performance of some known codes and compare with the bounds we have. The codes we have in mind are those found by Gold [12] and Kasami [13], which are frequently used in so called spread spectrum systems, [2]. In the usual application each user is assigned not a single codeword but a whole set of codewords, corresponding to some codeword and all its cyclic shifts. The reason for this is the assumed lack of synchronism between the different users in the system. In our present problem the question of synchronism does not arise (we assume perfect synchronization between all users). As a consequence we can assign codewords to a considerably larger set of users.

An m-sequence (maximum linear feedback shift register sequence, [2] p. 283) is the output sequence from a linear shift register with feedback connections determined by a primitive polynomial. Such a sequence together with all its cyclic shifts and the all-zero sequence form a linear code known as the simplex code ([8] ch. 14. p. 407). Let us assume the conventional isomorphic mapping from GF(2) into R ($0 \leftrightarrow n^{-1/2}$; $1 \leftrightarrow -n^{-1/2}$). Then in R^n the simplex code has absolute correlation $c = 1/n$. Gold [12] proved that for d odd it is possible to construct a set of 2^d+1 m-sequences of length $n = 2^d-1$ and with absolute correlation c (when represented in R^n) satisfying

$$n \cdot c = 2^{\frac{d+1}{2}} + 1 .$$

Taking all $n = 2^d-1$ codewords corresponding to each one of these 2^d+1 m-sequences and also adding the all-zero sequence (in $GF(2)^n$) gives us a linear code with

$$T = (2^d-1)(2^d+1) + 1 = 2^{2d}$$

codewords.

Applying Theorem 1 this gives us the bound

$$m \geq \left\lceil \frac{1}{2} \frac{2^d-1}{2^{\frac{d+1}{2}}+1} \right\rceil .$$

ACKNOWLEDGEMENT

Several discussions with Dr. Lazlo Györfi, Technical Institute, Budapest, and with Dr. Robert Schweikert, German Aerospace Research Establishment (DFVLR), Munich are gratefully acknowledged.

REFERENCES

[1] Mazo J.E. (1979): Some theoretical observations on spread spectrum communications. In: Bell Syst. J., Vol. 58, No. 9, 2013-2023.

[2] Simon M.K. (1985): Omura J.K. Scholtz R.A. Levitt B.K. Spread spectrum communications. Vol.I, Ch. 5, Computer science press, Rockvill, Maryland.

[3] Kautz W.H. (1964): Singleton R.C. Nonrandom binary superimposed codes. In: IEEE Trans on Inf. Theory, IT-10, No. 4, 363-377.

[4] Dyachkov A.G. (1982): Rykov V.V. Bounds on the length of disjunctive codes. In: Probl. Pered. Inf. Vol. 18, No. 3 (Engl. transl.), 166-171.

[5] Lindström B. (1972): On B_2-sequences of vectors. In: J. Number Theory, Vol. 4, No. 3, 261-265.

[6] Dyachkov A.G. (1981): Rykov V.V. A coding model for a multiple-access adder channel. In: Probl. Pered. Inf. Vol. 17, No. 2, (Engl.transl.), 94-104.

[7] Gallager R.G. (1968): Information and reliable communication. Wiley, New York.

[8] MacWilliams F.J. (1977): Sloane M.J.A. The theory of error-correcting codes. 3rd Ed., North Holland Publishing Co., New York.

[9] Sidelnikov V.M. (1974): Upper bounds for the number of points of a binary code with a specified code distance. In: Probl. Péred. Inf. Vol. 10, No. 2, (Engl. transl.) 124-131.

[10] Levenshtein V.I. (1975): Minimum redundancy of binary error-correcting codes. In: Probl. Pered. Inf. Vol. 10, No. 2 (Engl.transl.) 110-123.

[11] Graham R.L. (1980): Sloane N.J.A. Lower bounds for constant weight codes. In: IEEE Trans. on Inf. Theory, Vol. IT-26, 37-43.

[12] Gold R. (1967): Optimal binary sequences for spread spectrum multiplexing. In: IEEE Trans. on Inf. Theory, IT-13, 619-621.

[13] Kasami T. (1966): Weight distribution formula for some class of cyclic codes. Coordinated Science Lab., Univ. Illinois, report R-265.

University of Linköping
Dept. of Electrical Engineering
S-581 83 Linköping
Sweden

SOME NEW RESULTS IN THE NONPARAMETRICAL ESTIMATION OF FUNCTIONALS

Rafail Hasminskii, Ildar Ibragimov

Moscow, Leningrad

Key words: Gaussian Noise, minimax estimation, nonparametric estimation, nonlinear functionals

ABSTRACT

The solution of minimax estimation problem of the unbounded linear functional from observed in the Gaussian Noise signal is presented. Analogous problem is considered for nonlinear functionals with the bounded Hilbert-Schmidt norms of derivatives.

INTRODUCTION

The last decade is characterized by the intensive work in the area of nonparametrical estimation. We would like to present here two results in this area. The first one is the minimax estimation of the unbounded linear functionals from a signal S observed with an additive Gaussian Noise. The second result is an approach to the finding of asymptotical effective nonlinear estimators (AENE) from S. The first result is presented in §§1-3, and the second in §4.

§1. THE STATEMENT OF PROBLEM

We suppose that Y. A. Rozanov (1971) was the first who has formulated the minimax estimation problem of linear functional in the additive Gaussian Noise in a correct way. He has considered the observation X(t) having the form

$$X(t) = S(t) + n(t), \quad t \in T \tag{1.1}$$

where $S \in \Sigma$, Σ is a known (to a statistician) set, n(t) is a generalized Gaussian process with zero mean. The problem is to estimate the linear functional L(S). It is natural to restrict ourselves only to the linear (in suitable sense) estimators $\hat{L}$, $\hat{L} \in M$. Therefore we have two problems. The first one is to find minimax risk

$$\Delta^2(L,M) \triangleq \inf_{\hat{L}\in M} \sup_{S\in\Sigma} E_S|\hat{L} - L(S)|^2.$$

The other one is to find linear minimax estimator, i. e. estimator L^*, for which

$$\sup_{S\in\Sigma} E(L^* - L(S))^2 = \Delta^2(L,M)$$

The first solution of these problems was obtained also in Rozanov (1971), but for very special Σ.
The next step was made in the our paper (1984). We have assumed there that the set Σ is convex, and the process n is white Gaussian noise (WGN).

Later Dodunekova, Hasminskii (1987) have noticed that the approach of Ibragimov, Hasminskii (1984) is suitable for consideration of a similar problem for indirect observation. In more detail, we consider the following situation. Let

$$\dot{Y}(t) = S(t) + n_1(t)$$

be nonobservable, and let the observed process have the form

$$Z(t) = \int_0^t Y(u)du + n_2(t), \quad t \in [0,1].$$

Here $n_1(t)$ and $n_2(t)$ are independent WGN. As earlier, the problem is to estimate L(S), $S \in \Sigma$. We have treated this problem in the frame of Ito stochastic differentials.

Alternative approach is to consider the observation $\dot{Z}(t) = X(t)$. Here

$$X(t) = S(t) + n_1(t) + \dot{n}_2(t)$$

and we have the mentioned problem for generalized stochastic process $n = n_1 + \dot{n}_2$.

It is necessary to constitute a more precise mathematical model and to introduce a concept of linear estimator. We assume that n = n(t) in (1.1) is a generalized Gaussian process with zero mean and a correlation operator R. Concerning R, we assume only that it is a self-adjoined and strictly positive operator in a Hilbert space H, and $\mathcal{D}(R)$ is the range of R (as usually, $\mathcal{D}(R)$ is assumed to be everywhere dense in H).

Further, S is nonrandom signal and it is known a priori that $S \in \Sigma$, Σ is a known (to a statistician) set, $\Sigma \subset H$. The generalized process X is observable in the sense that the statistician can observe linear functionals

$$(X,\phi) = (S,\phi) + (n,\phi) \tag{1.2}$$

for all $\phi \in \Phi \subseteq H$ ($(.,.)$ is inner product in H). If $\phi_i \in \mathcal{D}(R)$, the vector $((n,\phi_1),\dots,(n,\phi_k))$ is Gaussian with zero mean and $\|(R\phi_i,\phi_j)\|$ is its covariance matrix. Particularly,

$$E(\phi,n)^2 = (R\phi,\phi)$$

It is very important to extend the set Φ as far as possible because it allows us to understand the observation X in the more rich sense (see (1.2)). The method of construction of this natural set Φ is well known (Rozanov (1983)), and we use it.

Let us start from $\mathcal{D}(R)$ and consider the new inner product in $\mathcal{D}(R)$: $(x,y)_1 = (Rx,y)$. Further we complete $\mathcal{D}(R)$ in the norm $\|x\|_1 = (x,x)_1^{1/2}$. Let H_1 denote this complete closure. If $H_1 \subset H$, then $H_1 = \mathcal{D}(R^{1/2})$. In the opposite case the operator $R^{1/2}$ may be extended to H_1 by continuity. Let $\tilde{R}^{1/2}$ be this extension. The operator $\tilde{R}^{1/2}$: $H_1 \to H$ has an inverse $\tilde{R}^{-1/2}$: $H \to H_1$, which is the extension of $R^{-1/2}$: $\mathcal{D}(R^{-1/2}) \to H$. Therefore, the operator $\tilde{R}^{-1} \triangleq (\tilde{R}^{-1/2})^2$: $\mathcal{D}(R^{-1/2}) \to H_1$, $\tilde{R} = (\tilde{R}^{-1})^{-1}$: $H_1 \to \mathcal{D}(R^{-1/2})$. Similar arguments allow us to assert that the inner product (x,y), $x \in H$, $y \in \mathcal{D}(R^{-1/2})$ can be extended by continuity to $x \in H_1$, $((x_n,y) = (R^{1/2}x_n, R^{-1/2}y))$.
Let us assume from now on that the condition

$$\Sigma \subset \mathcal{D}(R^{-1/2}) \tag{1.3}$$

is fulfilled.
This condition and the previous discussion allow us to consider equation (1.2) for $\phi \in H_1$; it corresponds to considering not only (X,ϕ), $\phi \in \mathcal{D}(R)$, but their limits in the mean square as well.

Definition 1. The class of statistics (X,ϕ), $\phi \in H_1$ is called the class of linear estimates and is denoted by M; ϕ is called the weight function of linear estimator.

Definition 2. The estimator $\hat{L}$ belongs to a class A, $\hat{L} \in A$, if it may be represented in a form

$$T((X,\phi_1),\dots,(X,\phi_k)), \qquad \phi_i \in H_1, \quad k = 1,2,\dots$$

or is l.m.s. of such estimators (T is measurable function).

We assume that a linear bounded functional L(S) is defined on Σ. Our purpose is to find the *linear minimax estimator*, i. e. estimator $\hat{L}_0$, for which

$$\sup_{S\in\Sigma} E|\hat{L} - L(S)|^2 = \Delta^2(L,M).$$

2. REVIEW OF THE RESULTS IBRAGIMOV, HASMINSKII (1984)

Let us assume first that operator R = I is a unit one, $H = L_2[0,1]$. Then our problem is equivalent to the following one (Ibraqimov, Hasminskii (1984)): estimate L(S), if

$$X(t) = \int_0^t S(u)du + w(t), \qquad t \in [0,1]$$

is observed.
Here w(t) is standard Wiener process. It is easy to see that, in this case, the class M consists of estimates of a form

$$\hat{L} = \int_0^1 m(t)dx(t); \qquad m \in L_2(0,1) = L_2.$$

A slight modification of the main result of Ibragimov, Hasminskii (1984) can be formulated in the following way.

Theorem 2.1. Assume that observation has the form (1.1) with unit operator R ($Rx \equiv x$) and Σ is a symmetric convex subset of H. Then for any linear functional L defined on Σ the equality

$$\Delta^2(L,M) = \sup_{S\in\Sigma} \frac{L^2(S)}{\|S\|^2+1} \tag{2.1}$$

is true. Assume moreover that the function $S \to L^2(S)/(\|S\|^2+1)$ takes on its maximum in the point $S_1 \in \Sigma$. Then the estimator

$$\hat{L}_0 = (m_0,X), \qquad m_0 = S_1 L(S_1)/(\|S_1\|^2 + 1) \tag{2.2}$$

is linear minimax. Any linear minimax estimator can be presented in the form (2.2) for the bounded set Σ.

The main step in the proof of this theorem is a verification of the fact that the corresponding minimax problem has a saddle point.

The following theorem is true in the general case.

Theorem 2.2. Assume that Σ is a symmetric convex subset of H and the condition (1.3) is fulfilled. Then for any linear functional L defined on Σ the equation

$$\Delta^2(L,M) = \sup_{S\in\Sigma} \frac{L^2(S)}{\|R^{-\frac{1}{2}}S\|^2+1} \tag{2.3}$$

is true.

Assume moreover that the function $S \to L^2(S)/(\|R^{-\frac{1}{2}}S\|^2+1)$ takes on its maximum in a point $S_1 \in \Sigma$. Then the weight function m_0 of the linear minimax estimator is the following

$$m_0 = \frac{L(S_1)}{\|R^{-\frac{1}{2}}S_1\|^2+1} \bar{R}^{-1}S_1 = \frac{L(S_1)}{(\bar{R}^{-1}S_1,S_1)+1} \bar{R}^{-1}S_1 \tag{2.4}$$

The weight function of any linear minimax estimator can be presented in the form (2.4), if the set $R^{-\frac{1}{2}}\Sigma$ is bounded.

Theorem 2.2 follows from theorem 2.1 almost immediately with help of so called "whitening", i. e. from the observation X going to $Y = R^{-\frac{1}{2}}X$.

See Ibragimov, Hasminskii (1987) for details.

The natural problem is to find minimax estimator in the class A. We cannot find this estimator now. However the following theorem is true.

Theorem 2.3. The inequality

$$\inf_{T\in A} \sup_{S\in\Sigma} E_S|T-L(S)|^2 \geq \kappa\Delta^2(L,M)$$

is true for any convex symmetric set $\Sigma \subset H$. Here $\kappa = \inf_{a>0} \frac{a^2+1}{a^2} h(a)$, h(a) is a value of a minimax quadratic risk in the problem of estimation $\theta \in [-a,a]$ over observation $X = \theta + \xi$, $L(\xi) = N(0,1)$.

Remark 1. It is easy to obtain an estimate $\kappa > 0,8$.

Remark 2. Theorem can be evidently reformulated for an observation in the form $X_\varepsilon = S + \varepsilon n$, for which

$$\Delta^2(L,M) = \varepsilon^2 \sup_{S\in\Sigma} L^2(S) / (\|R^{-\frac{1}{2}}S\|^2+\varepsilon^2)$$

3. EXAMPLES

1. Theorem 2.2 was stimulated by the corresponding estimation problem for indirect observation, see §1. In this case, $n(t) = n_1(t) + n_2'(t)$ is a generalized Gaussian process with the correlation operator

$$Ry = y - y''; \; \mathcal{D}(R) = \{y: y'' \in L_2[0,1], \; y'(0) = y(1) = 0\}$$

All nonasymptotic results of Dodunekova, Hasminskii (1987) can be treated from this point of view. Details see in Ibragimov,Hasminskii (1987).

2. Assume that Σ is a subspace of H and $\Sigma \subset \mathcal{D}(R^{-\frac{1}{2}})$. Then the functional L is bounded in Σ and can be written in the form (ℓ,S), $\ell \in \Sigma$. Let us denote by Q_Σ a bounded symmetric operator $\Sigma \to \Sigma$ which corresponds to the quadratic form $\|R^{-\frac{1}{2}}x\|^2$, $x \in \Sigma$.

The condition

$$\ell \in \mathcal{D}(Q_\Sigma^{-\frac{1}{2}}) \tag{3.1}$$

is necessary and sufficient for the finiteness of $\Delta^2(L,M)$, and the value of the minimax risk is

$$\Delta^2(L,M) = \|Q_\Sigma^{-\frac{1}{2}}\ell\|^2 .$$

The proof may be found in Ibragimov, Hasminskii (1987).

Minimax estimator is unbiased in this situation and therefore minimax estimator is the best linear unbiased estimator (BLUE). From this result it follows easily, that BLUE exists if R is bounded and the condition (1.3) is fulfilled. (It was known many years ago.) If R is an integral operator with the kernel R(s,t), and

$$\int_T R(t,t)dt < \infty$$

then the set $\mathcal{D}(R^{-\frac{1}{2}})$, with the norm $\|R^{-\frac{1}{2}}x\|$, is the same as a reproducing kernel Hilbert Space (RKHS) H_R which corresponds to the kernel R. This remark enables us to verify the most important condition $\Sigma \subset \mathcal{D}(R^{-\frac{1}{2}}) = H_R$ because in many cases $\mathcal{D}(H_R)$ is known. Some concrete examples may be found in Ibragimov, Hasminskii (1987).

3. Let B be operator in H with a property

$$(BS,S) \le \alpha\|S\|^2; \quad \alpha > 0 \; .$$

Then the set $\Sigma_B = \{S: (BS,S) < 1\}$ is an "ellipsoid". Assume that $L(S)$, $S \in \Sigma_B$ is a linear bounded functional. We would like to find a minimax estimator of $L(S)$, the observation (1.1) being known.

It is easy to see that the functional $L(S)$ in this case may be written in the form

$$L(S) = (\ell, B^{\frac{1}{2}}S) \ . \tag{3.2}$$

Assume the element ℓ to be defined by the equation (3.2), $\Sigma_B \subset \mathcal{D}(R^{-\frac{1}{2}})$. Then the minimax risk and the weight function of the minimax estimator can be found from the equations

$$\Delta^2(L,M) = (A^{-1}\ell,\ell); \ A = B^{-\frac{1}{2}}R^{-1}B^{-\frac{1}{2}}+I;$$

$$m_0 = \frac{(A^{-1}\ell,\ell)}{\|R^{-\frac{1}{2}}B^{-\frac{1}{2}}A^{-1}\ell\|^2+1} \tilde{R}^{-1} B^{-\frac{1}{2}} A^{-1} \ell \ .$$

Proof see in Ibragimov, Hasminskii (1987).

Let us moreover assume that operators B and R commutate and a spectrum of one of them is discrete. Then there exist a common basis $\phi_1,\dots,\phi_2,\dots$ of eigenelements: $B\phi_i = \lambda_i\phi_i$; $R\phi_i = \mu_i\phi_i$.
It is easy to find in this case that

$$\Delta^2(L,M) = \sum \frac{P_i^2}{1+(\lambda_i\mu_i)^{-1}} ,$$

$$m_0 = \frac{\Delta^2(L,M)}{1+\sum\frac{\ell_i(\lambda_i\mu_i)^{-1}}{(1+(\lambda_i\mu_i)^{-1})^2}} \sum \frac{\mu_i^{-1}\lambda_i^{-\frac{1}{2}}\ell_i}{1+(\lambda_i\mu_i)^{-1}} \phi_i \ .$$

Here $\ell_1,\dots,\ell_n,\dots$ are coordinates of ℓ in the basis $\phi_1,\phi_2,\dots$ More special examples see in Ibragimov, Hasminskii (1984), Dodunekova, Hasminskii (1987).

§4. ESTIMATION OF NONLINEAR FUNCTIONALS

An approach to an estimation of nonlinear bounded functionals of signal S, which is observed in a mixture with a Gaussian White Noise (GWN) was proposed in Hasminskii, Ibragimov (1980). Now, together with A. S. Nemirovskii, we improve this approach. It allows us not only to weaken the assumptions. We hope the results to be completed in a sense mentioned later.

Let us try to expose the essense of this approach (details see in Ibragimov, Nemirovskii, Hasminskii (1986)).

We assume that the observation has a form

(4.1) $$X(t) = S(t) + \varepsilon\, n(t), \quad 0 \leq t \leq 1$$

Here $S \in \Sigma$ is an unknown signal, n is GWN, ε is a small parameter. The problem consists in estimation of a sufficiently smooth bounded functional $\Phi(S)$, based on X.

"Coordinate" form of our problem is the following: observations $X(i) = S(i) + \varepsilon\, \xi(i)$, $i = 1,2,3,...$ are given. Here $L(\xi(i)) = N(0,1)$, $\xi(i)$ are i.i.d. $S = (S(1),...,S(n),...) \in \Sigma$, $\Sigma \subset \ell_2$ is compact. Denote S_n the projection of S onto an n-dimensional Euclidean space, generated by n initial coordinate vectors, $S^n = S - S_n$. We have, with help of Taylor formula,

$$\Phi(S) = \Phi(S_n+S^n) = \Phi(S_n)+\Phi'(S_n)[S^n] +...+ \frac{1}{k!}\Phi^{(k)}(S_n)[S^n,...,S^n] + R_k .$$

The remainder R_k is small, if some conditions of smoothness are fulfilled and n is chosen in a suitable way. The expressions $\Phi^{(j)}(S_n)$, $j = 0,...,k$, may be estimated by $\Phi^{(j)}(X_n)$ (here $X_n = (X(1),X(2),...$ $...,X(n),0,0,...)$. Therefore the problem is reduced to finding estimators of homogeneous forms $\Phi'(X_n)[S^n],...,\Phi^{(k)}(X_n)[S^n,...,S^n]$, based on $X(i)$, $i = n+1,n+2,...$ Dependence of $\Phi^j(X_n)$ on X_n is unessential in view of independence $X(i)$, $i > n$ of this value. Let us assume that the derivatives $\Phi^{(j)}(S)$, $i = 1,...,k$, have finite Hilbert -Smidt norms. Then the corresponding Hermitian polynomials of variables $X(n+1),X(n+2),...$ may be used for the estimation of the forms $\Phi^{(j)}(X_n)[S^n,...,S^n]$. The estimator Φ^*, which is constructed in this way, is the asymptotically effective nonparametrical estimator (AENE) in the sense Hasminskii, Ibragimov (1980), if some conditions, which are formulated in the theorem 4.1 (see below) are fulfilled. There are conditions of two types in this theorem: conditions concerning the smoothness $(k + \gamma)$ of the functional Φ and those concerning the quality of approximation of the set Σ by finite-dimensional projections, which is characterized by a number β. The main condition for $k > 2$ has the form

(4.2) $$(k + \gamma)\beta > \tfrac{1}{2}.$$

It means that more smooth functionals admit AENE for "less compact" Σ.

The mentioned above completeness of result means, that the change of the condition (4.2) to $(k + \gamma)\beta < \frac{1}{2}$ leads to an existence of functionals, for which the rest of the theorem conditions are fulfilled and AENE do not exist.

Theorem 4.1. Assume, that the following conditions are fulfilled:

I. The functional $\Phi(S)$ has Frechet derivatives up to the order $k - 1$ in the set $\{\|S\|<1\} = \mathcal{O}$ with bounded Hilbert-Schmidt (H.-S.) norms and for the $\Phi^{(k)}(S)[g]$ the inequality

$$\|\Phi^{(k)}(S)[g]\|_2 \le L\|g\|$$

is fulfilled. (Here $\|.\|_2$ denotes H.-S. norm, $\|.\|$ is the usual norm).

II. The inequality

$$\|\Phi^{(k)}(S_2)-\Phi^{(k)}(S_1)\| \le L\|S_2-S_1\|^\gamma$$

is true in $\mathcal{O}$.

III. The set $\Sigma\subset\text{int}\,\mathcal{O}$ is compact, and its n-dimensional Kolmogorov's diameters satisfy a condition $d_n(\Sigma) < c\, n^{-\beta}$.
Then there exists AENE of the functional $\Phi(S)$ for $k \ge 3$ if the condition (4.2) is true, too. The assertion of the theorem is true for $k = 1$ and $k = 2$, if the condition (4.2) is changed to a condition

$$(k + \gamma - 1)\beta > \tfrac{1}{2}.$$

The proof of this theorem is presented in Ibragimov, Nemirovskii, Hasminskii (1986).

REFERENCES

Rozanov Yu. A. (1971): On the Minimax Estimator of an Unknown Mean Value. J. of Multivariate Analysis, Vol. 1, Number 2, June 1971, pp. 158-166.

Ибрагимов И. А., Хасьминский, Р. З. (1984): О непараметрическом оценивании значения линейного функционала в гауссовском белом шуме. Теор. вероятн. и ее примен., т. 29, в. 1, с. 19-32.

Додунекова Р. Д., Хасьминский Р. З. (1987): Оценка значений линейного функционала по косвенным наблюдениям. Пробл. передачи информации, т. 23, в. 1.

Розанов Ю. Ф. (1983): Гауссовские бесконечномерные распределения. Тр. МИАН, т. 108, М.: Наука, с. 136.

Ибрагимов И. А., Хасьминский Р. З. (1987): Об оценивании линейных функционалов в гауссовском шуме.Теор. вероятн. и ее примен., т. 32, в. 1.

Hasminskii R. Z., Ibragimov I. A. (1980): Some estimation problems for stochastic differential equations. Lect. Notes Control Inform. Sci. B 25, s. 1-12.

Ибрагимов И. А., Немировский А. С., Хасьминский Р. З. (1986): Некоторые задачи непараметрического оценивания в гауссовском белом шуме. Теор. вероятн. и ее примен., т. 31, в. 3, с. 451-466.

IPPI AN-SSSR
Ul. Ermolovoj 19
Moskva 101447
USSR

LOMI AN SSSR
Fontanka 27,
Leningrad
USSR

ON INVARIANT PROBABILITY DENSITIES OF PIECEWISE MONOTONIC TRANSFORMATIONS

Marius Iosifescu

Bucharest

Key words: Frobenius-Perron operator, Ionescu Tulcea-Marinescu ergodic theorem

ABSTRACT

Differential properties of the invariant probability density of a piecewise monotonic transformation are established by applying the Ionescu Tulcea-Marinescu ergodic theorem to the Frobenius-Perron operator associated with the transformation.

PRELIMINARIES

Let $T : G \longrightarrow I = [0,1]$ be a continuous map, where $G \subset I$ is open and $\lambda(G) = 1$ (λ = Lebesgue measure). Then there exists a finite or countable collection $(I_a)_{a \in A}$ of closed intervals with disjoint interiors such that $\bigcup_{a \in A} I_a \supset G$, and for any $a \in A$ the set $I_a \cap (I \setminus G)$ consists exactly of the endpoints of I_a. Assume that for any $a \in A$ the restriction of T to $I_a \cap G$ is strictly monotonic and extends to a C^p-function T_a

on I_a with $T_a(I_a)=I$. A map T with the above properties is called a C^p piecewise monotonic transformation, $p \geq 0$.

For any $a \in A$ let f_a denote the function inverse to T_a, thus mapping I onto I_a. For any $a^{(n)}=(a_1,\ldots,a_n) \in A^n$, $n \geq 1$, put

$$f_{a^{(n)}} = f_{a_1} \circ \ldots \circ f_{a_n}$$

($\circ$ denotes composition of functions). Clearly, $f_{a^{(n)}}$ is the function inverse to $T_{a_n} \circ \ldots \circ T_{a_1}$ and maps I onto the closed interval $I_{a^{(n)}}$ with endpoints $f_{a^{(n)}}(0)$ and $f_{a^{(n)}}(1)$. For any $n \geq 1$, the $I_{a^{(n)}}$, $a^{(n)} \in A^n$, which are called fundamental intervals of rank n, exhaust I up to a set of λ-measure 0, and any two of them can have only an endpoint in common.

Consider the conditions

$$(C) : \sup_{x \in I} \left| f'_{a^{(n)}}(x) \right| \Big/ \inf_{x \in I} \left| f'_{a^{(n)}}(x) \right| \leq C$$

for any $a^{(n)} \in A^n$, $n \geq 1$, where C is a constant ≥ 1 ;

$$(E_m) : \sup_{x \in I} \left| f'_{a^{(m)}}(x) \right| \leq \theta$$

for any $a^{(m)} \in A^m$, where θ is a constant < 1.

Clearly, conditions (C) and (E_m) make sense for any C^0 piecewise monotonic transformation for which finite derivatives f'_a, $a \in A$, exists everywhere in I.

Proposition 1. Condition (C) implies condition (E_m) for some m.

This result is essentially due to Halfant (1977). See also Iosifescu (1986).

It is well known (Halfant (1977), Iosifescu (1986), Rényi(1957)) that, under condition (C), there exists a unique absolutely continuous (with respect to λ) T-invariant probability measure ρ , i.e., $\rho(T^{-1}B) = \rho(B)$ for any Borel set B in $[0,1]$. The probability density $r = d\rho/d\lambda$ of ρ satisfies almost everywhere on I the inequalities

$$1/C \leq r \leq C,$$

therefore ρ is equivalent to λ . Moreover, T is ergodic with respect to both ρ and λ .

Further assumptions on T imply further properties of r . For example, if besides (C) condition

$$(BV) : \sum_{a \in A} \text{var } f_a' < \infty$$

holds (here var stands for total variation), then r possesses a version which is a function of bounded variation. See Iosifescu (1986), Keller (1979), Lasota and Yorke (1973).

The aim of this paper is to show how r inherits from T differential properties. Such a study has been first made by Halfant (1977) in a framework of smaller generality as to the type of the transformations T considered. Our approach, which, unlike Halfant's , is theoretic-functional leads to stronger and more precise results. The main tool we use is the ergodic theorem of Ionescu Tulcea and Marinescu (1950). For the reader's convenience we state this theorem in the next section. The theorem is motivated by and in turn is of fundamental importance in the theory of dependence with complete connections.

THE IONESCU TULCEA-MARINESCU ERGODIC THEOREM

Consider two complex Banach spaces $(B, |\cdot|)$ and $(L, \|\cdot\|)$ with $L \subset B$. Assume that

(i) If $y_n \in L$, $\|y_n\| \leq c$ for all $n \geq 1$, $y \in B$, and

$\lim_{n\to\infty} | y_n - y | = 0$, then $y \in L$ and $\| y \| \leq c$.

Denote by $\mathcal{C}_k(L,B)$, $k \geq 1$, the set of all linear operators U from L into L which are bounded with respect to both $\|\cdot\|$ and $|\cdot|_L$, where the latter is the restriction of $|\cdot|$ to L, and in addition satisfy

(ii) $H = \sup_{n \geq 0} |U^n|_L < \infty$.

(iii) There exist two positive constants $d < 1$ and D such that

$$\| U^k y \| \leq d \, \| y \| + D \, | y | \quad , \; y \in L.$$

(iv) If L' is a bounded set in $(L, \|\cdot\|)$, then $U^k L'$ has compact closure in $(B, |\cdot|)$.

(Let us notice that (ii) and (iii) imply that $J = \sup_{n \geq 0} \| U^n \| < \infty$.)

For any complex number σ set

$$E(\sigma) = \{ y \in L \; : \; Uy = \sigma y \} ,$$

so that σ is an eigenvalue of U iff $E(\sigma) \neq \{0\}$.
If $|\sigma| = 1$ let

$$U_\sigma^n = \frac{1}{n} \sum_{j=0}^{n-1} \sigma^{-j} U^j .$$

<u>Theorem (C.T.Ionescu Tulcea - G.Marinescu).</u> Let $U \in \mathcal{C}_k(L,B)$.

(a) The set E of eigenvalues of U of modulus 1 is finite and $E(\sigma)$ is finite dimensional for each $\sigma \in E$.

(b) For every $\sigma, |\sigma| = 1$, and $y \in L$ there exists a $U_\sigma y \in L$ to which $U_\sigma^n y$ converges in B (i.e. $\lim_{n\to\infty} | U_\sigma^n y - U_\sigma y | = 0$). The linear operators U_σ have $|U_\sigma|_L \leq H$ and $\|U_\sigma\| \leq J$, and $U_\sigma U_{\sigma'} = 0$, $\sigma \neq \sigma'$, $U_\sigma^2 = U_\sigma$, $U_\sigma L = E(\sigma)$.

(c) The operator $V = U - \sum_{\sigma \in E} \sigma U_\sigma$ belongs to $\mathcal{L}_k(L,B)$, has spectral radius

$$r_L(V) = \lim_{n \to \infty} \|V^n\|^{1/n} < 1,$$

and $U_\sigma V = VU_\sigma$.

(d) $U^n = \sum_{\sigma \in E} \sigma^n U_\sigma + V^n$ for any $n \geq 1$.

A <u>proof</u> of this theorem can be found, e.g., in Norman (1972, pp.45–49).

Let us introduce the special spaces L and B we will deal with in this paper.

For any complex-valued function h defined on I put

$$u(h) = \sup_{x \in I} |h(x)| \quad , \; s(h) = \sup_{x_1 \neq x_2 \in I} \left| \frac{h(x_1)-h(x_2)}{x_1-x_2} \right| .$$

It is well known that the set $C^p(I)$, $p \geq 0$, of complex-valued functions defined on I which have continuous derivatives of order up to and including p is a Banach space under the norm

$$|h| = \sum_{i=0}^{p} u(h^{(i)})$$

with $h^{(0)} = h$, $h^{(i)}$ = i-th derivative of h , $i \geq 1$. The linear subspace $C^pL(I)$ of $C^p(I)$ consisting of the elements $h \in C^p(I)$ for which $s(h^{(p)}) < \infty$ is easily seen to be a Banach space under the norm

$$\|h\| = |h| + s(h^{(p)}).$$

It is a not difficult exercise to show that for any $p \geq 0$ the spaces $L = C^pL(I)$ and $B = C^p(I)$ satisfy condition (i) above.

THE MAIN RESULT

The fact that the probability density r (whose existence and uniqueness were discussed in the introductory section) is T-invariant amounts to the equation $Pr = r$ almost everywhere in I, where P is the Frobenius-Perron operator associated with T defined as

$$Ph(x) = \sum_{a\in A} \left| f_a'(x) \right| h(f_a(x)) , \quad x \in I ,$$

for any measurable, bounded, complex-valued function h on I. See Halfant (1977), Iosifescu (1986).

In this section we give sufficient conditions for r to be a Lipschitz function. This will follow from the Ionescu Tulcea-Marinescu theorem applied to P for the special case of spaces $L = CL(I) = C^0L(I)$, $B = C(I) = C^0(I)$.

Before proceeding we need some prerequisites. First note that since for any $n \geq 1$ the intervals $I_{a^{(n)}}$, $a^{(n)} \in A^n$, exhaust I up to a set of λ-measure 0, i.e.,

$$(1) \qquad \sum_{a^{(n)}\in A^n} \left| f_{a^{(n)}}(1) - f_{a^{(n)}}(0) \right| = 1 ,$$

condition (C) implies that

$$(2) \qquad \sup_{x\in I} \sum_{a^{(n)}\in A^n} \left| f'_{a^{(n)}}(x) \right| \leq C, \quad n \geq 1 .$$

Indeed, for any $x \in I$ and $n \geq 1$ we can write

$$\frac{\left|f'_{a^{(n)}}(x)\right|}{C} \leq \frac{\sup_{t\in I}\left|f'_{a^{(n)}}(t)\right|}{C} \leq \inf_{t\in I} \left| f'_{a^{(n)}}(t) \right| \leq \left| f_{a^{(n)}}(1) - f_{a^{(n)}}(0) \right| .$$

Next, we prove

Lemma 2. If condition (C) holds, then

$$\sup_{x\in I, a\in A} |f'_a(x)| < \infty. \tag{3}$$

Proof. Remark that on account of (C) one should have

$$i(a) = \inf_{x\in I} |f'_a(x)| > 0$$

for any $a \in A$.

By virtue of Proposition 1, condition (C) implies condition (E_m) for some $m \geq 1$. If $m=1$ then, clearly, (3) holds. Assume therefore that (E_m) holds with $m > 1$. Fix $a^{(m-1)} = (a_1, \ldots, a_{m-1})$ and assume (3) does not hold. Then there are $x_o \in I$ and $a \in A$ such that

$$|f'_a(x_o)| > 1/i(a_1) \ldots i(a_{m-1}).$$

Now, with $a_m = a$, $a^{(m)} = (a_1, \ldots, a_m)$, we have

$$f'_{a^{(m)}}(x) = f'_{a_1}(f_{(a_2,\ldots,a_m)}(x)) \ldots f'_{a_{m-1}}(f_{a_m}(x)) f'_{a_m}(x),$$

whence

$$|f'_{a^{(m)}}(x_o)| \geq i(a_1) \ldots i(a_{m-1}) |f'_a(x_o)| > 1,$$

thus contradicting condition (E_m). The proof is complete.

Finally, assuming that the derivatives f''_a, $a \in A$, exist everywhere in I, consider the condition

$$(BD^{(2)}) : \sup_{x\in I} \sum_{a\in A} |f''_a(x)| < \infty.$$

Clearly, condition $(BD^{(2)})$ implies condition (BV).

We have

Lemma 3. If conditions (C) and $(BD^{(2)})$ hold, then

$$\sup_{n\geq 1} \sup_{x\in I} \sum_{a^{(n)} \in A^n} |f''_{a^{(n)}}(x)| < \infty.$$

Proof. We use the formula

(4) $f''_{ij}(x) = f''_i(f_j(x))(f'_j(x))^2 + f'_i(f_j(x))f''_j(x)$,

which is valid for any u-tuple $i \in A^u$ and any v-tuple $v \in A^v$, $u,v \geq 1$. On account of (2) and Lemmas 2 and 3 it is easy to show that

$$D_n = \sup_{x \in I} \sum_{a^{(n)} \in A^n} \left| f''_{a^{(n)}}(x) \right| < \infty$$

for any $n \geq 1$. It remains to prove that the D_n are uniformly bounded.

We know that condition (C) implies condition (E_m) for some m . Choose an integer j such that $C\,\theta^j = d < 1$. Using (2), it follows from (4), with $u = (s-j)m+i$, $s \geq j$, $i < m$, and $v = jm$, that

(5) $D_{sm+i} \leq C\,\theta^j\,D_{(s-j)m+i} + CD_{jm}$,

and it is easily seen that (5) implies

(6) $D_n \leq D = \max_{1 \leq i < jm} D_i + \frac{C\,D_{jm}}{1-d}$, $n \geq 1$.

Now we are ready to prove

Theorem 4. Let T be a C^1 piecewise monotonic transformation such that the second derivatives f''_a , $a \in A$, exist everywhere in I . Assume that conditions (C) and ($BD^{(2)}$) hold. Then the invariant probability density r is a Lipschitz function on I and the equation $Pr = r$ is valid everywhere in I .

Proof. It is not difficult to see that the Frobenius-Perron operator P takes CL(I) into itself. (In fact, the proof of this assertion is similar to checking condition (iii) a few lines below.)

Since

$$P^n h(x) = \sum_{a^{(n)} \in A^n} \left| f'_{a^{(n)}}(x) \right| \, h(f_{a^{(n)}}(x)), n \geq 1 ,$$

on account of (2) we have $H \leq C$.

Let us check condition (iii). Choose m and j as in the proof of Lemma 3. Putting $k = jm$, for any $h \in CL(I)$ and $x_1, x_2 \in I$ we can write

$$P^k h(x_1) - P^k h(x_2) =$$

$$= \sum_{a^{(k)} \in A^k} \left(\left| f'_{a^{(k)}}(x_1) \right| - \left| f'_{a^{(k)}}(x_2) \right| \right) h(f_{a^{(k)}}(x_1)) +$$

$$+ \sum_{a^{(k)} \in A^k} \left| f'_{a^{(k)}}(x_2) \right| \left(h(f_{a^{(k)}}(x_1)) - h(f_{a^{(k)}}(x_2)) \right).$$

The first sum above is dominated by $D|h| \, |x_1-x_2|$, where D is given by (6). Indeed, that sum can be written as $\alpha(1) - \alpha(0)$ with

$$\alpha(t) = \sum_{a^{(k)} \in A^k} f'_{a^{(k)}}(tx_1 + (1-t)x_2) h(f_{a^{(k)}}(x_1)) \operatorname{sgn} f'_{a^{(k)}} ,$$

and the intermediate value formula and Lemma 3 justify our assertion. Similarly, the second sum is dominated by

$$C \, \theta^j \, s(h) \, |x_1-x_2| .$$

Therefore we have

$$s(P^k h) \leq d \, s(h) + D \, |h| ,$$

whence, on account of the fact that $|P^k h| \leq C \, |h|$,

$$\| P^k h \| \leq d \|h\| + (C-d+D) |h| , \; h \in CL(I) ,$$

i.e. condition (iii) .

To check condition (iv) we should note that if $L' \subset CL(I)$ is bounded in $CL(I)$, then the set $P^k L'$ is also bounded in $CL(I)$. It remains to use the Arzelà-Ascoli theorem.

To sum up, we have proved that $P \in \mathcal{O}_k(CL(I),C(I))$.

Now, using (1) it is immediate that

$$h_n = \frac{1}{n} \sum_{i=0}^{n-1} P^i(1) \geq 0 ,$$

and $\int h_n d\lambda = 1$, $n \geq 1$. Then the Ionescu Tulcea-Marinescu theorem implies that h_n converges as $n \to \infty$ in both $CL(I)$ and $C(I)$ to a probability density (a non-negative function with integral 1), which is invariant under P. Thus 1 is an eigenvalue of P, say $\sigma_1 = 1$, and, actually, h_n converges to $P_1(1) =$ $(= U_1(1))$, which should coincide with the invariant probability density r on account of its uniqueness.

The validity of the equation $Pr = r$ everywhere in I follows from the fact that both Pr and r are continuous functions on I.

The proof of Theorem 4 is complete.

A MORE GENERAL RESULT AND THE ANALYTIC CASE

Assuming that for a fixed integer $p \geq 2$ the derivatives $f_a^{(p)}$, $a \in A$, exist everywhere in I, consider the condition

$$(BD^{(p)}) \; : \; \sup_{x \in I} \sum_{a \in A} \left| f_a^{(p)}(x) \right| < \infty .$$

Theorem 4 can be easily generalized as follows .

Theorem 5. Let T be a C^p piecewise monotonic transformation, $p \geq 1$, such that the derivatives $f_a^{(p+1)}$, $a \in A$, exist everywhere in I. Assume that conditions (C) and $(BD^{(i)})$, $2 \leq i \leq p+1$, hold. Then the invariant probability density r belongs to the space $C^{p-1}L(I)$ and the equation $Pr = r$ is valid everywhere in I.

The proof is entirely similar to that of Theorem 4, using the Ionescu Tulcea-Marinescu theorem for the special case of spaces $L = C^{p-1}L(I)$ and $B = C^{p-1}(I)$.

Clearly, first of all, we have to prove the obvious generalization of Lemma 3, which is easy to do. Termwise differentiations needed in the proof for $p \geq 2$, are justified by dominated convergence ensured by conditions $(BD^{(i)})$ for $2 < i \leq p+1$.

Remark that the method used to prove Theorems 4 and 5 shows that the invariant probability density r is approached geometrically in $C^{p-1}L(I)$ by the iterated probability densities

$$\sum_{a^{(n)} \in A^n} \left| f'_{a^{(n)}}(x) \right| , \quad n \geq 1.$$

Theorem 5 and the above remark in conjunction with Theorem 10 of Halfant (1977) lead to

Theorem 6. Let T be an analytic piecewise monotonic transformation, i.e., the f_a, $a \in A$, are analytic on I. Assume that conditions (C) and $(BD^{(i)})$, $i \geq 2$, hold. Then the invariant probability density r is analytic on I and the equation $Pr = r$ is valid everywhere in I.

A CONJECTURE FOR THE SPECIAL CASE OF f-EXPANSIONS

Let f be a continuous non-negative strictly monotone function with inverse f^{-1}. Such a function can be used to associate with certain $x \in I$ an infinite integer sequence $(a_n(x))_{n \geq 1}$ for which

(7) $$x = f(a_1(x) + f(a_2(x) + \dots)).$$

Representation (7) is called an f-expansion. The 'digits' $a_n(x)$ and the 'remainders' $r_n(x)$ are defined recursively as follows

$$r_0(x) = x, \; a_n(x) = \left[f^{-1}(\{r_{n-1}(x)\})\right], \; r_n(x) = f^{-1}(\{r_{n-1}(x)\}), \; n \geq 1,$$

where $[\cdot]$ and $\{\cdot\}$ denote the integral part and the fractional part, respectively. S.Kakeya in 1924, B.H.Bissinger in 1944, C.J.Everett in 1946, and A.Rényi in 1957 gave sufficient conditions for the validity of representation (7). (For details, precise assumptions on f, and exact references see Grigorescu and Iosifescu (1982, pp.264-266), Iosifescu (1985), Rényi (1957).)

The setting here enters the general framework of piecewise monotonic transformations as follows. The map T is defined as $T(x) = \{f^{-1}(x)\}$, $x \in I$, the index set A is a segment (finite or infinite) of the non-negative integers, I_a is the closed interval with endpoints $f(a)$ and $f(a+1)$, $a \in A$, $T_a(x) = f^{-1}(x) - a$, $x \in I_a$, so that $f_a(x) = f(a+x)$, $x \in I$, $a \in A$. Under suitable assumptions (e.g. under only (C)) the existence of the digits $a_n(x)$ for any $n \geq 1$ and the validity of (7), meant as

$$\lim_{n \to \infty} f_{(a_1(x),\dots,a_n(x))}(0) = x,$$

are ensured. The above equation holds for all $x \in I$ not belonging to a certain countably infinite subset of I.

In particular, the case $f(t) = t/D$, $t \in [0,D]$, with D an integer > 1, leads to the D-adic expansion, while the case $f(t) = 1/t$, $t \geq 1$, leads to the continued fraction expansion.

The equation $Pr = r$, which in the present context can be written as

$$\sum_{a\in A} |f'(a+x)| \; r(f(a+x)) = r(x), \; x\in I ,$$

shows that under conditions (C) and $(BD^{(2)})$

$$q(a,x) = \frac{|f'(a+x)| \; r(f(a+x))}{r(x)} , \; a\in A,$$

is a probability distribution on A for <u>any</u> $x\in I$.

It was asserted in Iosifescu (1985) that

(8) $$q(a, f_{(a_n,\dots,a_1)}(0))$$

equals the conditional probability $\lambda(a_{n+1}=a \mid a_1,\dots,a_n)$.

(Note the reversal of order of digits in (8).) This is easily seen to be true for both the D-adic expansion and the continued fraction expansion. I was convinced to be in the possession of a proof of the above assertion but I have recently discovered a gap in it. Therefore, the things are at a conjecture level and I conclude by expressing my feeling that, however, this conjecture is true.

REFERENCES

Grigorescu S. and Iosifescu M.(1982) : Dependence with Complete Connections and Its Applications. Ed.ştiinţifică şi enciclopedică, Bucureşti. (In Romanian)

Halfant M.(1977) : Analytic properties of Rényi's invariant density. Israel J.Math. 27, 1-20.

Ionescu Tulcea C.T. and Marinescu G.(1950) : Théorie ergodique pour des classes d'opérations non complètement continues. Ann.of Math. (2) 52,140-147.

Iosifescu M.(1985) : f-Expansions : A result and a querry. Rev. Roumaine Math.Pures Appl.30, 749-750.

Iosifescu M.(1986) : Mixing properties for f-expansions. In : Proc.4th Vilnius Conference Probab.Theory and Math.Statist. (Vilnius, 1985). VNU Science Press, Utrecht. (To appear)

Keller,G.(1979) : Ergodicité et mesures invariantes pour les transformations dilatantes par morceaux d'une région bornée du plan. C.R.Acad.Sci.Paris Sér. A-B 289, A 625 - A 627.

Lasota A. and Yorke J.A.(1973) : On the existence of invariant measures for piecewise monotonic transformations. Trans.Amer.Math.Soc. 186, 481-488.

Norman M.F.(1972) : Markov Processes and Learning Models. Academic Press, New York.

Rényi,A.(1957): Representations for real numbers and their ergodic properties. Acta Math.Acad.Sci.Hungar.8,477-493.

Centre of Mathematical Statistics
174 Ştirbei Vodă St.
77104 Bucharest
Romania

ALGORITHMIC COMPLEXITY AND PSEUDO-RANDOM SEQUENCES

Ivan Kramosil

Prague

Key words: Algorithmic complexity, pseudo-random sequences, Turing machines

ABSTRACT

The notion of Kolmogorov algorithmic complexity of sequences of symbols is used to define pseudo-random sequences of stochastically independent and identically distributed samples from a finite set and pseudo-Markov chains with a finite set of states. Sufficient conditions are introduced and discussed, under which the ideas used in these two special cases can be generalized to obtain an appropriate complexity-based definition of pseudo-random sequences with a more complicated stochastical structure.

0. INTRODUCTION

For well-known reasons, the classical axiomatic probability theory ascribes the predicate of randomness rather to generators of sequences of results than to individualized sequences themselves. In fact, within this framework there is no immediate tool to classify a particular sequence of potential results as "random" and to separate it from the "non-random" ones. Various tests of randomness represent only a second-level and not quite sufficient remedy and the validity of their answers needs to be parametrized by a probability of error or significance level. So it is rather difficult, inside this classical probability calculus, to develop an appropriate theory of pseudo-random sequences (numbers), which would classify output sequences only from the viewpoint of their ability to simulate random inputs for some statistical computational or decision-making procedures, regardless of the, possibly deterministic, origin of such sequences. The demand for pseudo-random sequences has also a practical motivation, because of time and expenses savings following when such sequences replace the true-random inputs.

A. N. Kolmogorov (Kolmogorov (1965)) and some other authors have proposed and investigated an alternative approach to randomness revoking, in a sense, the von Mises' notion of collective (von Mises (1919)). Algorithmic complexity of a sequence of symbols (results, outputs) is defined as the length of the shortest program by which this sequence can be generated using a fixed computational device (a fixed universal Turing machine, in this paper). Adopting the most simple definition (cf. Fine (1973) for other alternatives), a finite sequence of the length n is defined to be pseudo-random, if its complexity is smaller than n only by the value of a fixed o(n)-function; for infinite pseudo-random sequences we need all initial segments to be pseudo-random. In this case pseudo-random sequences possess the main properties typical for i.i.d. samples from the uniform (equiprobable) distribution over a finite set, the aim of this paper is to define, using the terms of algorithmic complexity, pseudo-random sequences also for some stochastically more complicated sequences of random variables and to find some sufficient conditions under which this construction is possible.

1. INFINITE SEQUENCES OVER FINITE ALPHABETS

Let $A = \{a_1, a_2, \ldots, a_r\}$, $r \geq 2$, be a finite set (<u>alphabet</u>) of abstract symbols (<u>letters</u>) with A^n, A^∞, $A^* = \bigcup_{n=0}^{\infty} A^n$ ($A^0 = \{\Lambda\}$) taking their usual sense, let $\ell(x) = n$ iff $x \in A^n$. Each A^n is lexicographically ordered w.r. to the increasing indices of letters (i. e. $a_1 < a_2 < \ldots < a_r$); this ordering $<$ is extended to A^*, setting $x < y$, if $\ell(x) < \ell(y)$. If $n \in N = \{0,1,2,\ldots\}$, then n^* is the n-th element of A^* w.r. to $<$, hence, 0^* is Λ.

For $x = \langle x_1, x_2, \ldots, x_n \rangle \in A^n$, <u>(elementary) cylinder</u> $V(x) \subset A^\infty$

is defined by

(1) $V(x) = \{\tilde{x}: \tilde{x} \in A^{\infty}, \tilde{x}[n] = x\}, \quad \tilde{x}[n] = \langle \tilde{x}_1, \tilde{x}_2, \ldots, \tilde{x}_n \rangle$.

Let $F \subset P(A^{\infty})$ denote the minimal σ-field of subsets of A^{∞} containing all elementary cylinders. An infinite sequence $X = \langle X_1, X_2, \ldots \rangle$ of random variables, with each X_i taking an abstract probability space $\langle \Omega, S, P \rangle$ into A, defines a consistent system $\langle P_1^X, P_2^X, \ldots \rangle$ of probability measures on $A^1, A^2, \ldots$ by simply setting, for each $n \in N$, $x \in A^n$,

(2) $P_n^X(x) = P(\{\omega: \omega \in \Omega, \langle X_1(\omega), X_2(\omega), \ldots, X_n(\omega) \rangle = x\})$.

Due to the well-known assertion of probability theory, the sequence $\langle P_1^X, P_2^X, \ldots \rangle$ uniquely defines a probability measure P^X on F. Some more probabilistic constructions over A^* and A^{∞} will be taken into consideration below.

On the other hand, finite and infinite sequences of letters may be considered also from a purely syntactic point of view, i. e. as linear combinations of abstract symbols not endowed with any interpretation. As the most general formal apparatus for syntactical handling of such sequences may serve a <u>universal Turing machine</u> (UTM, c. f., e. g., Davis (1958), Rogers (1967), or elsewhere for definition and further references); UTM can be seen as an idealized abstraction of a computer, working beyond any space and time limitations. Let U be a fixed UTM over the alphabet A, let $p \in A^*$, $x \in (A \cup \{blank\})^*$, then $U(p) = x$ means: having written p on the (input) type of U and having conventionally initialized it, the machine U eventually stops its work with x written on the (output) tape. For $x \in A_0^{\infty}$, $U(p) = x$ means: for each $n \in N$ there is $m(n) \in N$ such that after $m(n)$ operations of U the initial segment $x[n]$ of the length n of x is on the output tape (strengthened version: and is never changed during further work of U). Evidently, the set

(3) $Rec = \{x: x \in A^{\infty}, (\exists p \in A^*) (U(p) = x)\}$

of recursive infinite sequences does not depend on U or on the version of $U(p) = x$.

In a sense, all syntactical properties of $x \in A^* \cup Rec$ are defined, up to U, by the set $\pi(x,U) = \{p: p \in A^*, U(p) = x\}$. More generally, given $S \in A^*$ and denoting by $*$ the concatenation, set

(4) $\pi(x,S,U) = \{p: p \in A^*, U(p*S) = x\}$.

The (Kolmogorov) algorithmic complexity of x given S and U is the length of the shortest element in $\pi(x,S,U)$ and is denoted by $K_U(x|S)$, hence,

(5) $K_U(x|S)=\min\{\ell(p):p\in\pi(x,S,U)\}=\min\{n:n\in N,(\exists p\in A^n)(U(p*S)=x)\}$

with the convention that $\min\{\emptyset\}=\infty$; c. f. Fine (1973), e. g., for more details and comments.
The three following properties of $K_U(x|S)$ hold; their proofs are very simple and can be found in Fine (1973) or elsewhere:

(6) $(\exists c_U\in N)(\forall x\in A^*)(K_U(x|(\ell(x))^*)\leq \ell(x)+c_U)$,

(7) $(\forall U_1,U_2\in \mathrm{UTM})(\exists c_{U_1,U_2}\in N)(\forall x,S\in A^*)(|K_{U_1}(x|S)-K_{U_2}(x|S)|\leq c_{U_1,U_2})$,

(8) $(\forall n,T\geq 0)(\forall S\in A^*)(\mathrm{card}\{x:x\in A^n,K_U(x|S)\geq n-T\}>r^n(1-(r-1)^{-1}r^{-T}))$.

2. SOME PARTICULAR TYPES OF PSEUDO-RANDOM SEQUENCES

Because of a more detailed explanation in Kramosil (1985), (1986a), (1986b), the presentation given below in this chapter is very brief and serves rather as an inspiration for the generalization investigated in Chapter 3. For a total function $f: N\to N$ we shall write $f(n)\in o(n)$, if $\lim_{n\to\infty} n^{-1}f(n)=0$, and $f(n)\in ML_r$ (f is a Martin-Löf function or f possesses the M.-L. property), if $\sum_{n=0}^{\infty} r^{-f(n)}<\infty$.

Given $T\in N$, a sequence $x\in A^*$ may be called T-(pseudo)-random, if $K_U(x|(\ell(x))^*)\geq \ell(x)-T$. Due to (8), the greatest part of finite sequences is T-random and this definition agrees with the intuition identifying "randomness" with the "lack of order". In fact, for each T fixed and for $\ell(x)$ increasing, relative frequency of occurrences of particular letters or blocks of letters in T-random sequences tends to the uniform (equiprobable) distribution over A or over the corresponding Cartesian power of A. These properties remain valid also if T replaced by $f(\ell(x))$ for $f(n)\in o(n)$.

Given $f(n)\in o(n)$, an infinite sequence $x\in A^\infty$ can be defined as f-(pseudo)-random, if $K_U(x[n]|n^*)\geq n-f(n)$ for almost all n's (i. e. up to a finite number of n's). Denote by D_f the set

(9) $D_f=\bigcup_{n=1}^{\infty}\bigcap_{m\geq n}\{x:x\in A^\infty,K_U(x[m]|m^*)\geq m-f(m)\}$.

Fact 1. (cf. Martin-Löf (1966), Fine (1973)). $D_f \neq \emptyset$ iff $f(n) \in ML_r$, and if it is the case, then $\tilde{P}_\infty(D_f) = 1$, where $\tilde{P}_\infty$ is defined on (A^∞, F) by an infinite i.i.d. sequence of random variables with uniform distribution on A. ∎

Hence, T-randomness cannot be extended to infinite sequences. If $f(n) \in o(n) \cap ML_r$, then f-random sequences satisfy the limit results mentioned above. If $F: N \to N$, $F(i) < F(i+1) < K(i+1)$ for $K \in N$ is a total recursive function and $x = \langle x_1, x_2, \ldots \rangle \in A^\infty$ is f-random, $f(n) \in o(n)$, then $x_F = \langle x_{F(1)}, x_{F(2)}, \ldots \rangle$ is f'-random for an $f'(n) \in o(n)$. The class $o(n) \in ML_r$ is non-empty, take $f(n) = \alpha \log_r n$, $\alpha > 1$.

The non-effectivity of the notion of T-randomness and f-randomness caused by the non-computability of $K_U(x|S)$ led to some approximations (cf. Chapter 5, Kramosil (1984)). As can be easily seen, f-random sequences can simulate i.i.d. samples ascribing the probability r^{-1} to each $a_j \in A$ and if applied, e. g., in Monte-Carlo methods, the estimates tend to the actual value in a stronger sense than that assured by the laws of large numbers (cf. Kramosil (1983)).

Given $x = \langle x_1, x_2, \ldots, x_n \rangle \in A^n$, set

(10) $$n_i(x) = \text{card}\{j: j \le n, x_j = a_i\},$$

(11) $$Q(x) = \bigcap_{i=1}^r \{y: y \in A^{\ell(x)}, n_i(y) = n_i(x)\} \subset A^{\ell(x)},$$

(12) $$\tilde{q}(x) = \text{card } Q(x) = \left(\sum_{i=1}^r n_i(x)\right)! \left(\prod_{i=1}^r (n_i(x)!)\right)^{-1},$$

let $q(x) \le \tilde{q}(x)$ denote the order number of x w.r. to the lexicographical ordering of Q(x). Set

(13) $$N(x) = \langle n_1^*(x), n_2^*(x), \ldots, n_r^*(x), q^*(x) \rangle.$$

Again, for $x = \langle x_1, x_2, \ldots, x_n \rangle \in A^n$, $j \le r$, denote by $x^j \in A^*$ the subsequence of those occurrences of symbols in x, which are immediately preceded by a_j, evidently, some x^j may be empty and x_1, $x^1, x^2, \ldots, x^r$ form a "disjoint covering of x". Set

(14) $$M(x) = \langle x_1, n_{11}^*(x), n_{12}^*(x), \ldots, n_{1r}^*(x), n_{21}^*(x), \ldots, n_{2r}^*(x), n_{31}^*(x), \ldots, n_{r1}^*(x), \ldots, n_{rr}^*(x), q^*(x^1), q^*(x^2), \ldots, q^*(x^r) \rangle,$$

where $n_{ij}(x) = n_j(x^i)$. Evidently, there are fixed programs π_1, π_2 independent of x such that $U(\pi_1 * \tilde{N}(x)) = U(\pi_2 * \tilde{M}(x)) = x$, $\tilde{N}(x)$ is a fixed encoding of $N(x) \in (A \cup \{\text{blank}\})^*$ into A^*, similarly for $\tilde{M}(x)$.

<u>Definition 1.</u> A sequence $x = \langle x_1, x_2, \ldots \rangle \in A^\infty$ is called <u>f-(pseudo)-independent and identically distributed sequence</u> (f-PIID-sequence), $f(n) \in o(n)$, if for almost all $n \in N$

(15) $$|K_U(x[n] \mid n^*) - K_U(\tilde{N}(x[n]) \mid \Lambda)| \leq f(n).$$

The sequence x is called <u>f-(pseudo)-Markov sequence</u> (f-PM-sequence), if for almost all $n \in N$

(16) $$|K_U(x[n] \mid n^*) - K_U(\tilde{M}(x[n]) \mid \Lambda)| \leq f(n).$$ ⌑

The non-emptiness of these definitions follows from these Facts (cf. Theorems 1 and 3 in Kramosil (1986a) for their proofs).

<u>Fact 2.</u> Let $p_i \in \langle 0,1 \rangle$, $i \leq r$, $\sum_{i=1}^{r} p_i = 1$, let $\varepsilon > 0$ be given, then there exist $f(n) \in o(n)$ and an f-PIID-sequence $x \in A^\infty$ such that the relative frequency $fr(a_i, x[n]) = n^{-1}\, n_i(x[n])$ converges and $\max\{|\lim_{n\to\infty} fr(a_i, x[n]) - p_i| : i \leq r\} < \varepsilon$. Moreover, for each $\alpha = \langle a_{i_1}, a_{i_2}, \ldots, a_{i_m} \rangle \in A^m$, $m \geq 1$

(17) $$\lim_{n\to\infty} fr(\alpha, B(m,x)[n]) = \Pi_{j=1}^{m} (\lim_{n\to\infty} fr(a_{i_j}, x[n])).$$

If x is f-PIID-sequence with rational limit values of relative frequencies, then there exist an alphabet A' and $f'(n) \in o(n)$ such that x results by factorization from an f'-random sequence $y \in A'^\infty$. ⌑

<u>Fact 3.</u> Let $\{p_{ij}\}_{i,j=1}^{r}$ be an $r \times r$ stochastic matrix $(\sum_{j=1}^{r} p_{ij} = 1)$, let $\varepsilon > 0$ be given. Then there exist $f(n) \in o(n)$ and an f-PM-sequence x such that, for all $i,j \leq r$, x^i is infinite, $r_{ij} = \lim_{n\to\infty} fr(a_j, x^i[n])$ exists, and

(18) $$\sum_{j=1}^{r} r_{ij} = 1, \quad \max(\{|r_{ij} - p_{ij}| : i,j \leq r\}) < \varepsilon.$$ ⌑

The generalization of this construction to chains with a fixed finite memory (degree of dependence) is obvious.

3. GENERAL PSEUDO-RANDOM SEQUENCES

In this chapter we shall try to generalize some ideas and methods from Chapter 2 with the aim to obtain a definition of pseudo-random sequences w.r. to more general sequences of random variables.

Given fixed <u>Acc</u> (accept), <u>Ref</u> (refuse), a set $B \subset A^*$ is called <u>recursive</u>, if there is a sequence $\underline{Ind}_B$ (indicator of B) such that, for all $x \in A^*$, $U(\underline{Ind_B}*x) = \underline{Acc}$, if $x \in B$, $U(\underline{Ind_B}*x) = \underline{Ref}$, if $x \in A^* - B$.

In Chapter 2 we used the idea that there are "typical" sequences w.r. to the given $X = \langle X_1, X_2, \ldots \rangle$, e. g. those with given limits of relative frequencies of letters or blocks of letters in the sequence itself or in some of its subsequences. Moreover, the set of "typical" sequences is of P_∞^X-measure one and generates, in each A^n, the set of "typical" n-tuples with P_n^X-measures increasing to one "rather quickly". To formalize this ideas, set, given $X = \langle X_1, X_2, \ldots \rangle$,

$$\mathcal{B}(X) = \{B: B \subset A^*,\ B \text{ is recursive},\ \textstyle\sum_{i=1}^\infty P_i^X(A^i - (A^i \cap B)) < \infty\}, \tag{19}$$

we shall write B_n for $A^n \cap B$, B_n^c for $A^n - B_n$, and similarly for other subsets of A^*. Evidently, $\mathcal{B}(X)$ is at most countable and closed w.r. to finite joints. Or, if $B^1, B^2 \in \mathcal{B}(X)$, the recursivity of $B^1 \cap B^2$ is trivial and

$$\textstyle\sum_{i=1}^\infty P_i^X((B_i^1 \cap B_i^2)^c) \le \sum_{i=1}^\infty P_i^X(B_i^{1c}) + \sum_{i=1}^\infty P_i^X(B_i^{2c}) < \infty. \tag{20}$$

On the other hand, $(\bigcap_{B \in \mathcal{B}(X)} B) \in \mathcal{B}(X)$ needs not hold, as this example demonstrates. Take $B^i \subset A^*$, $i \in N$, such that $B_n^i \supset B_n^{i+1}$ $i, n \in N$, and $P_n^X(B_n^{ic}) = \alpha(n^{1+i^{-1}})^{-1}$ for an $\alpha > 0$. Hence, for each i ,

$$\textstyle\sum_{n=1}^\infty P_n^X(B_n^{ic}) = \sum_{n=1}^\infty \quad (n^{1+i^{-1}})^{-1} < \infty, \tag{21}$$

but, setting $B^0 = \bigcap_{i=1}^\infty B^i$, so that $B_n^{0c} = \bigcup_{i=1}^\infty B_n^{ic}$,

$$\textstyle\sum_{n=1}^\infty P_n^X(B_n^{0c}) = \sum_{n=1}^\infty \sup_{i=1}^\infty P_n^X(B_n^{ic}) = \sum_{n=1}^\infty \sup_{i=1}^\infty \alpha(n^{1+i^{-1}})^{-1} = \sum_{n=1}^\infty (\alpha/n) = \infty. \tag{22}$$

<u>Definition 2.</u> A system $X = \langle X_1, X_2, \ldots \rangle$ is called <u>closed,</u> if $(\bigcap_{B \in \mathcal{B}(X)} B) \in \mathcal{B}(X)$. Given $B \in \mathcal{B}(X)$, X is called <u>B-uniform</u>, if there are ε_1^B, $\varepsilon_2^B > 0$ such that, for all $n \in N$, $x \in B_n$,

$$1 + \varepsilon_1^B \ge P_n^X(x)(P_n^X(B_n))^{-1} \operatorname{card} B_n \ge 1 - \varepsilon_2^B. \tag{23}$$ ⌑

For $C \subset A^n$, define the linear ordering on A^n, say $\ll_{c,n}$, in this way: if $x \in C$, $y \in A^n - C$, then $x \ll_{c,n} y$, if $x, y \in C$ or $x, y \in A^n - C$ and $x < y$, then $x \ll_{c,n} y$. We write $\ll_{X,B}$, if $C = B_n$, $B \in \mathcal{B}(X)$. Evidently, if C is recursive, $\ll_{c,n}$ is recursive, hence, given n and $m \le r^n$, a fixed program computes the m-th n-tuple $m^*(n,C)$ in A^n w.r.to $\ll_{c,n}$. We may define also a (recursive, if C is) mapping

$Y_{n,C}$: $A^n \to A^*$ in this way: $Y_{n,C}(x) = y$ iff $x = m^*(n,C)$ and $y = m^*$ (re-enumeration of n-tuples ordered w.r. to $<<_{C,n}$ by first r^n sequences from A^* w.r. to <), again, we write $Y_{X,B}$ if $C = B_n$, $B \in \mathcal{B}(X)$.

Given a recursive $C \subset A^*$, consider a program π_C (including $\underline{\text{Ind}}_C$) which, given $y \in A^*$, n^*, works in this way: computes $m \in N$ such that $y = m^*$ and tests, whether $m < r^n$. If it is not the case, $U(\pi_C * y * n^*) = \Lambda$, if $m < r^n$, $U(\pi_C * y * n^*) = m^*(n,C)$, the m-th n-tuple in A^n w.r. to $<<_{C_n,n}$. E. g., if $y = \Lambda$, then $y = 0^*$, $m = 0$, $0 < r^n$ trivially holds, hence, $m^*(n,C) = U(\pi_C * \Lambda * n)$ is either $a_1 a_1 \dots a_1$ (n-times), if $a_1 \dots a_1 \in C$, or the lexicographically first n-tuple in C_n, if $a_1 \dots a_1 \in A^* - C$. Evidently,

$$U(\pi_C * Y_{n,C}(x) * n^*) = x \tag{24}$$

and $Y_{n,C}(x)$ completely defines x, given U, π_C and n. Now,we are ready to propose the following generalization of the notion of pseudo-randomness.

Definition 3. Let $X = \langle X_1, X_2, \dots \rangle$ be a sequence of random variables, each of them taking $\langle \Omega, S, P \rangle$ into A, let $B \in \mathcal{B}(X)$, let $f: N \to N$ be a total o(n)-function. An infinite sequence $x \in A^\infty$ is called <f,X,B>-(pseudo)random, if there is $n(x) \in N$ such that, for all $n \geq n(x)$, $x[n] \in B_n$ and

$$|K_U(x[n] \mid n^*) - \ell(Y_X(x[n]))| < f(n). \tag{25}$$ ⌑

So, <f,X,B>-random sequence possesses, up to an initial segment, the property B (as can be easily proved, the P_∞^X-measure of the set of infinite sequences with this property equals one). This "typical" property B can be used to shorten the programs for initial segments in B by ascribing them smaller indices by $<<_{X,B}$, i.e. shorter codes or programs. At the same time, this way of using the B-property of $x \in A^\infty$ should be, up to tolerance limits given by an o(n)-function f(n), the optimal way how to shorten the programs for initial segments of x, supposing x is <f,X,B>-random. A sequence $x \in A^\infty$ is called <f,X>-random, if it is <f,X,B>-random for all $B \in \mathcal{B}(X)$, and is called X-random, if it is <f,X>-random for an o(n)-function f(n).

Again,the main problem is to find sufficient conditions for non-emptiness of Definition 3 as well as some quantitative characteristics of the extent of the set of <f,X,B>-random sequences;

both these problems are solved by the following assertion.

Theorem 1. Let $\mathcal{X}$ and B be as in Definition 3, let $\mathcal{X}$ be B-uniform, let $f(n) \in o(n) \cap ML_r$ be a total function taking N into itself, let $\mathcal{D}(f,\mathcal{X},B)$ be the set of all $\langle f,\mathcal{X},B\rangle$-random sequences, then $P_\infty^{\mathcal{X}}(\mathcal{D}(f,\mathcal{X},B)) = 1$, i. e., almost all (w.r. to $P_\infty^{\mathcal{X}}$) sequences in A^∞ are $\langle f,\mathcal{X},B\rangle$-random. ⌑

Proof. (We use the basic idea of Theorem 6 in Fine (1973) and omit $\mathcal{X}$ in $P^{\mathcal{X}}$.) (24) implies, with $C = B_n$, that for each $c < \ell(\pi_{\mathcal{X}})$, $x \in A^\infty$, $n \in N$,

$$(26) \qquad K_U(x[n] \mid n^*) < \ell(Y_{\mathcal{X},B}(x[n])) + c.$$

Set (the dependence on B will not be explicitly introduced)

$$(27) \qquad \begin{aligned} D_{n,1} &= \{x\colon x \in A^\infty,\ x[n] \in A^n - B_n\}, \\ D_{n,2} &= \{x\colon x \in A^\infty, x[n] \in B_n, K_U(x[n] \mid n^*) < \ell(Y_{\mathcal{X},B}(x[n])) - f(n)\}, \\ D_n &= D_{n,1} \cup D_{n,2}, \quad D_n^c = A^\infty - D_n. \end{aligned}$$

Evidently

$$(28) \qquad \mathcal{D}(f,\mathcal{X},B) = \bigcup_{n=1}^\infty \bigcap_{m=n}^\infty D_m^c,$$

hence

$$(29) \quad P_\infty(\mathcal{D}^c(f,\mathcal{X},B)) = \inf_{n=1}^\infty P_\infty\Big(\bigcup_{m=n}^\infty D_m\Big) \le \inf_{n=1}^\infty \sum_{m=n}^\infty P_\infty(D_m),$$

so, a sufficient condition for $P_\infty(\mathcal{D}(f,\mathcal{X},B)) = 1$ is that $\sum_{n=1}^\infty P_\infty(D_n) < \infty$.

It depends on $x[n]$, whether $x \in D_n$ or not, so D_n and D_n^c can be written as disjoint unions of cylinders of the length n. As P_∞ extends conservatively each P_n, $P_\infty(V(x)) = P_n(V(x))$ for each $n \in N$, $x \in A^n$. Evidently,

$$(30) \qquad \sum_{n=1}^\infty P_\infty(D_n) \le \sum_{n=1}^\infty P_\infty(D_{n,1}) + \sum_{n=1}^\infty P_\infty(D_{n,2}),$$

where

$$(31) \quad \begin{aligned} \sum_{n=1}^\infty P_\infty(D_{n,1}) &= \sum_{n=1}^\infty \sum_{x \in A^n - B_n} P_\infty(V(x)) = \sum_{n=1}^\infty \sum_{x \in A^n - B_n} P_n(x) = \\ &= \sum_{n=1}^\infty P_n(B_n^c) < \infty, \end{aligned}$$

as $B \in \mathcal{B}(\mathcal{X})$. The only we have to prove is that $\sum_{n=1}^\infty P_\infty(D_{n,2}) < \infty$.

Different n-tuples from A^n need different programs to be generated by U, so that the number of elementary cylinders contained in $D_{n,2}$ cannot exceed the number L_n of different sequences of the length at most

(32) $K = \max\{\ell(Y_{X,B}(x[n]))-f(n)-1: V(x[n]) \subset D_{n,2}\} < \log_r(\text{card } B_n)+1,$

as $Y_{X,B}(x[n])$ are r-adic codes of numbers from 0 to card B_n-1. Hence,

(33) $L_n = \sum_{i=0}^{K} r^i = (r-1)^{-1}(r^{K+1}-1) < (r-1)^{-1}r^{K+1} < r(r-1)^{-1}(\text{card}B_n)r^{-f(n)}.$

More precisely, there exist $L_{1,n} \le L_n$, $x^1,x^2,\ldots,x^{L_{1,n}} \in A^n$, such that $D_{n,2} = \bigcup_{i=1}^{L_{1,n}} V(x^i)$. But, due to the B-uniformity of X there is $\varepsilon < \infty$ such that, for all $x \in A^n$,

(34) $P_\infty(V(x)) = P_n(x) \le (1+\varepsilon)P_n(B_n)(\text{card } B_n)^{-1} \le (1+\varepsilon)(\text{card } B_n)^{-1},$

so that

(35) $\sum_{n=1}^{\infty} P_\infty(D_{n,2}) = \sum_{n=1}^{\infty} \sum_{V(x)\subset D_{n,2}} P_n(x) < \sum_{n=1}^{\infty} L_n(1+\varepsilon)(\text{card } B_n)^{-1} <$

$< \sum_{n=1}^{\infty} r(r-1)^{-1}(1+\varepsilon)r^{-f(n)} = r(r-1)^{-1}(1+\varepsilon) \sum_{n=1}^{\infty} r^{-f(n)} < \infty.$

The assertion is proved. Evidently, if X is B-uniform for all $B \in \mathcal{B}(X)$, then the P_∞-measure of the set of all $\langle f,X\rangle$-random and X-random sequences alsoequals one. ⌑

4. SOME FUNDAMENTAL PROPERTIES OF GENERAL PSEUDO-RANDOM SEQUENCES

The assertions presented below describe formally some basic properties of $\langle f,X,B\rangle$-random sequences. Theorem 2 shows that a typical sequence, which can be defined more precisely than by the set B, is not $\langle f,X,B\rangle$-random, Theorem 3 introduces sufficient conditions under which a subsequence of a pseudo-random sequence is itself pseudo-random. Finally, Theorem 4 presents some conditions under which $\mathcal{D}(f,X,B) = \mathcal{D}(f',X,B')$ for different $B, B' \in \mathcal{B}(X)$. Only the basic ideas of the proofs are given with all technical details being omitted.

<u>Theorem 2.</u> Let X and B be as in Theorem 1, let $f:N \to N$ be a total o(n)-function, let $C \subset A^*$ be a recursive set, let there exist $\varepsilon > 0$ and $n_0(C) \in N$ such that, for all $n \ge n_0$, $P_n^X(D_n) \le (1-\varepsilon)^{n-n_0(C)}$, where $D_n = C \cap B \cap A^n$. Let $x \in A^\infty$ be such that $x[n] \in D_n$ for all $n \ge n_1(x)$ and $\ell(Y_{X,B}(x[n])) = \lceil \log_r \text{card } B_n \rceil$ for finitely many n's, then x is not $\langle f,X,B\rangle$-random. ⌑

Proof. We may suppose that $x[n] \in B_n$ for all n large enough, take n such that $x[n] \in D_n$. The set D is recursive, so $<<_D$ on A^n and $Y_{n,D_n}(x)$ are defined as above and are recursive. If $n \geq n_0(C)$ and $P_n^X(B_n) \geq 1-\varepsilon$, then

(36) $(1-\varepsilon)^{n-n_0(C)} \geq P_n^X(D_n) = \sum_{x \in D_n} P_n^X(x) \geq$

$\geq (\text{card } D_n)(\text{card } B_n)^{-1}(1-\varepsilon_2)P_n^X(B_n) \geq (\text{card } D_n)(\text{card } B_n)^{-1}(1-\varepsilon_2)(1-\varepsilon)$

where ε_2 is the infimum of all ε_2^B, for which (23) holds. Hence,

(37) $$\frac{\text{card } D_n}{\text{card } B_n} \leq \frac{(1-\varepsilon)^{n-n_0-1}}{(1-\varepsilon_2)}$$

and

(38) $\log_r \text{card } D_n \leq \log_r \text{card } B_n - \log_r\left[(1-\varepsilon_2)(1-\varepsilon)^{n_0+1}\right] - n\log_r(1-\varepsilon)^{-1}.$

However, taking (24) for D_n and an appropriate π_D, we obtain that

(39) $K_U(x[n] \mid n^*) \leq \ell(Y_n, D_n(x[n])) + \text{const}_1 \leq \log_r(\text{card } D_n) + \text{const}_1 + 1 \leq$

$\leq \log_r(\text{card } B_n) + \text{const}_2 - Kn = \ell(Y_{X,B}(x[n])) + \text{const}_2 - Kn$

with the last equality holding for infinitely many n's, so that x cannot be $\langle f, X, B\rangle$-random. ¤

Theorem 3. Let X, f, B be as in Theorem 1, let $F: N \to N$ be a total increasing recursive function such that $P_n^X(B_n^{F,y}/B_n) < \alpha^n$ for some $\alpha < 1$ and for all $n \in N$, where $k(n) = \max\{j: F(j) \leq n\}$, and, for $y = \langle y_1, y_2, \ldots, y_{k(n)}\rangle \in A^{k(n)}$,

(40) $$B_n^{F,y} = \bigcap_{i=1}^{k(n)} \{x: x = \langle x_1, \ldots, x_n\rangle \in B_n, x_{F(i)} = y_i\}.$$

Then there exists a set $\mathcal{D}_F \subset \mathcal{D}(f, X, B)$ such that $P_\infty^X(\mathcal{D}_F) = 1$ and $\tilde{x}_F = \langle \tilde{x}_{F(1)}, \tilde{x}_{F(2)}, \ldots\rangle$ is $\langle f', X, B\rangle$-random for all $\tilde{x} = \langle \tilde{x}_1, \tilde{x}_2, \ldots\rangle \in \mathcal{D}_F$ and for an appropriate o(n)-function f'(n). ¤

Proof. Take into consideration (26) with x replaced by x_F, the assertion will be proved if we arrive at the contradiction supposing that there exist $0 < K < 1$ and infinitely many m's such that

(41) $$K_U(x_F[m] \mid m^*) < K(\ell(Y_{X,B}(x_F[m]))).$$

Take such an m, take n such that $m = k(n)$, set $B_n^F = B_n^{F,\tilde{y}}$, $\tilde{y} = \langle x_{F(1)}, x_{F(2)}, \ldots, x_{F(k(n))}\rangle = x_F[k(n)] \in A^{k(n)}$. Take n such that $m = k(n)$ satisfies (41) and $P_n^X(B_n) > 1 - \varepsilon_2^B$, this can be done. Then

$$(42)\quad \alpha^n > (\sum_{x \in B_n^F} P_n^X(x))(P_n^X(B_n))^{-1} = (P_n^X(B_n^F))(P_n^X(B_n))^{-1} \geq$$

$$\geq (\text{card } B_n^F)(\text{card } B_n)^{-1}(1-\varepsilon_2^B)P_n^X(B_n) \geq (\text{card } B_n^F)(\text{card } B_n)^{-1}(1-\varepsilon_2^B)^2,$$

so that

$$(43)\qquad (\text{card } B_n^F)(\text{card } B_n)^{-1} < (1-\varepsilon_2^B)^{-2}\,\alpha^n.$$

A^n can be recursively ordered in this way: first n-tuples from B_n^F, then those from $B_n - B_n^F$, finally those from $A^n - B_n$, inside each class the lexicographical ordering is preserved. Let $<<_{X,F}$ be the ordering of $A^{n-k(n)}$ resulting from the ordering just described by erasing the $F(i)$-th, $i \leq k(n)$, coordinates in n-tuples (in case some $(n-k(n))$-tuple is repeated, only its first occurrence is considered). Let $(x[n])^F$ be the "rest" of $x[n]$ when $<x_{F(1)}, x_{F(2)}, \ldots$ $\ldots, x_{F(k(n))}>$ erased, let $Y_{X,B}^F$ be defined as $Y_{X,B}$ but w.r. to the ordering defined above. $x_F[k(n)]$ and $(x[n])^F$ determine $x[n]$, so, due to (41),

$$(44)\quad K_U(x[n] \mid n^*) \leq \ell(Y_{X,B}^F((x[n])^F) + K\ell(Y_{X,B}(x_F[k(n)])) + \text{const}_1.$$

Different pairs $<(x[n])^F, x_F[k(n)]>$ yield different $x[n]$'s; the same idea as in the proof of Theorem 1 shows that there are $g \in o(n)$ and $\mathcal{D}g \subset \mathcal{D}(f,X,B)$ such that $P_\infty^X(\mathcal{D}g) = 1$ and for almost all initial segments of each $x \in \mathcal{D}g$,

$$(45)\quad |\ell(Y_{X,B}(x[n]) - (\ell(Y_{X,B}^F((x[n])^F)) + \ell(Y_{X,B}(x_F[k(n)])))| < g(n) + \text{const}_2.$$

The last two relations yield

$$(46)\quad K_U(x[n] \mid n^*) \leq \ell(Y_{X,B}^F((x[n])^F)) + K[\ell(Y_{X,B}(x[n])) - \ell(Y_{X,B}^F((x[n])^F)] +$$

$$+ g(n) + \text{const}_2 = K\ell(Y_{X,B}(x[n])) + (1-K)\,\ell(Y_{X,B}^F((x[n])^F)) + g(n) + \text{const}_2 \leq$$

$$\leq K\log_r(\text{card } B_n) + (1-K)\log_r(\text{card } B_n^F) + g(n) + \text{const}_3.$$

Due to (43)

$$(47)\qquad \log_r(\text{card } B_n^F) < \log_r(\text{card } B_n) + n\log\alpha + \text{const}_4,$$

and (46) yields, as $\log_r(\text{card } B_n) < n + 1$,

$$(48)\quad K_U(x[n] \mid n^*) \leq K\log_r(\text{card } B_n) + (1-K)[\log_r \text{card } B_n + n\log_r\alpha] + g(n) + \text{const}_5$$
$$\leq K_1 \log_r(\text{card } B_n) + \text{const}_6$$

for some $K_1 \in (1-(1-K)\log_r\alpha^{-1}, 1)$. Setting

(49) $\mathcal{E}_{K_2} = \bigcup_{n=1}^{\infty} \bigcap_{m=n}^{\infty} \{x \in A^{\infty}: x[m] \in B_m,\ \ell(Y_{X,B}(x[m])) \leq$
$\leq K_2 \log_r(\text{card } B_m)$,

we can prove, using the same idea as above, that $P_{\infty}^{X}(\mathcal{E}_{K_2}) = 0$ for each $K_2 < 1$. Hence, for all x from a set of P_{∞}^{X}-measure one, (48) contradicts the hypothetical <f',X,B>-randomness of x for each $f' \in o(n)$. The assertion is proved.

<u>Theorem 4.</u> Let X be as in Theorem 1, let $B^1, B^2 \in \mathcal{B}(X)$ satisfy these conditions: (i) X is B^1- and B^2 -uniform, (ii) $\lim_{n\to\infty} \text{card } B_n^1 = \lim_{n\to\infty} \text{card } B_n^2 = \infty$, (iii) there exist K, $n_0 \in N$ such that $n^{-K} \leq (\text{card } B_n^1)(\text{card } B_n^2)^{-1} \leq n^K$ for all $n \geq n_0$. Let $f(n) \in o(n)$, then there exists $f'(n) \in o(n)$ such that $P_{\infty}^{X}(\mathcal{D}(f',X,B^2)/\mathcal{D}(f,X,B^1)) = 1$ supposing that this conditional probability is defined. ⌑

<u>Proof.</u> Immediately from (iii),

(50) $|\log_r(\text{card } B_n^1) - \log_r(\text{card } B_n^2)| \leq K \log_r n + \text{const}_1$.

Let $\ell(Y_{X,B}(x[n])) \leq K \log_r(\text{card } B_n)$ for $B \in \mathcal{B}(X)$, $K < 1$ and $n \geq n_0$. $Y_{X,B}$ is 1-1 mapping, so the number of n-tuples satisfying this inequality does not exceed

(51) $$\sum_{i=0}^{L} r^i = (r-1)^{-1}(r^L-1) < (\text{card } B_n)^K$$

for $L = K \log_r(\text{card } B_n)$. If X is B-uniform, then

(52) $P_n^X(\{x[n]: \ell(Y_{X,B}(x[n])) \leq K \log_r(\text{card } B_n)\}) \leq (1+\varepsilon)(\text{card } B_n)^{K-1}$,

and this tends to 0 for $n\to\infty$ and $\text{card } B_n \to \infty$. Hence,

(53) $$|\ell(Y_{X,B}(x[n])) - \log_r(\text{card } B_n)| \in o(n)$$

with the P_{∞}^{X}-probability 1 for $B = B^1, B^2$. (50) then yields that

(54) $$|\ell(Y_{X,B^1}(x[n])) - \ell(Y_{X,B^2}(x[n]))| \in o(n),$$

again with the probability 1. The assertion is proved. ⌑

5. TIME- AND SPACE-RESTRICTED APPROXIMATIONS OF PSEUDO-RANDOMNESS

The general non-recursivity of algorithmic complexity $K_U(x|S)$ taken as a function of (Gödel numbers of) its arguments, and the resulting computational non-effectivity of pseudo-random sequences of various kinds introduced above, demonstrates, from the theoretical point of view, the limited abilities of mathematical formalisms when dealing with the notion of randomness and proves these limitations to be"Gödel type" principial ones. However, from a more practical viewpoint this situation may be taken as a weak point of the conception explained above and it is why we shall briefly mention, in this chapter,an approximation which may serve as a partial solution or remedy.

In Kolmogorov (1965) the notion of algorithmic complexity is defined not only for universal Turing machines, but for partially recursive functions $\Psi: A^* \to A^*$ in general, defining $K_\Psi(x|S)$ as the length of the shortest $p \in A^*$ such that, when Ψ applied to the concatenation $p * S$ as its argument, $\Psi(p*S)$ is defined and equals x, again with the convention that $K_\Psi(x|S) = \infty$ if no such p exists. Evidently, for some functions Ψ with weaker computational abilities than those of universal Turing machines, $K_\Psi(x|S)$ may be a recursive function. Here we shall apply this idea to a special and intuitively more transparent case. Suppose that each universal Turing machine U contains an oracle which counts the number of steps (applications of particular operations) and the number of boxes on tape (or tapes) which were empty in the initial state and which have been used, at least once, during the computation. This oracle terminates the work of U if at least one of these two numbers exceeds an a priori given treshold value and the instantaneous sequence on the (output) tape defines the result of the computation. So we write, for $p, x \in A^*$ and $m, n \in N$, $U(p,\langle m,n\rangle) = x$ iff U, applied to p, yields x and needs at most m steps and n initially empty boxes. The restricted algorithmic complexity $K_U^*(x|S,\langle m,n\rangle)$ is then defined by

$$(55) \qquad K_U^*(x|S,\langle m,n\rangle) = \min\{k: k \in N, (\exists p \in A^k)(U(p*S,\langle m,n\rangle) = x)\}$$

with $\min \emptyset = \infty$ again. Evidently, for each x, $S \in A^*$, $m, n, m', n' \in N$, $m' \geq m$, $n' \geq n$,

$$(56) \qquad K_U^*(x|S,\langle m,n\rangle) \geq K_U^*(x|S,\langle m',n'\rangle) \geq K_U^*(x|S,\langle\infty,\infty\rangle) = K_U(x|S),$$

as with m, n increasing some shorter potential candidates for programs giving x may occur. The following more general assertion holds.

<u>Theorem 5.</u> There exist functions $F, G: N \times A^* \to N$ such that, for all x, $S \in A^*$,

(57) $K_U(x|S) = K_U^*(x|S,\langle F(\ell(x),S),G(\ell(x),S)\rangle)$. ∎

Proof. Given p, S, $x \in A^*$, set m(p,S,x) to be the number of steps executed by U if U applied to p * S and U(p*S) = x , set $m(p,S,x) = \infty$ elsewhere. Similarly, set n(p,S,x) to be the number of initially empty boxes occupied if U works over p * S and U(p*S) = x, set $n(p,S,x) = \infty$ elsewhere. Set

(58) $m(S,x) = \min\{m(p,S,x): p \in \pi(x,S,U), \ell(p) = K_U(x|S)\}$,

(59) $n(S,x) = \min\{n(p,S,x): p \in \pi(x,S,U), \ell(p) = K_U(x|S)\}$,

cf. (4) for the definition of $\pi(x,S,U)$. Evidently, $m(S,x) < \infty$ for all x, $S \in A^*$. Finally, for each $k \in N$ take

(60) $F(k,S) = \max\{m(S,x): x \in A^k\}$, $G(k,S) = \max\{n(S,x): x \in A^k\}$.

Hence, within the time and space limitations given by F(k,S) and G(k,S) the shortest S-aided programs for all k-tuples from A^k can be executed, so (57) holds.

This result is still of rather theoretical sense, as the functions F and G above cannot be, in general, recursive (if they were, the halting problem for universal Turing machines would be positively solved). Nevertheless, there exist recursive functions F, G: $N \to N$ such that the corresponding modifications of pseudo-random sequences conserve certain characteristic and desirable limit properties of the original pseudo-random sequences. Let F, G, f: $N \to N$ be three total functions, a sequence $x \in A^\infty$ is called <u><F,G>-f-pseudo-random</u>, if it satisfies the definition of f-pseudo-randomness with $K_U(x[n]|n^*)$ replaced by $K_U^*(x[n]|n^*,\langle F(n),G(n)\rangle)$ and analogously for <u><F,G>-f-PIID-sequences</u> and <u><F,G>-f-PM-sequences</u>.

Theorem 6. (a) For each $k \in N$ there exist recursive functions $F_{1,k}, G_{1,k}: N \to N$ such that for each $f(n) \in o(n)$ and each $\langle F_{1,k},G_{1,k}\rangle$-f-pseudo-random sequence $x \in A^\infty$ the relative frequency of the blocks of the length at most k in x tends to the uniform distribution, i.e., for each $j \le k$, $\alpha \in A^j$,

(61) $\lim_{n\to\infty} fr(\alpha,B(x,j)[n]) = r^{-j}$,

where r = cardA, $B(x,j) = \langle\langle x_1,x_2,\dots,x_j\rangle,\langle x_{j+1},\dots,x_{2j}\rangle, \langle x_{2j+1},\dots,x_{3j}\rangle,\dots\rangle \in (A^j)^\infty$, and

(62) $fr(\alpha, B(x,j)[n]) = n^{-1}$ card $\{k: k \le n, \langle x_{(k-1)j+1}, \ldots, x_{kj} \rangle = \alpha\}$.

(b) There exist recursive functions F_2, G_2: $N \to N$ such that for each $f(n) \in o(n)$ and each $\langle F_2, G_2 \rangle$-f-PIID sequence x and each $j \le r$ the limit value $\lim_{n\to\infty} fr(a_j, x[n]) = r_j$ exists and $\sum_{j=1}^{r} r_j = 1$.
(c) There exist recursive functions F_3, G_3: $N \to N$ such that for each $f(n) \in o(n)$ and each $\langle F_3, G_3 \rangle$-f-PM sequence $x \in A^\infty$ the limit values $\lim_{n\to\infty} fr(a_j, x^i[n]) = r_{ij}$ exist for each $j \le r$ and each $i \le r$ such that the subsequence x^i of immediate successors of a_i in x is infinite, moreover, setting $r_{ij} = 0$ elsewhere, $\sum_{j=1}^{r} r_{ij} = 1$ for each $i \le r$. ⌑

<u>Proof</u> (a sketch of its basic idea). The proof uses the fact that if a sequence $x \in A^\infty$ does not possess a limit property mentioned in the assertion,there always exists a particular program (or "meta-program" dependent on the length of a segment) which enables to generate initial segments of x on the ground of a description which is substantially (i. e. by a multiplicative constant smaller than one and independent of the length of the segment) shorter than the "almost optimum" description allowed by the supposed kind of pseudo-randomness. Time and space complexity, in the sense defined above, of such a program are recursive functions of the segment in question, so the maximum time and space complexity, taken over all finite sequences of the same length and considered as functions of this length are recursive functions as well. Defining F_i and G_i by these maximum values we allow the "counter-example" programs to play their role in the sense that there are explicitly avoided if a sequence satisfies the corresponding definition of $\langle F_i, G_i \rangle$-f-pseudo--randomness. Take the most simple case as an illustration. Given $x \in A^*$, denote by $m_{k'}(x)$, $n_{k'}(x)$, $k' \in N$, the time and space complexities of the program which takes x as $B(x,k')$ and generates it given the frequencies $n_1(x), \ldots, n_{r^{k'}}(x)$ of letters from $A^{k'}$ in $B(x,k')$ and the order number of $B(x,k')$ w.r. to the lexicographical ordering of all sequences of k'-tuples with the same length and frequencies of letters as $B(x,k')$ (we omit the technical details in case $\ell(x)$ is not divisible by k'). Evidently, due to the Church-Turing thesis, $m_{k'}(x)$ and $n_{k'}(x)$ are recursive functions of x and k', hence, given s, $k \in N$ and setting

(63) $F_{1,k}(s) = \max\{\max\{m_{k'}(x): k' \leq k\}: x \in A^s\}$,

(64) $G_{1,k}(s) = \max\{\max\{n_{k'}(x): k' \leq k\}: x \in A^s\}$,

$F_{1,k}$, $G_{1,k}$ are recursive functions taking N into N and we can easily prove that (62) holds. The constructions in cases (b) and (c) of the assertion are similar. ◻

In spite of the analogies introduced in Theorem 6, time- and space-restricted pseudo-random sequences represent substantial weakening of their original versions supposing the time and space restrictions are recursive functions of the lengths of the sequences in question. E. g., for each pair <F,G> of recursive functions, the class of increasing recursive functions H: $N \to N$, such that the subsequence $x_H = \langle x_{H(1)}, x_{H(2)}, \dots \rangle$ of an <F,G>-f-pseudo-random sequence $x = \langle x_1, x_2, \dots \rangle$ A^∞ is again <F,G>-f-pseudo-random, is a proper subclass of the class of those increasing recursive functions H' for which $x_{H'}$ is f-pseudo-random supposing that x is f-pseudo--random.

The notion of <F,G>-<f,X,B>-pseudo-random sequence can be defined similarly as in the particular cases mentioned above, i.e. by a priori avoiding from consideration the programs the execution of which exceeds the given treshold values. However, the more detailed investigation of such time- and space-restricted general pseudo-random sequences is more complicated than in the introduced particular cases and will be postponed to another paper.

REFERENCES

Davis M. (1958): Computability and Unsolvability. McGraw-Hill Book Comp., New York.

Fine T. L. (1973): Theories of Probability - an Examination of Foundations. Academic Press, New York - London.

Kolmogorov A. N. (1965): Three Approaches to the Quantitative Definition of Information. Problemy peredači informacii vol. 1, no. 1, pp. 4-7.

Kramosil I. (1983): Monte-Carlo Methods from the Point of View of Algorithmic Complexity. In: Trans. of the 9-th

Prague Conference on Information Theory,..., 1982. Academia, Prague, pp. 39-51.

(1984): Recursive Classification of Pseudo-Random Sequences. Kybernetika, vol. 20 - supplement, pp. 1-34.

(1985): Pseudo-markovské posloupnosti (Pseudo-Markov Sequences - in Czech). Res. Rep. no. 1328, Inst. of Inf. Theory and Automation.

(1986a): Independent and Identically Distributed Pseudo-Random Samples. To appear in Kybernetika (Prague).

(1986b): On Some Types of Pseudo-Random Sequences. To appear in MFCS 86 Proceedings.

Kramosil I., Šindelář J. (1984): Infinite Pseudo-Random Sequences of High Algorithmic Complexity. Kybernetika, vol. 20, no. 6, pp. 429-437.

Martin-Löf P. (1966): The Definition of Random Sequences. Information and Control, vol. 9, no. 3, pp. 602--619.

von Mises R. (1919): Grundlangen der Wahrscheinlichkeitsrechnung. Math.Zeitschrift vol. 5, pp. 52-99.

Rogers H. (1967): Theory of Recursive Functions and Effective Computability. McGraw-Hill Book Comp., New York.

Schnorr C.-P. (1971): Zufälligskeit und Wahrscheinlichkeit. Lect. Notes in Math. 218, Springer-Verlag, Berlin--Heidelberg-New York.

Institute of Information Theory and Automation
Czechoslovak Academy of Sciences
Pod vodárenskou věží 4
182 08 Prague 8, Czechoslovakia

ON TRANSIENT PHENOMENA IN SELF-OPTIMIZING CONTROL SYSTEMS

Petr Mandl

Prague

Key words: Stochastic control, arcsine law

ABSTRACT

The asymptotic behaviour of the quadratic cost in self-optimizing linear systems is investigated from the view point of the arcsine law.

1. PROBLEM FORMULATION

Papers Mandl (1986), (1987) deal with n-dimensional linear controlled systems the trajectories of which fulfil the differential equation

(1) $$dX_t = fX_t\, dt + gU_t\, dt + dW_t, \qquad t \geq 0, \quad X_0 = x,$$

and to which the cost function

$$C_T = \int_0^T (X_t' c X_t + |U_t|^2)\, dt, \qquad T \geq 0,$$

is associated. $U = \{U_t, t \geq 0\}$ is the control input, $W = \{W_t, t \geq 0\}$ the n-dimensional Wiener process. f,g,c are matrices, c nonzero nonnegatively definite. Prime denotes the transposition.

The stabilizability of the pairs (f,g) and $(f', \sqrt{c})$ is assumed to guarantee that the steady state matrix Riccati equation

(2) $$wf + f'w - wgg'w + c = 0$$

has a unique nonnegatively definite solution see e.g. Kučera (1973). The optimal stationary control is then

(3) $$U_t = kX_t , \qquad t \geqq 0 ,$$

where

$$k = -g'w .$$

(3) yields the minimal average cost

$$\theta = \text{trace } w .$$

Let the matrices f,g,c depend in a continuously differentiable way on a parameter α the true value α_0 of which is unknown to the controller. Let $A \subset R^q$ denote the range of α. Write $f[\alpha]$, $g[\alpha]$, $c[\alpha]$, $w[\alpha]$, $k[\alpha]$, $\theta[\alpha]$. The controller computes an estimate α_t^* of α_0 from the observation of $\{X_s, s \leqq t\}$, and applying the certainty equivalence principle to (3) it uses the control

(4) $$U_t = k[\alpha_t^*] X_t , \qquad t \geqq 0 .$$

(4) is self-optimizing, if the estimate is strongly consistent, i.e.

(5) $$\lim_{t\to\infty} \alpha_t^* = \alpha_0 \quad \text{a.s.}$$

A transient phenomenon in the system is modelled by letting the parameter α depend on time,

$$\alpha = \beta(t), \qquad t \geqq 0,$$

with convergence to the limit value

(6) $$\lim_{t\to\infty} \beta(t) = \alpha_0 .$$

Set

$$f(t) = f[\beta(t)], \qquad k(t) = k[\beta(t)] \qquad \text{etc.}$$

Instead of (1) the nonautonomous equation

(7) $$dX_t = f(t)X_t dt + g(t) U_t dt + dW_t, \qquad t \geq 0, \quad X_0 = x,$$

holds, and the cost is

(8) $$C_T = \int_0^T (X_t' c(t) X_t + |U_t|^2) dt, \qquad T \geq 0.$$

The integrated minimal average cost

(9) $$\int_0^t \theta(s) ds, \qquad t \geq 0,$$

which becomes the product θt in the autonomous case, gives a curve around which C_t should fluctuate under good controls. The proportion of time spent by C_t above (9), namely

$$B_T = \frac{1}{T} \int_0^T \chi\{ C_t > \int_0^t \theta(s) ds\} dt,$$

measures the quality of the control in the sense made precise in the subsequent proposition.

Assume that $\beta(t)$ has a continuous derivative $\dot{\beta}(t)$. Propositions 2,3 of Mandl (1987) can be extended to the case of a transient phenomenon as follows. p lim refers to the convergence in probability.

<u>Proposition 1.</u> Let for a $\delta \geq 0$

$$\lim_{T\to\infty} \int_0^T |\dot{\beta}(t)| dt / T^{1/2-\delta} = 0.$$

Let U be any control under which

(10) $$E\chi_D |X_t|^2 \leq \text{const.} (1 + t^{\delta}), \qquad t \geq 0,$$

holds, where $P(D)$ can be made arbitrarily close to 1. Then

(11) $$\limsup_{T\to\infty} P(B_T \leq y) \leq \frac{2}{\pi} \arcsin \sqrt{y}, \qquad y \in [0,1].$$

If in addition to (10)

(12) $$\operatorname{p\,lim}_{T\to\infty} \frac{1}{\sqrt{T}} \int_0^T |U_t - k(t)X_t|^2 \, dt = 0 ,$$

then

(13) $$\lim_{T\to\infty} P(B_T \leq y) = \frac{2}{\pi} \arcsin \sqrt{y} , \qquad y \in [0,1] .$$

Under the hypotheses of Propo sition 1 the arcsine law applies to B_T for good controls. (11) states that the limiting arcsine distribution cannot be improved in the sense of stochastic ordering of probability distributions.

In this paper we present sufficient conditions for the validity of (10), (12), and consequently of (13), for controls having the form (4). Controlled Markov chains under transient phenomena were dealt with in Mandl and Hübner (1985).

2. CONDITIONS FOR OPTIMALITY

In this section we assume (6), (7) , (8) , and consider the controls satisfying (4), (5). $\{\alpha_t^*, t \geq 0\}$ is supposed to have piecewise continuous trajectories. We set

(14) $$K_t = k[\alpha_t^*] , \quad t \geq 0; \quad U_t = K_t X_t , \quad t \geq 0 .$$

Our aim is to prove (10), and to establish a sufficient condition for (12). We use $a > 0$ to express that matrix a is symmetric positively definite, $a \geq b$ to express that $a - b$ is nonegative definite. I denotes the unit matrix.

A matrix $q > 0$ can be found such that

(15) $$q(f(t) + g(t)\, k[\alpha]) + (f(t) + g(t)k[\alpha])' q + I \leq 0$$

holds for t sufficiently large and for α from a neighbourhood of α_0 in A, say

(16) $$t \geq T_0 , \qquad \alpha \in \omega .$$

To see this let $q > 0$ be the solution of

$$(17)\quad q(f[\alpha_0] + g[\alpha_0]\, k[\alpha_0]) + (f[\alpha_0] + g[\alpha_0] k[\alpha_0])' q + 2I = 0.$$

The theorem which we applied to (2) yields the existence of $q \geq 0$ fulfilling (17), and in fact $q > 0$. With regard to (6) and to the continuity of $k[\alpha]$ we obtain (15), (16).

Lemma 1. Let (4), (5) hold. Then for arbitrary $n > 0$, $\varepsilon > 0$,

$$(18)\quad \limsup_{T\to\infty} T^{-1} E\chi_D \int_0^T |X_t|^{2n} dt \leq \text{const.},$$

$$(19)\quad \lim_{T\to\infty} T^{-\varepsilon} E\, \chi_D |X_T|^{2n} = 0$$

hold with $P(D)$ arbitrarily close to 1. The constant in (18) does not depend on U or on the initial position $X_0 = x$.

Proof. For $S > 0$ let $\tau(S)$ be the time of the first exit of $\{\alpha_t,\ t \geq 0\}$ from ω after S. Let $\bar{X} = \{\bar{X}_t,\ t \geq 0\}$ be the solution of (7) corresponding to the control

$$\bar{U}_t = U_t, \qquad 0 \leq t < \tau(S), \quad \bar{U}_t = k[\alpha_0]\, X_t,\ t \geq \tau(S).$$

Introduce the events

$$D_0 = \{ |X_t| \leq M,\ t \in [0,S]\},\quad D_1 = \{\bar{U}_t = U_t,\ t \geq S\}, D = D_0 \cup D_1.$$

By choosing S and then M large enough $P(D)$ can be made arbitrarily near to 1.

Applying (15) and the Itô formula for $n = 1,2,\ldots$ we obtain

$$E\chi_{D_0}(\bar{X}_T' q \bar{X}_T - \bar{X}_S' q \bar{X}_S) = E\chi_{D_0}\Big(n \int_S^T (\bar{X}'q\bar{X})^{n-1} 2\bar{X}'q(f(t) + g(t)K)\bar{X}\, dt \;+$$

(20)

$$+\; n \operatorname{trace} q \int_S^T (\bar{X}'q\bar{X})^{n-1} dt + 2n(n-1)\int_S^T (\bar{X}'q\bar{X})^{n-2} |q\bar{X}|^2 dt\Big) \leq$$

$$\leq -nE\chi_{D_0} \int_S^T (\bar{X}'q\bar{X})^{n-1} |\bar{X}|^2 dt + n \operatorname{trace} q\, E\chi_{D_0} \int_S^T (\bar{X}'q\bar{X})^{n-1} dt \;+$$

$$+ \quad 2n(n-1)\, E\chi_{D_0} \int_S^T (\bar{X}\, q\bar{X})^{n-2} \quad |q\bar{X}|^2 \, dt \, .$$

From here successively for $n = 1,2,\ldots$ follows

(21) $$\limsup_{T\to\infty} \quad T^{-1}\, E\chi_{D_0} \int_S^T |\bar{X}_t|^{2n}\, dt \leqq \text{const.}$$

Noting that D_1 implies $\bar{X}_t = X_t$, $t \geqq 0$, we get form here

(22) $$\limsup_{T\to\infty} \quad T^{-1}\, E\chi_D \int_S^T |X_t|^{2n}\, dt \leqq \text{const.}$$

In (22) we can replace S by 0 to obtain (18).

Using (21) to estimate the right-hand side of (20) we get

$$\limsup_{T\to\infty} \quad T^{-1}\, E\chi_{D_0} |X_T|^{2n} \leqq \text{const.},$$

and hence

(23) $$\limsup_{T\to\infty} \quad T^{-1}\, E\chi_D |X_T|^{2n} \leqq \text{const.}$$

To prove (19) take m such that $0 < 1/m < \varepsilon$, and apply (23) to

$$\frac{E\chi_D |X_T|^{2n}}{T^{1/m}} \leqq \left(\frac{E\chi_D |X_T|^{2nm}}{T} \right)^{1/m} . \quad \square$$

Note that (19) implies (10) with $\delta > 0$ arbitrary.

<u>Proposition 2.</u> Let (4), (5) hold. Then (10) and (12) are valid, if for an $\varepsilon > 0$

(24) $$\underset{T\to\infty}{\text{p lim}} \int_0^T |K_t - k(t)|^2\, dt / T^{1/2-\varepsilon} = 0 \, .$$

Proof. The validity of (10) follows from Lemma 1. To show that (24) implies (12) we have for $m > 1$

$$\frac{1}{\sqrt{T}} \int_0^T |(K_t - k(t))\, X_t|^2\, dt \leq \frac{1}{\sqrt{T}} \int_0^T |K_t - k(t)|^2\, |X_t|^2\, dt \leq$$

$$\leq \frac{1}{\sqrt{T}} \left(\int_0^T |K_t - k(t)|^{2\frac{m}{m-1}}\, dt\right)^{\frac{m-1}{m}} \left(\int_0^T |X_t|^{2m}\, dt\right)^{\frac{1}{m}} \leq$$

$$\leq \left(\int_0^T |K_t-k(t)|^2 dt/T^{\frac{1}{2}(1-\frac{1}{m-1})}\right)^{\frac{m-1}{m}} \left(\sup_{t\geq 0} |K_t - k(t)|^{\frac{2}{m}}\right) .$$

$$\left(\int_0^T |X_t|^{2m}\, dt \,/\, T\right)^{\frac{1}{m}}$$

The last two terms are bounded in probability in virtue of (5), (14), (18) while the next term tends to 0 if $1/(m-1) < 2\varepsilon$. This yields (12). □

<u>Corollary 1.</u> Let

(25) $\beta(t) - \alpha_0 = \mathcal{O}(t^{-\frac{1}{4}-\delta}), \quad \alpha_t^* - \alpha_0 = \mathcal{O}(t^{-\frac{1}{4}-\delta})$ a.s.,

as $t \to \infty$. Then (12) is fulfilled.

Namely, from (25) and from the continuity of $k[\alpha]$ follows

$$|K_t - k(t)|^2 \leq 2|k[\alpha_t^*] - k[\alpha_0]|^2 + 2\,|k[\beta(t)] - k[\alpha_0]|^2 =$$
$$= \mathcal{O}(t^{-\frac{1}{2}-2\delta}), \qquad t \to \infty \quad \text{a.s.}$$

Integrating from 0 to T we get from here (24).

3. EXAMPLE

Let

$$f[\alpha] = f_0 + \alpha f_1 ,$$

where α is a one-dimensional parameter, and let g and c be independent of α. The controller, assuming $\beta(t) \equiv \alpha_0$, constructs its maximum likelihood estimate α_T^* of α_0 as follows. The logarithmic likelihood function is

$$L_T(\alpha) = \int_0^T (f[\alpha]X_t + gU_t)'dX_t - \frac{1}{2}\int_0^T |f[\alpha]X_t + gU_t|^2\, dt \quad .$$

L_T is the logarithm of the probability density of $\{X_t, t \in [0,T]\}$ with respect to the distribution of $\{x + W_t, t \in [0,T]\}$. Solving

$$\frac{d}{d\alpha} L_T(\alpha_T^*) = 0$$

he gets

$$\alpha_T^* = (\int_0^T X'f_1'dX - \int_0^T X'f_1' (f_0X + gU)\, dt)/Q_T$$

where

$$Q_T = \int_0^T |f_1X|^2\, dt \quad .$$

A treatment of maximum likelihood estimates of a multidimensional parameter is in Duncan and Pasik-Duncan (1985). Inserting from (7) for dX_t one obtains

$$(26) \qquad \alpha_T^* = \int_0^T \beta(t)|f_1X|^2\, dt/Q_T + \int_0^T X'f_1'dW/Q_T$$

Assume the validity of

$$(27) \qquad \beta(t) - \alpha_0 = O(t^{-\frac{1}{4}-\delta}) , \qquad t \to \infty ,$$

$$(28) \qquad Q_T \asymp T , \qquad T \to \infty , \qquad \text{a.s.}$$

Take $0 < \delta < \frac{1}{4}$ in (27).

We shall show that (27), (28) imply

$$(29) \qquad \alpha_T^{*} - \alpha_0 = \mathcal{O}(T^{-\frac{1}{4}-\delta}), \qquad T \to \infty, \qquad \text{a.s.}$$

To this purpose express the last term in (26) by means of a random time change in a Wiener process

$$\int_0^T X' f_1' \, dW/Q_T = \mathcal{W}_{Q_T}/Q_T \quad .$$

Applying (28) and the law of the iterated logarithm for the Wiener process we get

$$\mathcal{W}_{Q_T}/Q_T = \mathcal{O}\left(\sqrt{\frac{\log \log T}{T}}\right), \qquad T \to \infty, \qquad \text{a.s.}$$

Further we have the estimate

$$\left| \alpha_0 - \int_1^T \beta(t) |f_1 X|^2 \, dt/Q_T \right| \leq \text{const.} \int_1^T t^{-\frac{1}{4}-\delta} |f_1 X|^2 \, dt/T =$$

$$= \mathcal{O}(T^{-\frac{1}{4}-\delta}), \qquad T \to \infty, \qquad \text{a.s.}$$

Consequently, from (26) follows (29), and Corollary 1 applies.

REFERENCES

Duncan T.E. and Pasik-Duncan B. (1985): Adaptive control of continuous time linear systems. Preprint, University of Kansas.

Kučera V. (1973): A review of the matrix Riccati equation. Kybernetika (Prague) 9, No. 1, 42-61.

Mandl P. and Hübner G. (1985): Transient phenomena and self-optimizing control of Markov chains. Acta Univ. Carolinae - Mathem. et Phys. 26, No. 1, 33-49.

Mandl P. (1986): Asymptotic ordering of probability distributions for linear controlled systems with quadratic cost. In: Stochastic Differential Systems, Lecture Notes in Control and Inf. Sc. 78, Springer-Verlag, 277-283.

(1987): Limit theorems of probability theory and optimality in linear controlled systems with quadratic cost. To appear in Proc. of 5 th IFIP Working Conference on Stochastic Differential Sytems, Eisenach 1986, Springer - Verlag.

Charles University
Department of Probability
and Math ematical Statistics
Sokolovská 83
186 00 Prague 8
Czechoslovakia

LOCALLY STATIONARY COVARIANCES

Jiří Michálek

Prague

Key words: spectral decomposition, weakly stationary process, locally stationary process, self-adjoint operator, normal operator

ABSTRACT

The contribution deals with a spectral decomposition of locally stationary processes and studies random processes having normal covariances which generalize the notion of weakly stationary covariance.

The notion of a locally stationary covariance is due to Silverman who introduced these in Silverman (1957). The author of this contribution generalized Silverman's results in Michálek (1986a) and studied asymptotic behavior of locally stationary processes in Michálek (1986b). Following Silverman we shall say that a covariance $R(s,t)$, $s, t \in R_1$ is locally stationary if we can write for every pair s, t

$$R(s,t) = R_1(\frac{s+t}{2})\ R_2(s-t)$$

where $R_1 \geq 0$ and R_2 is a stationary covariance. A random process $x(t)$, $t \in R_1$ is said to be locally stationary if its covariance function is locally stationary. When $R_1 = \text{const} > 0$ then we see that a stationary covariance is locally stationary. A further example of locally stationary covariance is given by the product

$$R_1(\frac{s+t}{2})\ \Delta(s-t)$$

where R_1 is any nonnegative function and $\Delta(s-t)$ is 1 when $s = t$ and 0 otherwise. The example just given shows that the product $R_1.R_2$ can be a covariance without R_1 being a covariance. If both R_1 and R_2 are covariances the product $R_1.R_2$ is automatically a covariance. That implies the product of a nonnegative covariance of the form $R_1(\frac{s+t}{2})$ with a stationary covariance $R_2(s-t)$ is a locally stationary covariance. Covariances of the form $R_1(\frac{s+t}{2})$ were studied by Loève (1948) who calls them exponentially convex covariances. As noted in Loève (1948) any bilatelar Laplace transform of a probability distribution function is an exponentially convex covariance; conversely, any continuous exponential convex covariance R_1 can be represented by the bilatelar Laplace transform

$$R_1(s+t) = \int_{-\infty}^{+\infty} e^{\lambda(s+t)}\, dF(\lambda).$$

This fact enables to introduce a large class of locally stationary covariances. Let R_1 be the Laplace transform of a probability distribution function converging for every real number and let R_2 be any stationary covariance. Then their product

$$R(s,t) = R_1(\tfrac{s+t}{2})\, R_2(s-t)$$

is a locally stationary covariance. This type of covariances will be called an exponentially convex locally stationary covariance.

ELEMENTARY PROPERTIES OF LOCALLY STATIONARY COVARIANCES

From the definition of a locally stationary covariance immediately follows that in the case $R_1(0) = R_2(0) = 1$

$$R(s,s) = R_1(s); \quad R_2(s) = R(\tfrac{s}{2},-\tfrac{s}{2}).$$

This fact implies that the covariance function $R(s,t)$ is completely defined by its values on the diagonals $\{s=t\}$, $\{s=-t\}$ in the plane for

$$R(s,t) = R(\tfrac{s+t}{2}, \tfrac{s+t}{2})\, R(\tfrac{s-t}{2}, \tfrac{t-s}{2}). \tag{1}$$

By use of the new coordinates $\frac{s+t}{2} = u$, $\frac{s-t}{2} = v$ we can rewrite (1) in the form

$$R(u+v, u-v) = R(u,u).R(v,-v).$$

We see from (1) that the first part $R_1(\frac{s+t}{2})$ of R(s,t) is invariant with respect to any shift $S_\delta(s,t) = (s+\delta,\ t-\delta)$, $\delta \in R_1$, which is parallel to the diagonal $\{s=-t\}$, similarly, the second part $R_2(s-t)$ of R(s,t) is invariant with respect to any shift $M_\delta(s,t) = (s+\delta, t+\delta)$, $\delta \in R_1$ which is parallel to the diagonal $\{s=t\}$ in the plane.

EXPONENTIALLY CONVEX COVARIANCES

These are very important cases of locally stationary covariances. A covariance R(s,t) is exponentially convex if for every $s,\ t \in R_1$

$$R(s,t) = R_1(s+t).$$

It means that an exponentially convex covariance is invariant with respect to any shift S_δ described above. We see immediately that $R_1(s+t) \geq 0$ in every case because

$$R(s,t) = R_1(s+t) = R(\frac{s+t}{2}, \frac{s+t}{2}) \geq 0.$$

These facts lead us to describe a random process x(t), $t \in R_1$, having an exponentially convex covariance as *a symmetric* process for

$$E\{X_s \overline{X}_t\} = E\{X_{s+\delta} \overline{X_{t-\delta}}\} = E\{|X_{\frac{s+t}{2}}|^2\}$$

holds for every real δ. Conversely, every covariance function which is invariant with respect to every shift S_δ in the plane is exponentially convex. Under assumption of continuity every exponentially convex covariance defines a positive kernel $R_1(s+t)$ in the plane what means, see Widder (1946), the corresponding covariance function R(s,t) can be represented in the form

$$R(s,t) = \int_{-\infty}^{+\infty} e^{\lambda(s+t)}\, dF(\lambda) \tag{2}$$

where F is a probability distribution function. Thanks to the analycity of R_1 on the real line the transform (2) can be extended the complex plane, i. e. for complex s, t. The function R_1, as a bilatelar Laplace transform, is infinitely differentiable and $R_1(z)$, $z = x + iy$, is holomorphic in the whole complex plane for (2) is absolutely convergent for every real number. Any symmetric process is infinitely differentiable in the quadratic mean with

$$E\{X^{(n)}(s)\overline{X^{(m)}(t)}\} = \frac{\partial^{m+n}}{\partial s^n \partial t^m} R_1(s+t).$$

Analycity of R_1 on the real line insures analycity of X(t) in the quadratic mean, i. e.

$$X(s) = \sum_{n=0}^{\infty} \frac{(s-s_0)^n}{n!} X^{(n)}(s_0).$$

Without loss of generality we can consider $s_0 = 0$ and then define for arbitrary complex $z = x + iy$

$$X(z) = \underset{N\to\infty}{\text{l.i.m.}} \sum_{n=0}^{N} \frac{z^n}{n!} X^{(n)}(0) \tag{3}$$

with the covariance function

$$E\{X(z_1)\overline{X(z_2)}\} = R_1(z_1+\overline{z_2}) = \int_{-\infty}^{+\infty} e^{\lambda(z_1+\overline{z_2})}\, dF(\lambda) \tag{4}$$

where $\overline{z_2}$ is conjugate to z_2. The formula (4) and the Karhunen theorem enable us to express the process x(z), $z \in C$, in the form of a stochastic integral understood in the quadratic mean sense

$$x(z) = \int_{-\infty}^{+\infty} e^{z\lambda} d\xi(\lambda)$$

where $\xi(\lambda)$ is a second order random process with $E\{\xi(\lambda)\} = 0$ and with orthogonal incre

$$E\{d\xi(\lambda_1)\overline{d\xi(\lambda_2)}\} = dF(\min(\lambda_1,\lambda_2)).$$

Especially, any symmetric process x(s), $s \in R_1$, continuous in the quadratic mean can be represented as

$$x(s) = \int_{-\infty}^{+\infty} e^{s\lambda} d\xi(\lambda)$$

with the derivatives

$$x^{(n)}(s) = \int_{-\infty}^{+\infty} \lambda^n e^{s\lambda} d\xi(\lambda),\ n \geq 1.$$

INVERSION FORMULA FOR SYMMETRIC PROCESS

By use of the inverse formula for the Laplace transform we can express the probability distribution function F in (4) at its points of continuity by means of $R(s,t) = R_1(s+t)$ as

$$F(\lambda) = \lim_{T\to\infty} \frac{1}{2\pi i} \int_{c-iT}^{c+iT} \frac{e^{-\lambda z}}{z} R_1(z)dz$$

where $c > 0$ and $R_1(z) = \int_{-\infty}^{+\infty} e^{\lambda z} dF(\lambda)$.

This formula gives a hint for the inversion formula of symmetric processes.

Theorem 1. Let $x(s)$, $s \in R_1$, be a symmetric process continuous in the quadratic mean and hence expressible as

$$x(s) = \int_{-\infty}^{+\infty} e^{s\lambda} d\xi(\lambda).$$

Let λ be a point of continuity in the quadratic mean of $\xi(\lambda)$. Then

$$\xi(\lambda) = \underset{T\to\infty}{\mathrm{l.i.m.}} \frac{1}{2\pi i} \int_{c-iT}^{c+iT} \frac{e^{-\lambda z}}{z} x(z)dz$$

where $x(z)$ is the extension of $x(s)$ into the complex plane.

Remark. Before the proof of Theorem 1 we must mention a stochastic integral in the quadratic mean sense along a curve in the complex domain. Let $y(z)$ be a random process defined in a domain $\mathcal{O}$ in the complex plane with the covariance function R_y. Let K be a curve in $\mathcal{O}$. The stochastic integral in the quadratic mean sense of $y(z)$ along the curve K is defined as the limit in the quadratic mean of integral sums $\sum_{i=1}^{n} y(w_i)\Delta z_i$, $w_i \in K$, $z_i \in K$, $z_i < w_i < z_{i+1}$, $i = 1,2,\dots,n$ when the norm $\|\mathcal{D}\| = \max_{i=1,\dots,n} |\Delta z_i|$ of partition $\mathcal{D} = \{z_1,\dots,z_n\}$ tends to zero. The stochastic integral $\int_K y(z)dz$ exists if and only if the double integral $\int_K\int_K R_y(z_1,z_2)dz_1 d\bar{z}_2$ exists.

Proof of Theorem 1. For a fixed $T > 0$ the curve $K_T = \{z \in C: z = c+iv, -T\leq v\leq T, c>0\}$ is a one-to-one mapping of the interval $<-T,T>$ with finite variation. The covariance function $R_y(z_1,z_2)$ of $y(z) = \frac{e^{-\lambda z}x(z)}{z}$ on $K_T \times K_T$ is continuous and bounded; hence $\int_{K_T}\int_{K_T} R_y(z_1,z_2)dz_1 d\bar{z}_2$ exists. That means the stochastic integral $\int_{K_T} \frac{e^{-\lambda z}}{z} x(z)$ is well defined. The extension $x(z)$ in the complex plane belongs to the Hilbert space $L_2(x(.))$ generated by the values of the process $x(s)$, $s \in R_1$ with the scalar product $E\{\xi\bar{\eta}\} = <\xi,\eta>$. This fact immediately follows from (3).

We must prove that

$$\lim_{T\to\infty} E\{|\frac{1}{2\pi i}\int_{C_T}\frac{e^{-\lambda z}}{z}x(z)dz - \xi(\lambda)|^2\} = 0 \tag{5}$$

in the quadratic mean at every point of continuity of $\xi(\lambda)$. As $x(z) = \int_{-\infty}^{+\infty} e^{uz}d\xi(u)$, the change of order between the stochastic integral and the curve integral in $\int_{C_T}\frac{e^{-\lambda z}}{z}\left[\int_{-\infty}^{+\infty} e^{uz}d\xi(u)\right]dz$ would be suitable. The existence of the integrals $\int_{C_T}|\frac{e^{-\lambda z}}{z}|^2 dz$, $\int_{C_T}\int_{-\infty}^{+\infty}|e^{zu}|^2 dF(u)dz$ makes possible to change those integrals, hence almost surely $\int_{C_T}\frac{e^{-\lambda z}}{z}x(z)dz = \int_{-\infty}^{+\infty}\left[\int_{C_T}\frac{e^{-\lambda(z-u)}}{z}dz\right]d\xi(u)$, e. g. for details see Gichman, Skorochod (1965), Lemma 4, 5 in §3, Chapter V. Then the expected value in (5) can be written in the form

$$E\{|\frac{1}{2\pi i}\int_{-\infty}^{+\infty}\Big(\int_{C_T}\frac{e^{-z(\lambda-u)}}{z}dz-\psi_\lambda(u)\Big)d\xi(u)|^2\} = \int_{-\infty}^{+\infty}|F_T(\lambda,u)-\psi_\lambda(u)|^2 dF(u)$$

where $\psi_\lambda(u) = \begin{cases} 2\pi i & u > \lambda \\ \pi i & u = \lambda \\ 0 & u < \lambda \end{cases}$ and $F_T(\lambda,u) = \int_{C_T}\frac{e^{-z(\lambda-u)}}{z}dz$.

A simple computation gives

$$\lim_{T\to\infty} F_T(\lambda,u) = \lim_{T\to\infty} i\int_{-T}^{+T}\frac{e^{-(c+iv)(\lambda-u)}}{c+iv}dv = \psi_\lambda(u)$$

for every pair λ, u of reals. As the integral $\int_{-\infty}^{+\infty} e^{2cu}dF(u)$ is convergent there exists an integrable majorant function for the sequence $\{|F_T(\lambda,u)-\psi_\lambda(u)|^2\}_{T>0}$ and the Lebesgue theorem on dominated convergence finishes the proof. Q.E.D.

In the proof of Theorem 1 we mentioned the Hilbert space $L_2(x(.))$ generated by the values of the process x(s), $s \in R_1$. This space is the closure of the linear set $L(x(.))$ of all linear combinations $\xi = \sum_{i=1}^n \alpha_i x(s_i)$ with respect to the convergence in the quadratic mean. Let T_h be the shift-operator defined on $L(x(.))$ by $T_h\xi = \sum_{i=1}^n \alpha_i x(s_i+h)$, $h \in R_1$. As familiarly known, in case the process x(s), $s \in R_1$ is weakly stationary then the family $\{T_h\}_{h\in R_1}$ of operators forms a group of unitary operators in the space $L_2(x(.))$ and the famous Stone theorem, e. g. see Riesz, Nagy (1953), gives the spectral decomposition of the process $x(s)$, $s \in R_1$. We shall see that in the case of a symmetric process the corresponding family $\{T_h\}$ of the shift-operators forms a group of self-adjoint operators, in general unbounded. For these purposes let us consider a symmetric process $x(s)$, $s \in R_1$, which is continuous in the quadratic mean and hence its covariance function has the representation in the form of a bilatelar Laplace transform. The process $x(s)$, $s \in R_1$, can be expressed as the stochastic integral

$$x(s) = \int_{-\infty}^{+\infty} e^{s\lambda} d\xi(\lambda).$$

The inversion formula says that the corresponding Hilbert spaces $L_2(x(.))$ and $L_2(\xi(.))$ coincide. Let H_λ be the subspace of $L_2(\xi(.))$ generated by all $\xi(\mu)$, $\mu \le \lambda$ and let P_λ be the orthogonal projector in $L_2(\xi(.))$ generated by H_λ. That is simple to prove the family $\{P_\lambda\}_{\lambda \in R_1}$ of projectors forms a resolution of the identity in $L_2(\xi(.))$. Now, let us consider the operators $A_s = \int_{-\infty}^{+\infty} e^{s\lambda} dP_\lambda$ where $s \in R_1$. The operator A_s is defined on the subset $D(A_s) = \{\eta \in L_2(\xi(.)): \int_{-\infty}^{+\infty} e^{2s\lambda} d \langle P_\lambda \eta, \eta \rangle < +\infty\}$. As the process $\xi(\lambda)$, $\lambda \in R_1$ is a martingale in the quadratic mean and hence $P_\lambda\{\xi(+\infty)\} = \xi(\lambda)$ for every λ (the existence of $\xi(+\infty)$ is ensured by finite variation of the function $F(\lambda) = E|\xi(\lambda)|^2$), we immediately see $A_s\xi(+\infty) = x(s)$. Similarly, every $\xi(\mu)$ belongs to $D(A_s)$ because $\langle P_\lambda \xi(\mu), \xi(\mu) \rangle = F(\lambda)$ for $\lambda \le \mu$ and $\langle P_\lambda \xi(\mu), \xi(\mu) \rangle = F(\mu)$ for $\lambda > \mu$. Hence every operator A_s is defined on the linear set $L(\xi(.))$ which is everywhere dense in the space $L_2(\xi(.))$. Every $x(t)$, $t \in R_1$, belongs to $D(A_s)$ also because $dP_\lambda x(t) = e^{t\lambda} dP_\lambda \xi(+\infty) = e^{t\lambda} dP_\lambda x(0)$ if we put $\xi(-\infty) = 0$. This fact proves, at the same time, $A_s\, x(t) = x(t+s)$ and hence $A_s = T_s$ on $L(x(.))$. For every $s \in R_1$ the function $e^{s\lambda}$ is continuous and real-valued and hence the operator A_s is self-adjoint, in general unbounded. The properties $e^{s\lambda} = (s^\lambda)^s$ and $P_\lambda P_\mu = P_\mu P_\lambda = P_{\min(\lambda,\mu)}$ ensure that the family $\{A_s\}_{s \in R_1}$ forms a group of shift-operators on $L(x(.))$ because $A_0 = \int_{-\infty}^{+\infty} dP_\lambda = I$ on $L_2(\xi(.))$ and $A_{t+s}\eta = \int_{-\infty}^{+\infty} e^{(t+s)\lambda} dP_\lambda \eta =$

$$= \int_{-\infty}^{+\infty} e^{t\lambda} . e^{s\lambda} dP_\lambda \eta = \int_{-\infty}^{+\infty} e^{t\lambda} dP_\lambda \eta \int_{-\infty}^{+\infty} e^{s\lambda} dP_\lambda \eta \text{ for every } \eta \in L(x(.)).$$

Theorem 2. A process $x(s)$, $s \in R_1$, is symmetric if and only if the family of the shift-operators $\{T_h\}_{h \in R_1}$ in $L_2(x(.))$ forms a group of nonnegative symmetric operators on $L(x(.))$.

Proof. Let $x(s)$, $s \in R_1$, be symmetric, i. e. its covariance function $R(s,t)$ can be expressed as $R(s,t) = R_1(s+t)$ where R_1 is nonnegative. Let T_h be the shift-operator defined by $T_h\, x(s) = x(s+h)$ on the linear set $L(x(.))$ in $L_2(x(.))$. These operators form a group on $L(x(.))$ because $T_{h_1} T_{h_2}\, x(s) = T_{h_1}\, x(s+h_2) = x(s+h_1+h_2) = T_{h_1+h_2}\, x(s)$. Further, $E\{x(s)\overline{x(t+h)}\} = \langle x(s), T_h x(t)\rangle = R_1(s+t+h) = E\{x(s+h)\overline{x(t)}\} = \langle T_h x(s), x(t)\rangle$ and hence every operator T_h is symmetric. Because $\langle x(s), T_h x(s)\rangle = E\{|x(s+\frac{h}{2})|^2\} = \langle T_{h/2}x(s), T_{h/2}x(s)\rangle = \|T_{h/2}x(s)\|^2 \geq 0$, T_h is nonnegative.
Conversely, let $\{T_h\}$ corresponding to the process $x(s)$, $s \in R_1$, be symmetric and nonnegative; hence $E\{x(s)\overline{T_h x(t)}\} = E\{T_h x(s)\overline{x(t)}\}$. As $x(s) = T_s x(0)$, $x(t) = T_t x(0)$ then $R(s,t) = \langle T_s x(0), T_t x(0)\rangle = \langle T_t T_s x(0), x(0)\rangle = \langle T_{s+t} x(0), x(0)\rangle = R(s+t,0) = R_1(s+t)$. The function R_1 is nonnegative because $\{T_h\}_{h \in R_1}$ are nonnegative. Q.E.D.

Remark. If the covariance function $R(s,t)$ of $x(s)$, $s \in R_1$, is continuous then, as proved above, there exists an extension of every T_h into a self-adjoint operator in $L_2(x(.))$.

Theorem 3. Let $x(s)$, $s \in R_1$ be a symmetric process continuous in the quadratic mean. Then the corresponding group of the symmetric operators $\{T_h\}_{h \in R}$ is a group of bounded operators if and only if at least one pair (T_h, T_{-h}), $h > 0$, is bounded.

Proof. In one direction the proof is trivial. Let us assume that T_1, T_{-1} are bounded. As for every $\xi \in L(x(.))$ ($L(x(.))$ is everywhere dense in $L_2(x(.))$,

$$\|T_{\frac{1}{2}}\xi\|^2 = \langle T_{\frac{1}{2}}\xi, T_{\frac{1}{2}}\xi\rangle = \langle T_1\xi, \xi\rangle \leq \|T_1\xi\|\,\|\xi\| \leq \|T_1\|.\|\xi\|^2$$

then $\sup_{\xi \neq 0} \frac{\|T_{\frac{1}{2}}\xi\|}{\|\xi\|} \leq \|T_1\|^{\frac{1}{2}}$ and hence $T_{\frac{1}{2}}$ must be bounded, too.
Further, $\|T_1\xi\| = \|T_{\frac{1}{2}}T_{\frac{1}{2}}\xi\| \leq \|T_{\frac{1}{2}}\|^2\|\xi\|$ what implies, $\|T_1\| \leq \|T_{\frac{1}{2}}\|^2$. We obtain together $\|T_1\| = \|T_{\frac{1}{2}}\|^2$. In a similar way we can prove $\|T_1\| = \|T_{\frac{1}{4}}\|^4 = \|T_{\frac{1}{8}}\|^8 = \ldots = \|T_{\frac{1}{2^n}}\|^{2^n}$ for every natural n. Let h be any number between 0 and 1. Let $h = \sum_{i=0}^{\infty} \frac{\alpha_i}{2^i}$ in the dyadic expan-

sion and let us consider the sequence of the operators

$T_{(n)} = T_{\sum_{i=0}^{n} \frac{\alpha_i}{2^i}}$. As $T_{\sum_{i=1}^{n} \frac{\alpha_i}{2^i}} = T_{\frac{\alpha_n}{2^n}} T_{\frac{\alpha_{n-1}}{2^{n-1}}} \cdots T_{\frac{\alpha_0}{2^0}}$ every $T_{(n)}$ is bounded and for every $\xi \in L(x(.))$, $\lim_{n\to\infty} \|T_{(n)}\xi - T_h\xi\| = 0$ because $T_{(n)}x(s) = x(s + \sum_{i=0}^{n} \frac{\alpha_i}{2^i})$ and we assume the continuity of $x(s)$, $s \in R_1$, in the quadratic mean. As $\sup_{n\in N} \|T_{(n)}\| < \infty$, then the Banach--Steinhaus theorem yields that T_h is bounded also. Further, for every $\xi \in L(x(.))$ the inequality $\langle T_2\xi, T_2\xi\rangle^{\frac{1}{2}} = \|T_2\xi\| = \|T_1(T_1\xi)\| \leq \|T_1\|^2 \|\xi\|$ holds and hence T_2 is bounded too. By means of the mathematical induction we can immediately prove that for every natural n the operator T_n is bounded. As $T_s = T_{[s]+h} = T_{[s]}T_h$, $s > 0$, where $[s]$ is natural and $0 \leq h < 1$, we proved that every T_s, $s \geq 0$ is bounded. For every $\xi \in L(x(.))$ the equality $\|\xi\|^2 = \langle T_1\xi, T_{-1}\xi\rangle$ holds. That yields, if $T_1\xi = 0$ or $T_{-1}\xi = 0$ then $\xi \equiv 0$ must hold. As T_1 is bounded, T_1 can be in the unique way extended onto the whole space $L_2(x(.))$. Let $T_1\xi = 0$ for some $\xi \in L_2(x(.))$. Then $\|T_{\frac{1}{2}}\xi\|^2 = \langle T_1\xi, \xi\rangle = 0$ and hence $T_{\frac{1}{2}}\xi = 0$; similarly, for every natural n one can prove $T_{\frac{1}{2^n}}\xi = 0$. As $\lim_{n\to\infty} \|T_{\frac{1}{2^n}}\xi - \xi\| = 0$ we proved $\xi = 0$. That means T_1 is a one-to-one mapping defined on $L_2(x(.))$. A simple consideration yields that the range of T_1 is in general everywhere dense in $L_2(x(.))$. Then there exists an inverse operator T_1^{-1} and need not be bounded. T_1^{-1} is bounded if and only if $R(T_1) = L_2(x(.))$. As T_1 is symmetric, $R(T_1) = L_2(x(.))$ if and only if there exists a constant $c > 0$ that for every $\xi \in L_2(x(.))$ $\|T_1\xi\| \geq c\|x\|$. The spectral representation of T_1 and T_{-1} gives $R(T_1) = D(T_{-1})$, $R(T_{-1}) = D(T_1)$ and $T_1^{-1} = T_{-1}$. We assume the boundedness of T_{-1} and hence, in a similar way for T_s, $s > 0$, one can prove that every T_s, $s < 0$ must be bounded too. Further, the symmetric property of T_s yields

$$\|T_{\frac{t+s}{2}}\| \leq \|T_t\|^{\frac{1}{2}} \|T_s\|^{\frac{1}{2}} \leq \tfrac{1}{2}(\|T_s\| + \|T_t\|)$$

and hence the function $\|T_s\|$ is convex. Q.E.D.

Problems concerning the group $\{T_h\}_{h\in R_1}$ of shift-operators in $L_2(x(.))$ can be solved sometimes better in the corresponding reproducing kernel Hilbert space (RKHS) than in $L_2(x(.))$. As known, every covariance function $R(s,t)$ forms a kernel for the Hilbert space of functions $m(t) = E\{\xi\overline{x(t)}\}$, $\xi \in L_2(\overline{x(.)})$ with the reproducing property

$$m(s) = \langle m(.),R(s,.)\rangle$$

where $\langle m(.),n(.)\rangle = E\{\xi\bar{\eta}\}$. The group $\{T_h\}$ of the shift-operators defined on $L(x(.))$ can be in an easy way transformed into RKHS

$$T_h m(t) = E\{T_h\xi.\overline{x(t)}\}$$

As $R(s,t)$ for every fixed s belongs to RKHS, $R(s,t) = E\{s(x)\overline{x(t)}\}$ then

$$T_h R(s,t) = E\{x(s+h)\overline{x(t)}\} = R(s+h,t)$$ and hence the operator T_h in RKHS is generated by the shift $R(s,.) \xrightarrow{T_h} R(s+h,.)$. Now, let us assume that the covariance function $R(s,t)$ of a symmetric process $x(s)$, $s \in R_1$, is continuous. Then

$$R(s,t) = \int_{-\infty}^{+\infty} e^{\lambda(s+t)} dF(\lambda)$$ and hence for every

$\xi = \sum_{i=1}^{N} \alpha_i x(s_i) \in L(x(.))$ the corresponding $m(t) \in$ RKHS has the form
$m(t) = \int_{-\infty}^{+\infty} e^{\lambda t} \sum_{1}^{N} \alpha_i e^{\lambda s_i} dF(\lambda)$ with $\|m(t)\|^2 = \int_{-\infty}^{+\infty} |\sum_{1}^{N} \alpha_i e^{\lambda s_i}|^2 dF(\lambda)$.

It implies that RKHS consists of elements in the form

$$m(t) = \int_{-\infty}^{+\infty} e^{\lambda t} f(\lambda) dF(\lambda)$$

where $f(\lambda)$ belongs to $L_2\{e^{\lambda t}, t \in R_1\}$. In case the family $\{e^{\lambda t}, t \in R_1\}$ spans the whole space $L_2(F)$, then RKHS is isomorphic and isometric to $L_2(F)$. Conversely, if there exists an isomorphism between RKHS and $L_2(F)$ then the family of the function $\{e^{\lambda t}, t \in R_1\}$ is complete in $L_2(F)$. If $\xi = \sum_1^n \alpha_i x(s_i) \in L(x(.))$, then $T_h\xi = \sum_1^N \alpha_i x(s_i+h)$ and hence in RKHS

$$m_\xi(t) = \int_{-\infty}^{+\infty} e^{\lambda t} \sum_1^N \alpha_i e^{\lambda s_i} dF(\lambda)$$

$$T_h m_\xi(t) = m_{T_h\xi}(t) = \int_{-\infty}^{+\infty} e^{\lambda t} \sum_{1}^{N} \alpha_i e^{\lambda s_i} e^{\lambda h} dF(\lambda).$$

We obtain that there exists a one-to-one correspondence between the operator T_h in RKHS and the multiplying by $e^{\lambda h}$ in $L_2(F)$.

Theorem 4. The operator T_h is bounded on $L(x(.))$ if and only if the subspace corresponding to RKHS in $L_2(F)$ is closed with respect to multiplying by $e^{\lambda h}$, i. e. if $f \in L_2\{e^{\lambda t}, t \in R_1\}$ then $f(\lambda)\, e^{\lambda h} \in L_2\{e^{\lambda t}, t \in R_1\}$ also. In other words if $f(\lambda) \in L_2\{e^{\lambda t}, t \in R_1\}$ then $e^{\lambda h} f(\lambda)$ must be in $L_2\{e^{\lambda t}, t \in R_1\}$ too, i. e. $\int_{-\infty}^{+\infty} e^{2\lambda h} |f(\lambda)|^2 dF(\lambda) < +\infty$.

Proof. As proved before there exists a one-to-one mapping between RKHS and the subspace $L_2\{e^{\lambda t}, t \in R_1\} \subset L_2(F)$ given by the relation

$$m(t) = \int_{-\infty}^{+\infty} e^{\lambda t} f(\lambda) dF(\lambda)$$

where $m(t) \in$ RKHS and $f(\lambda) \in L_2\{e^{\lambda t}, t \in R_1\}$. The shift-operator T_h is characterized via multiplying by $e^{\lambda h}$ and let us suppose that T_h is defined on the whole RKHS, i. e. when $f \in L_2\{e^{\lambda t}, t \in R_1\}$ then $e^{\lambda h} f(\lambda) \in L_2\{e^{\lambda t}, t \in R_1\}$, too. Such an operator is symmetric because $e^{\lambda h}$ is real and hence T_h must be bounded. Conversely, T_h is defined on the linear subset of all linear combinations $\sum_{i=1}^{N} \alpha_i e^{\lambda s_i}$ and surely $e^{\lambda h} \sum_1^n \alpha_i e^{\lambda s_i}$ belongs to $L_2\{e^{\lambda t}, t \in R_1\}$, too; let T_h be bounded there. As these all linear combinations are everywhere dense in $L_2\{e^{\lambda t}, t \in R_1\}$, T_h can be in the unique way extended on the whole $L_2\{e^{\lambda t}, t \in R_1\}$ and this extension must be bounded and hence with $f(\lambda) \in L_2\{e^{\lambda t}, t \in R_1\}$, $e^{\lambda h} f(\lambda) \in L_2\{e^{\lambda t}, t \in R_1\}$, too. Q.E.D.

NORMAL COVARIANCES

Let $x(s), s \in R_1$ be a locally stationary process with the covariance function

$$R(s,t) = R_1\left(\frac{s+t}{2}\right) R_2(s-t)$$

where both R_1, R_2 are continuous covariances. Then, thanks to the Bochner theorem, $R_2(s-t) = \int_{-\infty}^{+\infty} e^{i\mu(s-t)} dF_2(\mu)$ and the Laplace transform yields that

$$R_1\left(\frac{s+t}{2}\right) = \int_{-\infty}^{+\infty} e^{\lambda(s+t)} dF_1(\lambda).$$

Hence we can write

$$R(s,t) = \int_{-\infty}^{+\infty} e^{\lambda(s+t)} dF_1(\lambda) \int_{-\infty}^{+\infty} e^{i\mu(s-t)} dF_2(\mu) =$$

$$= \int_{-\infty}^{+\infty} \int_{-\infty}^{+\infty} e^{s(\lambda+i\mu)}\, e^{t(\lambda-i\mu)}\, dF_1(\lambda)\, dF_2(\mu) =$$

$$= \int_{-\infty}^{+\infty} \int_{-\infty}^{+\infty} e^{sz} e^{t\bar{z}} \, dF_1(\mathrm{Re}z) dF_2(\mathrm{Im}z)$$

In this way we obtain the spectral decomposition of R(s,t) in the form of a double integral with respect to a two-dimensional probability distribution function. Hence we obtain a class of random processes with covariance functions expressible in the form

$$R(s,t) = \iint_{-\infty}^{+\infty} e^{sz} e^{t\bar{z}} dF(z)$$

where F(z) = F(Rez,Imz) is a two-dimensional probability distribution function. We shall call such covariances as *normal* covariances.

The following Theorem 5 expresses a relation between locally stationary and normal covariances.

Theorem 5. A normal covariance is locally stationary if the corresponding probability function in its spectral decomposition is a product probability distribution function. If a locally stationary covariance is normal then the corresponding probability distribution function is a product distribution function.

Proof. Let R(s,t) be normal, i. e. $R(s,t) = \iint_{-\infty}^{+\infty} e^{sz} e^{t\bar{z}} dF(z)$ with

$$F(z) = F_1(\mathrm{Re}z).F_2(\mathrm{Im}z).$$

Then by the Fubini theorem $R(s,t) = \int_{-\infty}^{+\infty} e^{\lambda(s+t)}.e^{i\mu(s-t)} dF_1(\lambda) dF_2(\mu) =$

$$= \int_{-\infty}^{+\infty} e^{\lambda(s+t)} dF_1(\lambda) \int_{-\infty}^{+\infty} e^{i\mu(s-t)} dF_2(\mu) = R_1(\tfrac{s+t}{2}) R_2(s-t)$$

and hence R(s,t) is locally stationary. Conversely, if R(s,t) is locally stationary and normal, i. e. $R(s,t) = R_1(\frac{s+t}{2}) R_2(s-t)$ and $R(s,t) = \iint_{-\infty}^{+\infty} e^{sz} e^{tz} dF(z)$, too, then

$R_1(\frac{s+t}{2}) = R(\frac{s+t}{2}, \frac{s+t}{2})$ and $R_2(s-t) = R(\frac{s-t}{2}, \frac{t-s}{2})$. Hence

$$R_1(\tfrac{s+t}{2}) = \iint_{-\infty}^{+\infty} e^{\frac{s+t}{2}z} e^{\frac{s+t}{2}\bar{z}} dF(z) = \iint_{-\infty}^{+\infty} e^{\lambda(s+t)} ddF(\lambda,\mu) =$$

$= \int_{-\infty}^{+\infty} e^{\lambda(s+t)} dF_1(\lambda)$ where F_1 is the marginale of F; similarly

$$R_2(s-t) = \iint_{-\infty}^{+\infty} e^{\frac{s-t}{2}z}.e^{\frac{t-s}{2}\bar{z}} ddF(\lambda,\mu) = \iint_{-\infty}^{+\infty} e^{i\mu(s-t)} ddF(\lambda,\mu) =$$

$= \int_{-\infty}^{+\infty} e^{i\mu(s-t)} dF_2(\mu)$ where F_2 is the other marginale of F.

Together we obtain

$$R(s,t) = \int_{-\infty}^{+\infty} e^{\lambda(s+t)} dF_1(\lambda) . \int_{-\infty}^{+\infty} e^{i\mu(s-t)} dF_2(\mu) =$$

$$= \int_{-\infty}^{+\infty} \int_{-\infty}^{+\infty} e^{sz}.e^{t\bar{z}} dF_1(\lambda) dF_2(\mu) = \int_{-\infty}^{+\infty}\!\!\int e^{sz}.e^{t\bar{z}} ddF(\lambda,\mu)$$

and properties of the two-dimensional Laplace transform give

$$F_1(\lambda).F_2(\mu) = F(\lambda,\mu). \quad \text{Q.E.D.}$$

Any random process having a normal covariance can be expressed in the form of a stochastic integral as the following Theorem 6 states.

Theorem 6. A random process $x(s)$, $s \in R_1$, can be written as

$$x(s) = \int_{-\infty}^{+\infty}\!\!\int e^{sz} d\xi(z), \; z = \lambda + i\mu,$$

where $\xi(z) = \xi(\lambda,\mu)$ is a plane-martingale in the quadratic mean sense, $E\{\xi(z)\} = 0$, $E\{\xi(z_1)\overline{\xi(z_2)}\} = F(\min(z_1,z_2))$, if and only if its covariance function $R(s,t)$ is normal,

$$R(s,t) = \int_{-\infty}^{+\infty}\!\!\int e^{sz} e^{t\bar{z}} dF(z).$$

Proof. The probability distribution function F induces a measure on the Borel sets in the plane and hence the assertion of Theorem 6 is an application of the Karhunen theorem in a general form quoted e.g. in Gichman, Skorochod (1965). In case the system of functions $\{e^{s\lambda}e^{i\mu s}, s \in R_1\}$ is complete in the Hilbert space $L_2(F)$ the values of the martingale $\xi(z)$ can be expressed by means of $x(s)$, $s \in R_1$.
Q.E.D.

Now, let $x(s)$, $s \in R_1$ be a random process with a normal covariance and hence we can write

$$x(s) = \int_{-\infty}^{+\infty}\!\!\int e^{sz} d\xi(z).$$

Let $L_2(\xi(.))$ be the Hilbert space generated by the values of $\xi(z)$, $z \in C$, with the scalar product $<\xi(z_1),\xi(z_2)> = E\{\xi(z_1)\overline{\xi(z_2)}\}$. Let H_z be the subspace of $L_2(\xi(.))$ generated by all random variables $\xi(w)$ where $w \le z$, i. e. $\mathrm{Re}w \le \mathrm{Re}z$, $\mathrm{Im}w \le \mathrm{Im}z$. The system $\{H_z, z \in C\}$ of these subspaces induces a complex resolution of the identity in

$L_2(\xi(.))$. Let P_z be the orthogonal projector onto H_z; then $\{P_z, z \in C\}$ forms a resolution of the identity in $L_2(\xi(.))$. Let $s \in R_1$ be fixed and consider the operator

$$A_s = \int_{-\infty}^{+\infty}\!\!\int e^{sz} dP_z$$

which is defined on the everywhere dense subset $D(A_s) =$
$= \{\eta \in L_2(\xi(.)) : \int_{-\infty}^{+\infty}\!\!\int e^{sz} e^{s\bar{z}} d\langle P_z \eta, \eta\rangle$ exists$\}$ in $L_2(\xi(.))$. As follows from the spectral theory of unbounded operators in a Hilbert space the operator A_s is a maximal normal operator, i. e. A_s is defined on an everywhere dense subset, A_s is closed, $A_s = A_s^{**}$ and $A_s A_s^* =$ $= A_s^* A_s$ where A_s^* is the adjoint operator, A_s^{**} the second adjoint operator; for details see Rudin (1973).

As the function F is of finite variation and $\xi(\underset{\frown}{z})$ is a plane-martingale, there exists the limit $\xi(+\infty) = \underset{z \to \infty}{\text{l.i.m.}}\ \xi(z)$ and the spectral decomposition of $x(s)$ gives $x(s) = \xi(+\infty)$. For any $\xi(z_0)$

$$A_s\ \xi(z_0) = \int_{-\infty}^{\lambda_0}\!\!\int^{\mu_0} e^{sz} d\xi(z)$$

where $z_0 = \lambda_0 + i\mu_0$ because $P_z \xi(z_0) = \xi(z)$ for $z \le z_0$ and $P_z \xi(z_0) =$ $= \xi(z_0)$ for $z > z_0$. Very important is the following relation

$$A_s\ x(0) = A_s \xi(+\infty) = \int_{-\infty}^{+\infty}\!\!\int e^{sz} dP_z \xi(+\infty) =$$
$$= \int_{-\infty}^{+\infty}\!\!\int e^{sz} d\xi(z) = x(s)$$

and hence A_s is the shift-operator on the linear subset $L(x(.))$. Properties of the function e^{sz} and of the resolution of the identity enable to prove that the operators A_s, $s \in R_1$, form a group on $L(x(.))$ for $A_0 \eta = \eta$ $A_{t+s}\eta = A_t A_s \eta$ for every $\eta \in L(x(.))$. We see that a normal covariance function induces a group of shift-operators on $L(x(.))$ which can be extended into maximal normal operators defined in $L_2(\xi(.))$. In general, the operator A_s need not be bounded nevertheless a random process with a normal covariance function is continuous in the quadratic mean and hence its covariance function is continuous, too. That follows from the spectral decomposition of normal covariances.

In case we consider the restriction T_s of A_s onto $L(x(.))$ in $L_2(\xi(.))$ only the situation with the boundedness of $\{T_s\}_{s \in R_1}$

is similar as in the case of symmetric operators. The spectral decomposition of $\{A_s\}_{s\in R_1}$ gives further that $\mathcal{D}(A_s) = \mathcal{R}(A_{-s})$, $\mathcal{R}(A_s) = \mathcal{D}(A_{-s})$ and A_s, A_{-s} are mutually inverse.

Theorem 6. All restrictions $\{T_s, s\in R_1\}$ are bounded on $L(x(.))$ if at least one pair $(T_h\ T_{-h})$ is bounded on $L(x(.))$.

Proof. Without loss of generality we can suppose T_1, T_{-1} are bounded on $L(x(.))$ and hence in the unique way T_1, T_{-1} can be extended onto the whole $L_2(x(.))$ which is a subspace in $L_2(\xi(.))$. As A_s is normal in $L_2(\xi(.))$ then $S_s = A_sA_s^* = A_s^*A_s$ is a nonnegative self-adjoint operator with $\mathcal{D}(S_s) = \{y\in\mathcal{D}(A_s): A_sy\in\mathcal{D}(A_s)\}$ because $\mathcal{D}(A_s) = \mathcal{D}(A_s^*)$. That implies S_s is defined on $L(x(.))$ and its quadratic root H_s is defined on $L(x(.))$, too. $\|A_sy\| = \|H_sy\|$ for every $y\in L(x(.))$ and hence the restrictions of H_1, H_{-1} onto $L(x(.))$ are bounded because $A_1 = T_1$, $A_{-1} = T_{-1}$ on $L(x(.))$. $\{H_s, s\in R_1\}$ forms a group of symmetric operators on $L(x(.))$ and at this moment we can use the method presented in the proof of Theorem 3. Q.E.D.

A very important case occurs when the functions $\{e^{sz}, s\in R_1\}$ span the whole space $L_2(F)$ and hence $L_2(x(.)) = L_2(\xi(.))$. In this situation the operators $\{A_s, s\in R_1\}$ form a group on $L(x(.))$ which is everywhere dense in $L_2(\xi(.))$. If A_1, A_{-1} are bounded on $L(x(.))$ then we obtain a group of bounded normal operators on $L_2(x(.))$ satisfying the condition $\lim_{s\to 0}\|A_sy-y\| = 0$ for every $y\in L_2(x(.))$ because the underlying process $x(s)$, $s\in R_1$, is continuous in the quadratic mean. Then the group $\{A_s, s\in R_1\}$ can be expressed as

$$A_s = e^{sA},\ s\in R_1,$$

where A is the infinitesimal generator of $\{A_s, s\in R_1\}$. A is normal, too, however may be unbounded. In every case the generator A can be expressed as

$$A = \int_{-\infty}^{+\infty}\!\!\int z\,dP_z$$

and hence $\mathcal{D}(A) = \{y\in L_2(x(.)): \int_{-\infty}^{+\infty}\!\!\int |z|^2 d\langle P_zy,y\rangle \text{ exists}\}$.

This generator A is closely connected with the derivative of the process $x(s)$, $s\in R_1$, in the quadratic mean sense because if $x(s)\in \mathcal{D}(A)$, i. e. $\int_{-\infty}^{+\infty}\!\!\int |z|^2 e^{zs}.e^{\bar{z}s}dF(z)$ exists,

$$Ax(s) = \underset{\varepsilon \to 0}{\text{l.i.m.}} \frac{A_\varepsilon x(s) - x(s)}{\varepsilon} = x'(s).$$

A similar situation sets even in a general case. Let us consider the normal operator $A = \int_{-\infty}^{+\infty}\int zdP_z$ where $\{P_z\}$ is the complex resolution of the identity defining the operators $\{A_s,\ s \in R_1\}$. If $x(s) \in \mathcal{D}(A)$ then

$$Ax(s) = x'(s)$$

and hence there exists the derivative $x'(s)$ of $x(s)$ in the quadratic mean sense.

INVERSE FORMULA FOR NORMAL COVARIANCES

Let $R(s,t) = \int_{-\infty}^{+\infty}\int e^{sz}e^{t\bar{z}}dF(z)$ be a normal covariance. We see immediately that for $s = t$ $R(\frac{s}{2}, \frac{s}{2}) = \int_{-\infty}^{+\infty}\int e^{s(\frac{z+\bar{z}}{2})} dF(z)$ is an exponentialy convex function and for $s = -t$

$$R(\tfrac{s}{2}, \tfrac{-s}{2}) = \int_{-\infty}^{+\infty}\int e^{s(\frac{z-\bar{z}}{2})} dF(z)$$

we obtain a stationary covariance function. The spectral decomposition of $R(s,t)$ gives further $R(s,t)$ is an exponentially convex covariance if and only if

$$F(z) = F_1(\lambda)\ H(\mu),\ z = \lambda + i\mu,$$

where F_1 is any one-dimensional probability distribution function, $H(\mu) = 0$ for $\mu \le 0$, $H(\mu) = 1$ otherwise. Similarly, $R(s,t)$ is a stationary covariance if and only if

$$F(z) = H(\lambda)\ .\ F_2(\mu)$$

where F_2 is any one-dimensional probability distribution function. To obtain an inverse formula for normal covariances we start with an inverse formula for a random process $x(s)$, $s \in R_1$, having a normal covariance function and hence expressed as a stochastic integral

$$x(s) = \int_{-\infty}^{+\infty}\int e^{sz}\ d\xi(z).$$

In general, the subspace $L_2(x(.))$ generated by the values of $x(s)$, $s \in R_1$ is a proper subspace in $L_2(\xi(.))$. As long as the functions $\{e^{sz},\ s \in R_1\}$ form a complete system in $L_2(F)$ in the plane then there exists a one-to-one-mapping between $L_2(x(.))$ and $L_2(F)$ and hence for every $f \in L_2(F)$ there exists a random variable $u(f) \in L_2(x(.))$, namely $u(f) = \int_{-\infty}^{+\infty}\int f(z)d\xi(z)$, and $u(f)$ can be expressed by means of $x(s)$, $s \in R_1$. Let us suppose that the functions $\{e^{sz},\ s \in R_1\}$ span the space $L_2(F)$. Under this condition the random variable $x(u,v)$

$$x(u,v) = \int_{-\infty}^{+\infty}\int e^{\lambda u} e^{i\mu v} d\xi(z)$$

belongs to $L_2(x(.))$ and by means of these variables we express the martingale $\xi(z)$. For analytic expression of an inverse formula an extension of $x(u,v)$ onto the complex plane is necessary. Let us put $u = u_1 + iu_2$, $v = v_1 + iv_2$ and let us consider, meanwhile formally, the integral

$$x(u,v) = \int_{-\infty}^{+\infty}\int e^{\lambda(u_1+iu_2)}\ e^{i\mu(v_1+iv_2)}\ dd\xi(\lambda,\mu).$$

This integral makes sense if and only if the function $e^{\lambda u}\ e^{i\mu v}$ belongs to $L_2(F)$, i. e. the integral

$$\int_{-\infty}^{+\infty}\int e^{\lambda u}\ e^{i\mu v}\ e^{\lambda\bar{u}}.e^{-i\mu\bar{v}}\ ddF(\lambda,\mu)$$

must be convergent. As $\lambda(u+\bar{u}) + i\mu(v-\bar{v}) = 2\lambda u_1 - 2\mu v_2$ and as $E\{|x(s)|^2\}$ is finite for a real s we must put $v_2 = 0$. That means, we can extend $x(u,v)$ onto the complex domain for any complex u and any real v, i. e.

$$x(u,v) = \int_{-\infty}^{+\infty}\int e^{\lambda u_1}.e^{i\lambda u_2}.e^{i\mu v} dd\xi(\lambda,\mu).$$

Let v be fixed and consider the process $x(u,v) = y(u)$, u real, then

$$E\{y(u_1)\overline{y(u_2)}\} = \int_{-\infty}^{+\infty}\int e^{\lambda(u_1+u_2)} ddF(\lambda,\mu) \quad \text{and hence}$$

the covariance function of $y(u)$, $u \in R_1$, is exponentially convex. Under this fact the process $y(u)$, $u \in R_1$, can be extended onto the complex plane because $y(u)$, u R_1 is a symmetric process.

Theorem 7. Let (λ,μ), $(\lambda,\mu+h_2)$, $(\lambda+h_1,\mu)$ and $(\lambda+h_1,\mu+h_2)$, $h_1 > 0$, $h_2 > 0$, be continuity points of $\xi(z)$ in the quadratic mean sense. Then

$$\Delta_{h_1}\Delta_{h_2}\xi(\lambda,\mu) = \underset{\substack{T_1\to\infty\\ T_2\to\infty}}{\text{l.i.m.}} \frac{1}{4\pi^2}\int_{c-iT_1}^{c+iT_1}\int_{-T_2}^{T_2} \frac{e^{-(\lambda+h_1)z}-e^{-\lambda z}}{z} \times$$

$$\times \frac{e^{-i(\mu+h_2)v}-e^{-i\mu v}}{-iv}\, x(z,v)\,dz\,dv \quad \text{where } c>0.$$

Proof. We have to prove that

$$(6)\quad E\{|\Delta_{h_1}\Delta_{h_2}\xi(\lambda,\mu)-\frac{1}{4\pi^2}\int_{c-iT_1}^{c+iT_1}\int_{-T_2}^{T_2}\frac{\Delta_{h_1}e^{-\lambda z}}{z}\cdot\frac{\Delta_{h_2}e^{-i\mu v}}{-iv}x(z,v)\,dz\,dv|^2\} \underset{T_1,T_2\to\infty}{\to} 0.$$

We see that

$$\Delta_{h_1}\Delta_{h_2}\xi(\lambda,\mu) = \int_{-\infty}^{+\infty}\!\!\int \psi_{<\lambda,\lambda+h_1>}(\alpha)\psi_{<\mu,\mu+h_2>}(\beta)\,dd\xi(\alpha,\beta) \quad \text{and}$$

$$\int_{c-iT_1}^{c+iT_1}\int_{-T_2}^{T_2}\frac{\Delta_{h_1}e^{-\lambda z}}{z}\cdot\frac{\Delta_{h_2}e^{-i\mu v}}{-iv}x(z,v)\,dz\,dv =$$

$$= i\int_{-T_1}^{T_1}\int_{-T_2}^{T_2}\frac{\Delta_h\, e^{-\lambda(c+iu)}}{c+iu}\cdot\frac{\Delta_{h_2}e^{-i\mu v}}{-iv}\,x(c+iu,v)\,du\,dv.$$

The spectral decomposition of $x(u,v)$ gives

$$x(c+iu,v) = \int_{-\infty}^{+\infty}\!\!\int e^{\alpha(c+iu)}e^{i\beta v}dd\xi(\alpha,\beta) \quad \text{and}$$

at this moment that is suitable to change order between a stochastic integral understood in the quadratic mean sense and a usual integral in the Riemann sense. On the base of analogy with Lemmas 4, 5 in Gichman, Skorochod (1965) under the square-integrability of the functions in the inverse formula and in the spectral decomposition such a change of order is admitted. Hence we can write

$$i\int_{-T_1}^{T_1}\int_{-T_2}^{T_2}\frac{\Delta_{h_1}e^{-\lambda(c+iu)}}{c+iu}\cdot\frac{\Delta_{h_2}e^{-i\mu v}}{-iv}x(c+iu,v)\;du\,du =$$

$$= i\int_{-\infty}^{+\infty}\!\!\int\{\int_{-T_1}^{T_1}\int_{-T_2}^{T_2} e^{\alpha(c+iu)}e^{i\beta v}\;\frac{\Delta_{h_1}e^{-\lambda(c+iu)}}{c+iu}\cdot\frac{\Delta_{h_2}e^{-i\mu v}}{-iv}\,du\,dv\}\;dd\xi(\alpha,\beta) =$$

$$= i\int_{-\infty}^{+\infty}\!\!\int\{\int_{-T_1}^{T_1}\int_{-T_2}^{T_2}\frac{\Delta_{h_1}e^{(\alpha-\lambda)(c+iu)}}{c+iu}\cdot\frac{\Delta_{h_2}e^{-iv(\mu-\beta)}}{-iv}\,du\,dv\}\,dd\xi(\alpha,\beta) =$$

$$= i \int_{-\infty}^{+\infty}\int \{\int_{-T_1}^{T_1} \frac{\Delta_{h_1} e^{(\alpha-\lambda)(c+iu)}}{c+iu} du \int_{-T_2}^{T_2} \frac{\Delta_{h_2} e^{-iv(\mu-\beta)}}{-iv} dv\} dd\xi(\alpha,\beta) =$$

$$= \int_{-\infty}^{+\infty}\int \{\int_{-T_1}^{T_1} e^{c(\alpha-\lambda)} \left[\frac{c\Delta_{h_1}\cos(u(\alpha-\lambda)) + u\Delta_{h_2}\sin(u(\alpha-\lambda))}{c^2+u^2}\right] du \times$$

$$\times \int_{-T_2}^{T_2} \frac{\Delta_{h_2}\sin(v(\mu-\beta))}{v} dv\} \; dd\xi(\alpha,\beta).$$

By use of these conclusions and of the properties of stochastic integral the expected value (6) can be calculated as

$$\int_{-\infty}^{+\infty}\int (\psi_{<\lambda,\lambda+h_1>}(\alpha)\psi_{<\mu,\mu+h_1>}(\beta) -$$

$$- \int_{-T_1}^{T_1} \frac{e^{c(\alpha-\lambda)}\left[c\Delta_{h_1}\cos(u(\alpha-\lambda)) + u\Delta_{h_2}\sin(u(\alpha-\lambda))\right]}{c^2+u^2} du \times$$

$\times \int_{-T_2}^{T_2} \frac{\Delta_{h_2}\sin(v(\mu-\beta))}{v} dv)^2 \, ddF(\alpha,\beta)$. The Lebesque theorem on the dominated convergence finishes the proof. Q.E.D.

To prove the inverse formula for a normal covariance function we need to realize that firstly

$$E\{|\Delta_{h_1}\Delta_{h_2}\xi(\lambda,\mu)|^2\} = \Delta_{h_1}\Delta_{h_2}F(\lambda,\mu)$$

and secondly the convergence in the quadratic mean implies the convergence of the corresponding norms.

Theorem 8. Let $R(s,t) = \int_{-\infty}^{+\infty}\int e^{\lambda(s+t)} e^{i\mu(s-t)} ddF(\lambda,\mu)$ be a normal covariance, let (λ,μ), $(\lambda,\mu+h_2)$, $(\lambda+h_1,\mu)$, $(\lambda+h_1,\mu+h_2)$ be a continuity points of $F(\lambda,\mu)$. Then

$$\Delta_{h_1}\Delta_{h_2}F(\lambda,\mu) = \lim_{\substack{T_1,T_2,T_3,T_4\\ \to\infty}} \frac{1}{16\pi^2} \int_{c-iT_1}^{c+iT_1}\int_{c-iT_2}^{c+iT_2}\int_{-T_3}^{T_3}\int_{-T_4}^{T_4} \frac{\Delta_{h_1} e^{-\lambda z_1}}{z_1} \frac{\Delta_{h_2} e^{-\lambda \bar{z}_2}}{\bar{z}_2}$$

$$\frac{\Delta_{h_1} e^{-i\mu v_1}}{-iv_1} \frac{\Delta_{h_2} e^{i\mu v_2}}{iv_2} R\left(\frac{z_1+\bar{z}_2}{2}+\frac{v_1-v_2}{2}, \frac{\bar{z}_1+z_2}{2}+\frac{v_2-v_1}{2}\right) dz_1 dz_2 dv_1 dv_2.$$

Proof. As mentioned above the proof follows immediately from Theorem 7. Properties of stochastic integrals imply that

$$E\{x(z_1,v_1)\overline{x(z_2,v_2)}\} = \int_{-\infty}^{+\infty}\int e^{\lambda(z_1+\bar{z}_2)} e^{i\mu(v_1-v_2)} ddF(\lambda,\mu)$$

and the covariance function R(s,t) defined originally for real s, t only can be extended onto the complex domain,too,because the process x(z,v) can be defined for z complex. Q.E.D.

CONCLUSIONS

Investigation on the spectral decomposition of locally stationary processes discovered a very important class of second order random processes whose covariance functions are normal. Spectral analysis was successfull to prove,that the theory of weakly stationary processes is closely connected with a group of unitary shift-operators as symmetric processes are connected with a group of self--adjoint shift-operators, the both cases are covered by the class of processes having normal covariances,whose spectral theory is connected with a group of normal shift-operators. A continuous stationary covariance function is completely characterized by the Bochner theorem, a continuous exponentially convex covariance is completely monotone function and hence it would be desirable to describe in a similar way the class of normal covariances.

REFERENCES

Silverman R. A. (1957): Locally Stationary Processes. IRE Transactions on Information Theory, Vol IT-3, 183-187.

Michálek J. (1986a): Spectral Decomposition of Locally Stationary Random Processes, Kybernetika (in print).

Michálek J. (1986b): Ergodic Properties of Locally Stationary Processes, Kybernetika (in print).

Loève M. (1948): Fonctions aléatoires du second order;in P. Lèvy: Processus Stochastiques et mouvement brownien, Gauthier-Villars, Paris.

Widder D. V. (1946): The Laplace Transform, Princeton University Press.

Gichman I. I., Skorochod A. V. (1965): Introduction to Theory of Random Processes, Nauka, Moscow (in Russian).

Riesz F., Sz.-Nagy B. (1953): Leçons d'analyse fonctionnalle, Budapest.

Rudin W. (1973): Functional Analysis, McGraw-Hill Book Company, New York.

Czechoslovak Academy of Sciences
Institute of Information Theory and Automation
182 08 Prague
Pod vodárenskou věží 4
Czechoslovakia

ON OPTIMAL SET-VALUED ESTIMATORS

H. Meister and O. Moeschlin

D-5800 Hagen 1, West Germany

Key words: Bochner-integral, unbiasedness of set-valued estimators with minimal risk, set-valued Bayesian estimators, method of directional derivatives

ABSTRACT

Set-valued estimators are studied in a non-Bayesian as well as in a Bayesian framework. In the first case, we base on the Bochner-integral in order to introduce the notion of unbiased set-valued estimators with minimal risk.

In the second case we introduce a Bayes risk for set-valued estimators. In both cases optimal estimators are characterized by the method of directional derivatives.

1. INTRODUCTION

The estimation of the values of a parameter function in the sense of the usual estimation theory leads to estimates which often with probability one do not coincide with the value that should be estimated. This fact, of course, lies in the nature of point estimation theory.

One way to overcome this difficulty is to appeal to concepts of confidence estimation. We try to overcome this difficulty here - not only in the classical, non-Bayesian , but also in the Bayesian case - studying set-valued estimators (estimation correspondences) for set-valued parameter functions (parameter correspondences). (With these parameter functions we determine those values the statistician tends to accept.)

For obvious reasons we first have to establish an adequate space of estimates, which enjoys a vector space structure and allows integration. The integral to be used is the one of Bochner, which - under certain conditions - coincides with the one of Aumann. This fact is not only interesting from a theoretical point of view but also allows an intuitive interpretation of unbiasedness of set-valued estimators in the non-Bayesian theory.

In both cases, non-Bayesian and Bayesian resp., we use the method of directional derivatives in order to give characterizations for optimal estimators, i.e. for unbiased estimators with minimal risk and Bayesian estimators, resp.

2. PRELIMINARIES

Let E be a separable real Banach space. The norm of an element $x \in E$ is denoted by $|x|$. For two nonempty subsets A and B of E, we define the Hausdorff distance δ of A and B by

$$\delta(A,B) := \max\{\sup_{x\in A}\inf_{y\in B}|x-y|, \sup_{y\in B}\inf_{x\in A}|x-y|\}.$$

Let C(E) be the system of all nonempty convex-compact subsets of E. Then, δ is a metric on C(E). The space C(E) will henceforth be endowed with the corresponding metric space structure.

In order to introduce a convex structure on the space C(E) we define:

$$\sum_{i=1}^{r} t_i C_i := \{\sum_{i=1}^{r} t_i c_i \mid c_1 \in C_1,\dots,c_r \in C_r\}$$

for $t_1,\dots,t_r \in \mathbb{R}$ and $C_1,\dots,C_r \in C(E)$.

By a famous theorem of Rådström (1952), the space C(E) endowed with this linear structure may be embedded as a closed convex cone into a real Banach-space V with norm $\|.\|$ s.t.

$$\|C_1 - C_2\| = \delta(C_1,C_2) \text{ for } C_1,C_2 \in C(E),$$

where the convex structure in V induces the one in C(E). With this linear structure in C(E) convexity can be introduced for real functions $f: C(E) \to \mathbb{R}$ in a canonical way. The function $f: C(E) \to \mathbb{R}$ is said to be convex, iff

$$f(tC_1 + (1-t)C_2) \leq tf(C_1) + (1-t)f(C_2)$$

holds for all $C_1, C_2 \in C(E)$ and $t \in [0,1]$.

With regard to future applications we first cite a result of convex optimization theory. Let W be any convex subset of a real vector space, and let $f : W \to \mathbb{R}$ be any convex function. Then, the directional derivative of f at the point $w \in W$ w.r.t. $w' \in W$ is defined by

$$D_{w'} f(w) := \lim_{t \downarrow 0} \frac{f((1-t)w+tw') - f(w)}{t} .$$

The minima of f in W can be characterized by means of directional derivatives.

Lemma 2.1: *Let $f : W \to \mathbb{R}$ be convex, and let $w \in W$. Then, the following two statements are equivalent:*

(1) $$f(w) = \min_{w' \in W} f(w') ;$$

(2) $$D_{w'} f(w) \geq 0 \qquad (w' \in W) .$$

The space E will be endowed with its Borel-σ-field $\mathcal{B}(E)$, i.e. the σ-field generated by the open subsets of E.

The Borel-σ-field induced by the metric space structure of C(E) will be denoted by $\mathcal{B}(C(E))$.

Furthermore, let $(\mathbb{H}, H, P)$ be a probability space. By a correspondence Z from $\mathbb{H}$ to E we understand a set-valued mapping, which associates with every element $x \in \mathbb{H}$ a nonempty subset Z(x) of E. On the other hand, the correspondence Z may be seen as a function from $\mathbb{H}$ to the set of all nonempty subsets of E. These two aspects of correspondences have led to different concepts of measurability. We base on the measurability concept of the later aspect of Z. The measurability of Z regarded as a function from $\mathbb{H}$ to C(E) is defined in the usual sense.

The notion of integral to be used in the sequel for mappings $Z : \mathbb{H} \to V$ will be the Bochner-integral (Bochner (1933)), as introduced for instance, too, in Dunford and Schwartz (1958), chapter IV, 10.7 , Def. 7 . We denote it by

$\int Z\,dP$. In addition to this notion of integral we refer to the integral introduced by Aumann (1965) : Thereby let

$$\Phi(Z) := \{f \mid f \text{ being } \mathcal{H}\text{-}\mathcal{B}(E)\text{-measurable, } f(x) \in Z(x) \quad (x \in \mathbb{H})\}$$

be the set of measurable selectors of Z. (For the existence of a measurable selector f within a correspondence Z compare Castaing and Valadier (1977), Theorem III. 6, p. 65). The Aumann-integral is defined by

$$\int^{A} Z\,dP := \{z \in E \mid z = \int f\,dP \text{ for some } f \in \Phi(Z)\} .$$

Following Hiai and Umegaki (1977), Theorem 4.5, we have:

If E has the Radon-Nikodym-Property (RNP, then

(2.2) $$\int Z\,dP = \int^{A} Z\,dP .$$

The merit of having stated the above equality first is due to Debreu (1967), Theorem 6.5; but this equality does not hold true without any assumption on E. Compare the Remark following Theorem 4.5 in Hiai and Umegaki (1977).

3. UNBIASED SET-VALUED ESTIMATORS WITH MINIMAL RISK

An (non-Bayesian) estimation problem with set-valued parameter function is determined by an eight-tuple

$$(\mathbb{H}, \mathcal{H}, \Gamma, \mathcal{W}, E, \mathcal{S}, L, v) ,$$

what we call an estimation experiment; thereby $(\mathbb{H}, \mathcal{H})$ (sample space) is a measurable space. Γ (parameter space) is a nonempty set and $\mathcal{W} := \{P_\gamma(.) \mid \gamma \in \Gamma\}$ (set of possible sample distributions) is a set of probability measures P_γ on $\mathcal{H}$, such that $\gamma \to P_\gamma(.)$ is one-to-one. E denotes a real separable Banach space, $\mathcal{S}$ is a nonempty closed, convex subcone of $C(E)$ within $\mathcal{V}$ and $L : \Gamma \to \mathcal{S}$ (set-valued parameter function, parameter correspondence) is a mapping. $v : \Gamma \times \mathcal{S} \to \mathbb{R}_+$ ((convex) loss function) denotes a mapping, such that $v(\gamma,.) : \mathcal{S} \to \mathbb{R}_+$ is (convex) and $\mathcal{B}(C(E))$-$\mathcal{B}$-measurable for all $\gamma \in \Gamma$.

The given set-up follows the line of the usual (point-)estimation problem with the alteration that L is now set-valued - with values in the subcone $\mathcal{S}$

of $C(E)$ - instead of being point-valued as in the case of a point-estimation problem. In the intuitive interpretation the convex and compact set $L(\gamma)$ determines those points in E, which the statistician tends to accept, provided that γ is the accurate parameter.

If for $S := C(E)$ a function $v(.,.) : \Gamma \times S \to \mathbb{R}_+$ is defined by

$$v(\gamma,C) := \delta(L(\gamma),C) \qquad (\gamma\in\Gamma,\ C\in C(E)),$$

then function v is a loss function. Moreover it is a convex loss function, for a proof see [4].

For $S := E = \mathbb{R}^n$ any (convex) loss function to a (usual) point-estimation experiment is a (convex) loss function in the sense of the set-up given here.

Definition 3.1: *Let an estimation experiment* $(\mathbf{H},H,\Gamma,W,E,S,L,v)$ *be given. Then any* H-$B(C(E))$ *- measurable mapping* $Z : \mathbf{H} \to S$ *is called an* <u>*estimation correspondence*</u> *(to the given estimation experiment).*

Z_S *denotes the set of all estimation correspondences.*

By the concept laid down, the values of the estimation correspondence and the ones of the parameter correspondence lie in the same closed and convex subcone S of $C(E)$. It is clear, that the point-estimation problem is covered with this set-up, too. But even if one is interested in true set-valued estimators, it is clear that for practical purposes the subcone S is kept as small as possibel.

Remark 3.2: As Z has values in the closed, convex cone S, it follows that $\int Z\,dP_\gamma \in S$ for all $\gamma \in \Gamma$.

Definition 3.3: *An estimation correspondence* Z *is called* <u>*unbiased,*</u> *iff*

$$\int Z\,dP_\gamma = L(\gamma) \qquad (\gamma\in\Gamma)\ .$$

Definition 3.4: *Let an estimation experiment be given.*

1. *For* $Z \in Z_S$ *the mapping*

$$\rho(.,Z) : \Gamma \to [0,\infty],$$

defined by

$$\rho(\gamma,Z) := \int v(\gamma,Z)\, dP, \qquad (\gamma\in\Gamma)$$

is called the *risk* *of* Z.

2. $Z \in Z_S$ *is called of* *finite risk*, *iff*

$$\rho(\gamma,Z) < \infty \quad (\gamma\in\Gamma) .$$

The set of all unbiased correspondences $Z \in Z_S$ *with finite risk is denoted by* $Z_S^{u\rho}$.

3. $\bar{Z} \in Z_S \subset Z_S$ *is called of minimal risk within* Z_S', *iff*

$$\rho(\gamma,\bar{Z}) \leq \rho(\gamma,Z) \qquad (\gamma\in\Gamma,\ Z\in Z_S') .$$

Theorem 3.5: *The space of all unbiased estimation correspondences is convex.*

Proof: The proof is an immediate consequence of the linearity of the Bochner-integral. □

The set of all unbiased estimation correspondences to a given estimation problem is denoted by Z_S^u .

Theorem 3.6: *Let* E *have the* RNP. *Then an estimation correspondence* Z *is unbiased, iff*

$$L(\gamma) = \{z\in E \mid z = \int f\, dP_\gamma,\ f \in \Phi(Z)\} \qquad (\gamma\in\Gamma).$$

Proof: The proof follows from 2.2 □

If we assume E to be the $\mathbb{R}^n$, then by the well-known Theorem of Radon-Nikodym $E = \mathbb{R}^n$ has the RNP.

In case that the elements of S have a special geometric structure, for instance if they are closed balls $B(m,r)$ with center $m \in \mathbb{R}^n$ and radius $r \geq 0$ the unbiasedness of the now ball-valued estimators may be characterized by the unbiasedness of the respective center and radius function.

Denote the space $\{B(m,r) | m \in \mathbb{R}^n, r \in \mathbb{R}_+\}$ by $C_0(\mathbb{R}^n)$.

Corollary 3.7: *Let an estimation experiment with* $S = C_0(\mathbb{R}^n)$ *and* L *having the center- and the radius function* $m^L : \Gamma \to \mathbb{R}^n$ *and* $r^L : \Gamma \to \mathbb{R}_+$ *be given; then the estimation correspondence* Z, $m : \mathbf{H} \to \mathbb{R}^n$ *and* $r : \mathbf{H} \to \mathbb{R}_+$ *being its center- and radius function is unbiased, iff*

$$\int m \, dP_\gamma = m^L(\gamma), \int r \, dP_\gamma = r^L(\gamma) \qquad (\gamma \in \Gamma) .$$

The proof is similar to the one given in Meister and Moeschlin (1986).

Applying Lemma 2.1 to the estimation experiment $(\mathbf{H}, H, \mathcal{W}, E, S, L, v)$ we obtain necessary and sufficient conditions for an estimation correspondence to be of minimal risk. These conditions may be seen as a generalization of the RAO-Covariance-Method well-known from point estimation theory, where the loss function is given by the square of the norm of the $\mathbb{R}^n$, being the space of estimates.

Note, however, that we do not require here the Fréchet-differentiability of the loss function, which makes our approach applicable even for (separable) Banach space-valued statistics, where the norm in general need not be Fréchet-differentiable.

Theorem 3.8: *Let* Z' *be any convex subset of* Z_S^{up}; *and let* v *be a convex loss function. Then for given* $Z, Z' \in Z'$ *the function* $x \mapsto D_{Z'(x)} v(\gamma, Z(x))$ *defined on* $\mathbf{H}$ *is* P_γ*-quasiintegrable for every* $\gamma \in \Gamma$. *A necessary and sufficient condition for an estimation correspondence* $\overline{Z} \in Z'$ *to be of minimal risk is given by*

$$\int D_{Z'(x)} v(\gamma, \overline{Z}(x)) \, dP_\gamma(x) \geq 0 \qquad (Z' \in Z', \gamma \in \Gamma) .$$

For a proof see Meister and Moeschlin (1986); it follows the line of a proof given in Eberl and Moeschlin (1982) for the point-estimation case.

4. SET-VALUED BAYESIAN ESTIMATORS

A Bayesian estimation experiment is given by a nine-tuple

$$(\mathbb{H},H,\Gamma,G,P,E,S,L,v)$$

where $(\mathbb{H},H)$, Γ,E,S,L and v have the same meaning as in the non-Bayesian estimation experiment. We now assume additionally that $\mathbb{H}$ and Γ are Polish spaces, H is the completed Borel-σ-field of $\mathbb{H}$ and G is the Borel-σ-field of Γ, while L is now a G-$B(C(E))$-measurable correspondence and $v:\Gamma\times C(E)\to\mathbb{R}_+$ is measurable w.r.t. the product σ-field $G\otimes B(C(E))$. Finally let P be a probability distribution on $\mathbb{H}\times\Gamma$, which represents the beliefs of the statistician on the samples and parameters of the Bayesian estimation experiment. The marginal distributions of P on $\mathbb{H}$ and Γ are denoted by $P_{\mathbb{H}}$ and P_Γ, resp.; P_Γ is called the prior distribution. The conditional distribution of the projection on $\mathbb{H}$ w.r.t. the projection on Γ defines the family

$$W_{\mathbb{H}} := \{P(\cdot|\gamma)\ |\gamma\in\Gamma\}$$

of possible sample distributions on $\mathbb{H}$. Obviously, $W_{\mathbb{H}}$ corresponds to the set W in chapter 3. On the other hand, the conditional distribution of the projection on Γ w.r.t. the projection on $\mathbb{H}$ yields the family

$$W_\Gamma := \{P(\cdot|(x)\ |\ x\in\mathbb{H}\}$$

of posterior distributions on Γ.

Definition 4.1: *The (Bayes)risk of the estimation correspondence Z is defined by*

$$R(Z) := \int v(\gamma,Z(x))dP(x,\gamma)\ .$$

An estimation correspondence Z is called a Bayes estimation correspondence (BEC), iff it minimizes the Bayes risk among all estimation correspondences.

Hence, the Bayesian statistician has to solve the following optimization problem: minimize R(Z) among all estimation correspondences Z. The following lemma shows that the global minimization of R is equivalent to local minimization a.e., that means, minimization of the posterior risk $r(x,.):C(E)\to[0,\infty]$

defined by

$$r(x,C) := \int v(\gamma,C)P(d\gamma|x) \qquad (C\in C(E))$$

subject to given sample $x\in\mathbf{H}$. Obviously, r is a $H\otimes\mathcal{B}(C(E))$-measurable function and

$$R(Z) = \int r(x,Z(x))\,dP_{\mathbf{H}}(x)$$

holds for all estimation correspondences Z.

Lemma 4.2: *Suppose, there exists an estimation correspondence* Z' *with finite risk* R(Z'). *Then, the following two statements are equivalent:*

(1) Z *is a Bayes estimation correspondence.*

(2) Z *is an estimation correspondence s.t.*

$$r(x,Z(x)) = \min_{C\in S} r(x,C)$$

is satisfied $P_{\mathbf{H}}$-a.e.

A proof is given in Meister (1985). It is an application of a measurable selection theorem as stated in Castaing and Valadier (1977), Theorem III. 6, p. 65.

A characterization of Bayes estimation correspondences can be given by Lemma 2.1 and 4.2 .

Theorem 4.3: *Let* v *be a convex loss function. For* $x\in\mathbf{H}$ *let*

$$\mathcal{D}(x) := \{C\in S \mid r(x,C) < \infty\} .$$

Then, $\mathcal{D}(x)$ *is convex and for given* $C,C' \in \mathcal{D}(x)$ *the function* $\gamma \mapsto D_{C'}v(\gamma,C)$ *defined on* Γ *is* $P(\cdot|x)$-*quasiintegrable. A necessary and sufficient condition for an estimation correspondence* Z *to be a Bayes estimation correspondence is given by*

$$\int D_{C'}v(\gamma,Z(x))\,P(d\gamma|x) \geq 0 \qquad (C'\in\mathcal{D}(x))\ P_{\mathbf{H}}\text{- a.e.}$$

As an application of this result one obtains necessary and sufficient conditions for confidence intervals to be of minimal expected Hausdorff distance.

Example 4.4: Let $\Gamma = E = \mathbb{R}$ and $S = C(\mathbb{R})$. Further, let the correspondence L be defined by

$$L(\gamma) := [l(\gamma), u(\gamma)] \qquad (\gamma \in \mathbb{R})$$

for some measurable functions $l, u : \mathbb{R} \to \mathbb{R}$ with $l \leq u$; and let the loss function v be defined by

$$v(\gamma, C) := \delta(L(\gamma), C) = \max(|l(\gamma)-a|, |u(\gamma)-b|)$$
$$(\gamma \in \mathbb{R}, C = [a,b] \in C(\mathbb{R})) .$$

Suppose that the distributions $Q_+(x)$ of $l+u$ and $Q_-(x)$ of $u-l$ w.r.t. $P(\cdot|x)$ are atomless $(x \in \mathbf{H})$.

Then, a calculation as carried out in Meister (1985), Theorem 5.1, (using 4.3), shows that an estimation correspondence Z with

$$Z(x) =: [a(x), b(x)] \qquad (x \in \mathbf{H})$$

is a Bayes estimation correspondence, iff

$$a(x) + b(x) \in \operatorname{med} Q_+(x), \; b(x) - a(x) \in \operatorname{med} Q_-(x)$$

holds $P_{\mathbf{H}}$ - a.e., where med denotes the set of medians.

REFERENCES

AUMANN R.J. (1965): Integrals of set-valued functions, J.Math.Anal. Appl. 12, 1-12

BOCHNER S. (1933): Integration von Funktionen, deren Werte die Elemente eines Vektorraumes sind, Fund. Math. 20, 262-276

CASTAING C., M. VALADIER (1977): Convex analysis and measurable multifunctions, Lecture Notes in Mathematics, Vol. 580, Berlin

DEBREU G. (1967): Integration of correspondences, Proc. Fifth Berkeley Sympos. on Math. Statist. and Probability, Vol. II, Part I, 351-372

DUNFORD N., J.T. SCHWARTZ (1958): Linear Operators, Part I; General Theory, New York

EBERL W., O. MOESCHLIN (1982): Mathematische Statistik, Verlag de Gryuter, Berlin

HIAI F., H. UMEGAKI (1977): Integrals, Conditional Expectations and Martingales of Multivalued Functions J. Multivariate Anal. 7, 149-182

MEISTER H. (1985): A Bayesian risk concept and optimality conditions for confidence regions, to appear in Statistics and Decisions

MEISTER H., O.MOESCHLIN (1986): Unbiased set-valued estimators with minimal risk Preprint, FernUniversität Hagen

RÅDSTRØM H. (1952): An embedding theorem for spaces of convex sets, Proc. Amer. Math. Soc. 3, 165-169

FernUniversität Hagen
Fachbereich Mathematik und Informatik
Postfach 940

D-5800 Hagen 1, FRG

ON THE NON-ASYMPTOTIC DISTRIBUTION OF THE M.L. ESTIMATES IN CURVED EXPONENTIAL FAMILIES

Andrej Pázman

Bratislava

Key words: nonlinear regression, exponential families, maximum likelihood

ABSTRACT

The paper presents an approximative expression for the non-asymptotic probability density of the maximum likelihood estimates of parameters in a curved exponential family dominated by the Lebesgue measure. A typical example is the gaussian non-linear regression. A matrix in this expression is considered a possible alternative to the estimate of the Fisher information matrix in nonlinear models. The approach is differential-geometrical and the level of the approximation of the probability density is expressed in geometrical terms.

INTRODUCTION

If we are interested in models which are nonlinear in the expected values of the observed variables and which can be smoothly parametrized, curved exponential families investigated in Efron (1978) or in Amari (1985) can help us to understand and solve the underlying statistical problems on a clear geometrical basis. However, the majority of statistical results obtained so far by geometric methods have been asymptotic. It seems that the geometric approach should be especially useful when non-asymptotic statistical problems are considered, since geometrical terms like the radius of curvature, can help to express how exact the obtained non-asymptotic but approximative formulas are.

In this paper we present formulas for the non-asymptotic probability density of the maximum likelihood (M.L.) estimates in curved exponential families. A similar approach has been used in the particular case of the gaussian nonlinear regression (cf. Pázman (1984) and (1986b)). Results concerning the asymptotic probability density of the M.L. estimates have been presented in Barndorf-Nielsen (1983), Field (1982), Skovgaard (1981) and Hougaard (1985).

PROPERTIES OF CURVED EXPONENTIAL FAMILIES

Let $\mathcal{P} := \{P_\theta : \theta \in \Theta\}$ be an exponential family of probability distributions supported by a common open set $\mathcal{X} \subset R^N$ and defined by the densities

$$(1) \qquad f(x|\theta) := \frac{dP_\theta(x)}{d\lambda} = \exp\{-\varphi[t(x)] + \theta' t(x) - \varkappa(\theta)\}$$

with respect to the Lebesgue measure λ on $\mathcal{X}$ (cf. Barndorff-Nielsen (1979)). We shall make the following assumptions:

(A) The set Θ is an open subset of R^k $(k \leq N)$.

(B) The matrix of the first-order derivatives $\partial t / \partial x'$ of the mapping $t: \mathcal{X} \mapsto R^k$ (the sufficient statistic) is continuous on $\mathcal{X}$ and has the rank k.

The family $\mathcal{P}$ is only auxiliary. What we are really interested in is a curved exponential subfamily of $\mathcal{P}$.

Let $m < k$ and let Γ be an open subset of R^m. Let $\theta(.)$: $\gamma \in \Gamma \mapsto \theta(\gamma) \in \Theta$ be a one-to-one mapping which has continuous second-order derivatives $\partial^2\theta/\partial\gamma_i\partial\gamma_j$ and which has a full rank matrix of first-order derivatives $\partial\theta/\gamma'$. The family

$$\mathcal{P}_\Gamma := \{P_{\theta(\gamma)} : \gamma \in \Gamma\}$$

is a curved exponential family. The family $\mathcal{P}$ is the embedding family.

By definition, the M.L. estimate of γ is equal to $\hat{\gamma} := \hat{\gamma}(x)$ $:= \arg\max\, l(\gamma;x)$, where $l(\gamma;x) := \ln f(x|\theta(\gamma))$.

The families $\mathcal{P}$ and $\mathcal{P}_\Gamma$ have the following properties.

(P1). The function $\varkappa: \Theta \mapsto R$ is infinitely many times differentiable, and the mean and the variance of $t(x)$ are equal to

$$(2) \qquad E_\theta(t) = \frac{\partial \varkappa(\theta)}{\partial \theta}, \qquad D_\theta(t) = \frac{\partial^2 \varkappa(\theta)}{\partial \theta \, \partial \theta'}$$

(cf. Barndorff-Nielsen (1979)).

(P2). As a consequence of (A), (B) and of (P1), the matrix $D_\theta(t)$ is positive definite for every $\theta \in \Theta$, and the mapping $\theta \in \Theta \mapsto E_\theta(t) \in R^k$ is one-to-one and differentiable. The set

$$\mathcal{H} := \{ E_\theta(t) : \theta \in \Theta \}$$

is open.

(P3). The sample space of the sufficient statistic

$$\mathcal{T} := \{ t(x) : x \in \mathcal{X} \}$$

is open in R^k, and the family induced from $\mathcal{P}$ by the mapping t is dominated by the Lebesgue measure on $\mathcal{T}$ (cf. Pázman (1986a)). The corresponding densities are again exponential

$$(3) \qquad \exp\{ -\Psi(t) + \theta' t - \varkappa(\theta) \} \quad ; \quad (\theta \in \Theta)$$

where Ψ is a function on $\mathcal{T}$.

(P4). The M.L. estimate $\hat{\gamma}(x)$ is unique with probability one, i.e. $P_\theta \{ x : \hat{\gamma}(x) \neq \tilde{\gamma}(x) \text{ are two M.L. estimates} \} = 0$ for every $\theta \in \Theta$ (cf. Pázman (1986a)).

EXAMPLES. A typical example of $\mathcal{P}_\Gamma$ satisfying the assumptions (A) and (B) is the gaussian nonlinear regression. $\mathcal{P}$ is defined by $f(x|\theta) = \exp\{ [-(1/2) \ln(2\pi \det \Sigma) - (1/2) x' \Sigma^{-1} x] + \theta' x - (1/2) \theta' \Sigma \theta \}$ where Σ is a given p.d. matrix, $\mathcal{X} = \Theta = R^N$. Here $k=N$ and $t(x)=x$.

Normal densities multiplied by a polynomial are often used to approximate probability densities. Hence another example is given by $\mathcal{X} = \Theta = R^N$, $f(x|\theta) = \exp\{-(1/2)x'x\} \, p(x) \exp\{x'\theta - \varkappa(\theta)\}$ where $p(x) = p(x_1, \dots, x_N)$ is a positive polynomial. It can be verified that

$$\exp\{\varkappa(\theta)\} = (2\pi)^{N/2} \exp\{\tfrac{1}{2}\theta'\theta\} \, p[(\theta_1 + \frac{\partial}{\partial \theta_1}), \dots, (\theta_N + \frac{\partial}{\partial \theta_N})] 1$$

where $1 : \Theta \mapsto R$ is identically equal to 1.

Other examples satisfying the assumptions (A), (B) are given e.g. in Pázman (1986a).

THE GEOMETRY OF $\mathcal{P}_\Gamma$.

A nice exposition of the geometry of curved exponential families is presented in Efron (1978) and (1980), other (asymptotic) aspects are emphasized in Amari (1985). Here we give a short exposition with the stress on geodesic curves.

The I-divergence in the embedding family $\mathcal{P}$ is defined by

$$I(\theta,\bar{\theta}) := E_\theta[\ln(dP_\theta/dP_{\bar{\theta}})]$$

It relates any two points $\theta,\bar{\theta}$ of the set Θ , but due to the one-to-one correspondence between θ and $\mu := E_\theta(t)$, the I-divergence can be considered to be defined also on $\mathcal{H}$: $I(\mu,\bar{\mu}) := I(\theta,\bar{\theta})$. As well known, in exponential families the I-divergence is equal to

$$(4) \qquad I(\theta,\bar{\theta}) = (\theta-\bar{\theta})'\mu - [\varkappa(\theta) - \varkappa(\bar{\theta})]$$

(cf. e.g. Efron (1978)). Other divergences are available (c.f. Vajda (1982)), but the I-divergence is closely related to the maximum likelihood estimates.

The curved family $\mathcal{P}_\Gamma$ can be represented equivalently by two manifolds. The "canonical manifold"

$$\Theta_\Gamma := \{\theta(\gamma) : \gamma \in \Gamma\}$$

and the "expectation manifold"

$$\mathcal{H}_\Gamma := \{\mu(\gamma) : \gamma \in \Gamma\}$$

Consider the geometry of Θ_Γ induced by the I-divergence on Θ. By differentiation of Eq. (4) we obtain

$$\frac{\partial I(\theta(\gamma),\bar{\theta})}{\partial \gamma_i} = \frac{\partial I(\bar{\theta},\theta(\gamma))}{\partial \gamma_i} = 0$$

$$\frac{\partial^2 I(\theta(\gamma),\bar{\theta})}{\partial \gamma_i \partial \gamma_j} = \frac{\partial^2 I(\bar{\theta},\theta(\gamma))}{\partial \gamma_i \partial \gamma_j} = \frac{\partial \theta'(\gamma)}{\partial \gamma_i} \frac{\partial \mu(\gamma)}{\partial \gamma_j}$$

when we set $\theta(\gamma)$ for $\bar{\theta}$. Thus, if we require that the squared distance of two neighbourhood points $d[\theta(\gamma), \theta(\gamma + d\gamma)]$ should be proportional to $I(\theta(\gamma),\theta(\gamma + d\gamma))$, resp. to $I(\theta(\gamma + d\gamma), \theta(\gamma))$, we obtain

(5) $$d[\theta(\gamma),\ \theta(\gamma + d\gamma)] \approx [\, d\gamma' M(\gamma)\, d\gamma\,]^{1/2}$$

where

$$M(\gamma) := \frac{\partial \theta'(\gamma)}{\partial \gamma}\frac{\partial \mu(\gamma)}{\partial \gamma'}$$

is the Fisher information matrix. Denote $\Sigma_\gamma := D_{\theta(\gamma)}(t)$. According to Eq. (2) we have

(6) $$\frac{\partial \mu(\gamma)}{\partial \gamma_i} = \Sigma_\gamma \frac{\partial \theta(\gamma)}{\partial \gamma_i}$$

Eq. (5) leads to the following inner product in the tangent space to Θ_Γ at the point γ:

(7) $$\left\langle \frac{\partial \theta}{\partial \gamma_i}, \frac{\partial \theta}{\partial \gamma_j} \right\rangle_\gamma := \frac{\partial \theta'}{\partial \gamma_i} \Sigma_\gamma \frac{\partial \theta}{\partial \gamma_j} = \{M(\gamma)\}_{ij}$$

Similarly, the inner product in the tangent space to $\mathcal{M}_\Gamma$ is given by

$$\left\langle \frac{\partial \mu}{\partial \gamma_i}, \frac{\partial \mu}{\partial \gamma_j} \right\rangle^\gamma := \frac{\partial \mu'}{\partial \gamma_i} \Sigma_\gamma^{-1} \frac{\partial \mu}{\partial \gamma_j} = \{M(\gamma)\}_{ij}$$

On a surface S which is in a Euclidean space we define a geodesics as a curve $l: u \in (a,b) \mapsto S$ having its vector of curvature $d^2 l(u)/du^2$ always orthogonal to S (i.e. to the tangent vectors to S at the point $l(u)$). By analogy we define a θ-geodesics on Θ_Γ as a mapping $z:(a,b) \mapsto \Gamma$ which is twice continuously differentiable and which satisfies the differential equations

(8) $$\left\langle \frac{d^2\theta[z(u)]}{du^2}, \frac{\partial \theta}{\partial \gamma_i} \right\rangle_{z(u)} = 0 \ ; \qquad (i=1,\ldots,m)$$

for every $u \in (a,b)$. The vector $d^2\theta[z(u)]/du^2$ is the vector of curvature and

(9) $$\rho_{\theta,z}(u) := \{\|\, d\theta[z(u)]/du \|_{z(u)}\}^2 / \| d^2\theta[z(u)]/du^2 \|_{z(u)}$$

is the radius of curvature at $\gamma = z(u)$. (Cf. Efron (1978), eq. (3.8))

In fact, the numerical computation of the radius of curvature at a given point γ does not require the solution of any differential equation but only of some linear algebraic equations (cf. e.g. the appendix in Pázman (1984)).

Denote by

$$\rho_\theta(\gamma) := \inf_z \rho_{\theta,z}(u) \tag{10}$$

the infimum being taken over all θ-geodesics z with $z(u)=\gamma$.

From $\partial l(\hat{\gamma};x)/\partial\gamma_i = 0$; $(i=1,\dots,m)$ we obtain that $\hat{\gamma}$ is the solution of the normal equation

$$[t(x) - \mu(\gamma)]' \Sigma_\gamma^{-1} \frac{\partial\mu}{\partial\gamma'} = 0. \tag{11}$$

A solution of (11) is the M.L. estimate only if the matrix with entries

$$\frac{\partial^2 l(\gamma;x)}{\partial\gamma_i\,\partial\gamma_j} = \{M(\gamma)\}_{ij} + [t(x) - \mu(\gamma)]' \frac{\partial^2\theta}{\partial\gamma_i\,\partial\gamma_j}$$

is positive definite (=p.d.). This is certainly true if the sample point $t(x)$ is close to $\mathcal{H}_\mu$, because $M(\gamma)$ is p.d. For more distant points $t(x)$ we generalize eq. (5.4) in Efron (1978) as follows.

Let z be a θ-geodesics such that $z(0)=\gamma$. We have

$$\frac{dz'(0)}{du}\frac{\partial^2 l(\gamma;x)}{\partial\gamma\partial\gamma'}\frac{dz(0)}{du} = \frac{d\mu'[z(0)]}{du}\frac{d\theta[z(0)]}{du} - [t(x)-\mu(\gamma)]'\frac{d^2\theta[z(0)]}{du^2}$$

hence $\partial^2 l(\gamma;x)/\partial\gamma\partial\gamma'$ is p.d. iff

$$[t-\mu(\gamma)]'\frac{\partial^2\theta[z(0)]}{du^2} < \left|\left|\frac{d\theta[z(0)]}{du}\right|\right|_\gamma^2$$

for every θ-geodesics z(.). Thus from the Schwarz inequality and from Eqs. (9) and (10) we obtain that the condition

$$||t(x)-\mu(\gamma)||^2 < \rho_\theta(\gamma) \tag{12}$$

is sufficient to ensure that $\partial^2 l(\gamma;x)/\partial\gamma\partial\gamma'$ is p.d..

THE PROBABILITY DENSITY OF $\hat{\gamma}$.

Take $\bar{\gamma} \in \Gamma$ and denote $\bar{\theta} := \theta(\bar{\gamma})$. We shall consider the probability density of $\hat{\gamma}$ under the fixed probability density of t :

$$\exp\{-\Psi(t) + \bar{\theta}' t - \varkappa(\bar{\theta})\}.$$

With every $\gamma \in \Gamma$ we associate an auxiliary affine set

$$\Theta_\gamma := \{\bar{\theta} + \frac{\partial \theta(\gamma)}{\partial \gamma'} \delta \quad : \delta \in R^m\}.$$

Denote $\theta_\gamma(\delta) := \bar{\theta} + \frac{\partial \theta(\gamma)}{\partial \gamma'} \delta$, $\mu_\gamma(\delta) = E_{\theta_\gamma(\delta)}(t)$. Denote by δ^* the solution of the equation

$$[\mu(\gamma) - \mu_\gamma(\delta)]' \frac{\partial \theta(\gamma)}{\partial \gamma'} = 0$$

which is an analogue to the normal equation (11). Finally denote $\theta^*(\gamma) := \theta_\gamma(\delta^*) = \bar{\theta} + [\partial\theta(\gamma)/\partial\gamma'] \delta^*$, $\mu^*(\gamma) := \mu_\gamma(\delta^*)$, $\Sigma^*_\gamma := D_{\theta^*(\gamma)}(t)$. Using this notation we have

$$(13) \qquad [\mu(\gamma) - \mu^*(\gamma)]' \frac{\partial \theta(\gamma)}{\partial \gamma'} = 0$$

Take k-m vectors $w_1(\gamma), \ldots, w_{k-m}(\gamma) \in R^k$ which are differentiable in a neighbourhood of γ and are such that

$$(14) \qquad w_l'(\gamma) \Sigma_\gamma^{-1} \frac{\partial \mu}{\partial \gamma'} = 0, \qquad w_l'(\gamma) \Sigma_\gamma^{*-1} w_r(\gamma) = \delta_{lr}.$$

With every $t \in \mathcal{T}$ associate new coordinates $\gamma_1, \ldots, \gamma_m$, $a_1, \ldots, a_{k-m}$ as follows: $\gamma := (\gamma_1, \ldots, \gamma_m)$ is a solution of Eq. (11) and $a_i := w_i'(\gamma) \Sigma_\gamma^{*-1}(t - \mu^*(\gamma))$. That means

$$(15) \qquad t = \mu^*(\gamma) + \sum_{l=1}^{k-m} a_l w_l(\gamma).$$

Let $\mathcal{T}_0$ be an open subset of $\mathcal{T}$ containing the point $\mu(\bar{\gamma})$ and such that for every $t \in \mathcal{T}_0$ there is a unique solution of Eq.(11) and that the inequality (12) holds for this solution. We do not specify $\mathcal{T}_0$ explicitly in the general case, but in the gaussian nonlinear regression it is a sphere centred in $\mu(\bar{\vartheta})$, (cf. Pázman (1986b)). Obviously, the coordinates $\vartheta_1,\ldots,\vartheta_m$, $a_1,\ldots,a_{k-m}$ are defined uniquely on $\mathcal{T}_o$, and $\vartheta_i = \hat{\vartheta}_i$; $(i=1,\ldots,m)$. Let g be a mapping defined by

$$(16)\qquad g(\vartheta,a) := \mu^*(\vartheta) + \sum_{l=1}^{k-m} a_l w_l(\vartheta)$$

Denote by ∇g the matrix $\nabla g := (\partial g/\partial\vartheta', \partial g/\partial a')$. Obviously, $(\partial g/\partial a') = W := (w_1(\vartheta),\ldots,w_{m-k}(\vartheta))$. We have

$$\frac{\det^2 \nabla g}{\det \Sigma^*_\vartheta} = \det(\nabla g' \Sigma^{*-1}_\vartheta \nabla g)$$

$$(17)\qquad = \det \begin{pmatrix} \frac{\partial g'}{\partial\vartheta}\Sigma^{*-1}_\vartheta \frac{\partial g}{\partial\vartheta'}, & \frac{\partial g'}{\partial\vartheta}\Sigma^{*-1}_\vartheta W \\ W'\Sigma^{*-1}_\vartheta \frac{\partial g}{\partial\vartheta'}, & I \end{pmatrix}$$

$$= \det\left\{\frac{\partial g'}{\partial\vartheta}\Sigma^{*-1}_\vartheta (I-P^\vartheta)\frac{\partial g}{\partial\vartheta'}\right\}$$

where $P^\vartheta := WW'\Sigma^{*-1}_\vartheta$ and where we used eq. (IIb) in Gantmacher (1966), chapter II, §5. It can be easily verified that P^ϑ is the projector onto the set $B_\vartheta := \{t : t \in R^k,\ t'\Sigma^{-1}_\vartheta(\partial\mu/\partial\vartheta') = 0\}$, orthogonal with respect to the inner product $a'\Sigma^{*-1}_\vartheta b$; $(a,b \in R^k)$. We have $P^\vartheta\Sigma^*_\vartheta(\partial\theta/\partial\vartheta_i) = WW'(\partial\theta/\partial\vartheta_i) = 0$ (see Eq. (14)), hence the vectors $\Sigma^*_\vartheta(\partial\theta/\partial\vartheta_i)$; $(i=1,\ldots,m)$ constitute a linear basis of the subspace othogonal to B_ϑ. It follows that the projector orthogonal to P^ϑ is equal to

$$I-P^\vartheta = \Sigma^*_\vartheta \frac{\partial\theta}{\partial\vartheta'}\left[\frac{\partial\theta'}{\partial\vartheta}\Sigma^*_\vartheta\frac{\partial\theta}{\partial\vartheta'}\right]^{-1}\frac{\partial\theta'}{\partial\vartheta}$$

After setting this expression into Eq. (17) we obtain

$$(18)\qquad \frac{\det^2 \nabla g}{\det \Sigma^*_\vartheta} = \frac{\det^2\left[\frac{\partial g'}{\partial\vartheta}\frac{\partial\theta}{\partial\vartheta'}\right]}{\det\left[\frac{\partial\theta'}{\partial\vartheta}\Sigma^*_\vartheta\frac{\partial\theta}{\partial\vartheta'}\right]}$$

From Eq. (16) we obtain that

$$\frac{\partial g'}{\partial \gamma}\frac{\partial \theta}{\partial \gamma'} = \frac{\partial \mu^{*\prime}}{\partial \gamma}\frac{\partial \theta}{\partial \gamma'} + \sum_l a_l \frac{\partial w_l'}{\partial \gamma}\frac{\partial \theta}{\partial \gamma'},$$

Applying $\partial/\partial\gamma$ to both sides of Eq. (13) we obtain

$$\frac{\partial \mu^{*\prime}}{\partial \gamma}\frac{\partial \theta}{\partial \gamma'} = M(\gamma) + [\mu(\gamma) - \mu^*(\gamma)]' \frac{\partial^2 \theta}{\partial \gamma \partial \gamma'}$$

Similarly we obtain from $w'(\gamma)(\partial\theta/\partial\gamma') = 0$ (Eq.(14))

$$\frac{\partial w_l'}{\partial \gamma}\frac{\partial \theta}{\partial \gamma'} = -w_l' \frac{\partial^2 \theta}{\partial \gamma \partial \gamma'}$$

Hence

$$\frac{\partial g'}{\partial \gamma}\frac{\partial \theta}{\partial \gamma'} = Q(\gamma,\bar{\gamma}) - \sum_{l=1}^{k-m} a_l D(w_l(\gamma)) \tag{19}$$

where

$$D(b) := D(b,\gamma) := b' \frac{\partial^2 \theta}{\partial \gamma \partial \gamma'}; \qquad (b \in R^k) \tag{20}$$

and

$$Q := Q(\gamma,\bar{\gamma}) := M(\gamma) + D[\mu(\gamma) - \mu^*(\gamma)] \tag{21}$$

We have expressed the dependence of Q on $\bar{\gamma}$ since $\mu^*(\gamma)$ depends on $\bar{\gamma}$.

From Eq.(12) it follows that the left-hand side of Eq.(19) is p.d. on $\mathcal{T}_0$. Hence putting the expression (19) into Eq.(18) we obtain the following lemma:

Lemma 1. The Jacobian of the mapping g is equal to

$$|\det \nabla g| = \frac{\det^{1/2}\Sigma_\gamma^* \det Q(\gamma,\bar{\gamma})}{\det^{1/2}[\frac{\partial \theta'}{\partial \gamma}\Sigma_\gamma^* \frac{\partial \theta}{\partial \gamma'}]} \det[I - \sum_{l=1}^{k-m} a_l D(w_l) Q^{-1}(\gamma,\bar{\gamma})]$$

Lemma 2. We have the equality

$$\exp\{-\Psi(t) + \bar{\theta}' t - \varkappa(\bar{\theta})\}$$

$$= \exp\{-I[\mu^*(\gamma), \mu(\bar{\gamma})]\} \exp\{-\Psi(t) + \theta^{*\prime}(\gamma) t - \varkappa[\theta^*(\gamma)]\}$$

Proof. From Eq.(5) we obtain

$$-\psi(t)+\theta^{*\prime}(\gamma)t-\varkappa[\theta^{*}(\gamma)] - I[\mu^{*}(\gamma),\bar{\mu}]$$
$$= -\psi(t)+\theta^{*\prime}(\gamma)t+\mu^{*\prime}(\gamma)\bar{\theta} -\mu^{*\prime}(\gamma)\theta^{*}(\gamma)-\varkappa(\bar{\theta}).$$

From Eqs.(11) and (13) we have $[\bar{\theta}-\theta^{*}(\gamma)]'[t-\mu^{*}(\gamma)]=0$, hence $\bar{\theta}'t=\theta^{*\prime}(\gamma)t+\bar{\theta}'\mu^{*}(\gamma)-\theta^{*\prime}(\gamma)\mu^{*}(\gamma)$. We put this expression into Eq. (22) and obtain the required equality. □

For any $m\times m$ matrix A denote by $A^{(s)}$ the $\binom{m}{s}\times\binom{m}{s}$ matrix with entries equal to the $s\times s$ minors of A. Consequently, $\mathrm{tr}A^{(s)}$ is the sum of all $s\times s$ principal minors of A. The following rule holds (cf. Fiedler (1981))

$$A^{(s)}B^{(s)}=(AB)^{(s)} \tag{23}$$

We have the following equality (cf. Gantmacher (1966),III, §7,eq.(5.7))

$$\det[I-\sum_{l=1}^{k-m} a_l D(w_l)]=1+\sum_{s=1}^{m}(-1)^s \mathrm{tr}[\sum_{l} a_l D(w_l)Q^{-1}]^{(s)}$$

Denote by $r(a;\gamma,\bar{\gamma})$ the right-hand side of this equality. It is a polynomial in the variables $a_1,\dots,a_{k-m}$. From Lemma 1 and Lemma 2 we obtain the following statement.

Theorem 1. The contribution of the "samples" belonging to to the probability density of the M.L. estimate $\hat{\gamma}$ is equal to

$$q(\hat{\gamma}|\bar{\gamma})[\det^{1/2}\Sigma^{*}_{\hat{\gamma}}]\int_{A_{\hat{\gamma}}}\exp\{-\psi[g(\hat{\gamma},a)]+\theta^{*\prime}(\hat{\gamma})g(\hat{\gamma},a)-\varkappa[\theta^{*}(\hat{\gamma})]\}\, r(a;\hat{\gamma},\bar{\gamma})\, da \tag{24}$$

where

$$q(\hat{\gamma}|\bar{\gamma}):=\frac{\det Q(\hat{\gamma},\bar{\gamma})}{\det^{1/2}[\frac{\partial\theta'(\hat{\gamma})}{\partial\gamma}\Sigma^{*}_{\hat{\gamma}}\frac{\partial\theta(\hat{\gamma})}{\partial\gamma'}]}\exp\{-I[\mu^{*}(\hat{\gamma}),\mu(\bar{\gamma})]\} \tag{25}$$

and where

$$A_{\hat{\gamma}}:=\{a_1,\dots,a_{k-m}:\ g(\hat{\gamma},a)\in\mathcal{T}_{\circ}\}.$$

The obtained results need some comments.

i) The expression $q(\hat{\gamma}|\bar{\gamma})$ can be considered the leading term in Eq.(24), since, as shown in Hougaard (1985) based on previous results in Field (1982) and Skovgaard (1981), the asymptotic probability density of $\hat{\vartheta}$, when the whole experiment is repeated n times, is equal to

$$(n/2\pi)^{m/2}\exp\{-nI[\mu^*(\hat{\vartheta}),\mu(\bar{\vartheta})]\}\{\frac{\det Q(\hat{\vartheta},\bar{\vartheta})}{\det^{1/2}[\frac{\partial\theta'(\hat{\vartheta})}{\partial\vartheta}\Sigma^*_{\hat{\vartheta}}\frac{\partial\theta(\hat{\vartheta})}{\partial\vartheta'}]}+o(\frac{1}{n})\}$$

(We note that stronger regularity conditions than ours are required in Hougaard (1985)).

ii) We can give bounds to the coefficients of the polynomial $r(a;\hat{\vartheta},\bar{\vartheta})$ expressed in geometrical terms (see Theorem 2). The function $g(\vartheta,a)$ is linear in $a_1,\dots,a_{k-m}$, hence the evaluation of the integral in Eq. (24) leads to the computation of higher order moments in an exponential family, which itself can yield to the differentiation of a certain function.

iii) The situation is especially favourable in the case of the gaussian nonlinear regression (cf. Pázman (1984) and (1986b)). In this case the approximative density is given by

$$(2\pi)^{-m/2}q(\hat{\vartheta}|\bar{\vartheta})$$

and the expressions in $q(\hat{\vartheta}|\bar{\vartheta})$ are simplified:

$$Q(\hat{\vartheta},\bar{\vartheta})=M(\hat{\vartheta})+[\mu(\hat{\vartheta})-\mu(\bar{\vartheta})]'[I-P^{\hat{\vartheta}}](\partial^2\hat{\theta}/\partial\vartheta\partial\vartheta')$$

(26)
$$\frac{\partial\theta'(\hat{\vartheta})}{\partial\vartheta}\Sigma^*_{\hat{\vartheta}}\frac{\partial\theta(\hat{\vartheta})}{\partial\vartheta'}=M(\hat{\vartheta})$$

$$I[\mu^*(\hat{\vartheta}),\mu(\bar{\vartheta})]=(1/2)||P^{\hat{\vartheta}}[\mu(\hat{\vartheta})-\mu(\bar{\vartheta})]||^2$$

Moreover, the bounds for the difference between $q(\hat{\vartheta}|\vartheta)$ and the formula given in Theorem 1 can be expressed in terms of the radius of curvature.

Let us return to the general case, and denote $e:=\sum_l a_l w_l/||\sum_l a_l w_l||_{\vartheta}$. We can write

(27)
$$r(a;\hat{\vartheta},\bar{\vartheta})=\sum_{s=1}^{k-m}(-1)^s[||\sum_l a_l w_l(\hat{\vartheta})||_{\hat{\vartheta}}]^s\ \mathrm{tr}[D(e,\hat{\vartheta})Q^{-1}(\hat{\vartheta},\bar{\vartheta})]^{(s)}$$

Theorem 2. Under the assumption that $||\mu(\hat{\vartheta})-\mu^*(\hat{\vartheta})||_{\hat{\vartheta}}<(1/2)\,\rho_\theta(\hat{\vartheta})$ we have the inequality

$$|\ \mathrm{tr}[D(e,\hat{\vartheta})Q^{-1}(\hat{\vartheta},\bar{\vartheta})]^{(s)}|\le\binom{m}{s}[\frac{2}{\rho_0(\hat{\vartheta})}]^s$$

Proof. Let $\hat{\gamma}=\gamma$ be fixed. We shall omit to write it when no confusion is likely to arise.
There is a matrix U such that $U'MU=I$. Denote

$$\bar{D}(e):=U'D(e)U$$

$$\bar{Q}:=U'QU$$

Using the rule (23) we obtain

$$\mathrm{tr}[D(e)Q^{-1}]^{(s)} = \mathrm{tr}[\bar{D}(e)\bar{Q}^{-1}]^{(s)}$$

Denote by $c^{(1)},\dots,c^{(m)}$ the orthogonal eigenvectors of $\bar{D}(e)$ (i.e. $c^{(i)\prime}c^{(j)}=\delta_{ij}$), and by $\lambda_1,\dots,\lambda_m$ the corresponding eigenvalues.

Let z be a θ-geodesics such that $z(0)=\gamma$, $dz(0)/du = Uc^{(1)}$. Obviously

$$\left\|\frac{d\theta[z(0)]}{du}\right\|_\gamma = c^{(1)\prime}U'MUc^{(1)}=1$$

$$e'\frac{d^2\theta[z(0)]}{du^2} = c^{(1)\prime}\bar{D}(e)c^{(1)} = \lambda_1$$

Hence, using the Schwarz inequality and Eqs. (9) and (10) we obtain the important inequality

$$|\lambda_1| \le \frac{1}{\varrho_\theta(\gamma)} \tag{28}$$

Denote $C:=(c^{(1)},\dots,c^{(m)})$, $\Lambda:=\mathrm{diag}\,(\lambda_1,\dots,\lambda_m)$. From $\bar{D}(e)=C\Lambda C'$ we obtain using the rule (23)

$$\mathrm{tr}[\bar{D}(e)\bar{Q}^{-1}]^{(s)} = \mathrm{tr}\,\{\Lambda^{(s)}[[C'\bar{Q}C]^{-1}]^{(s)}\}.$$

The matrix $\Lambda^{(s)}$ is diagonal having on the diagonal all products of the form $\lambda_{i_1}\dots\lambda_{i_s}$; $(i_1<\dots<i_s)$. Hence from Eqs. (28) we obtain

$$\mathrm{tr}[D(e)Q^{-1}]^{(s)} \le [1/\varrho_\theta(\gamma)]^s\ \mathrm{tr}\{(C'QC)^{-1}\}^{(s)}$$

$$=(1/\varrho_{\theta(\gamma)})^s\ \mathrm{tr}(\bar{Q}^{-1})^{(s)} \tag{29}$$

To evaluate $\mathrm{tr}(\bar{Q}^{-1})^{(s)}$ denote $\tilde{e}:=[\mu(\gamma)-\mu^*(\gamma)]/\|\mu(\gamma)-\mu^*(\gamma)\|_\gamma$. From Eq. (21) it follows that

(30) $\bar{Q}=I+||\mu-\mu^*||_\gamma \bar{D}(\tilde{e})$

Let $\tilde{\pi}_1,\ldots,\tilde{\pi}_m$ be the eigenvalues of $\bar{D}(\tilde{e})$, and let F be the matrix of the corresponding eigenvectors. In exactly the same way as in Eq. (28) we obtain

(31) $|\tilde{\pi}_1| \leq 1/\rho_\theta(\gamma).$

According to Eq. (30), the matrices $\bar{Q}$ and $\bar{D}(e)$ have the same eigenvectors, but the eigenvalues of $\bar{Q}$ are equal to

$$1+||\mu-\mu^*||_\gamma \pi_1$$

From the inequality (31) it follows that

$$1-\frac{||\mu-\mu^*||_\gamma}{\rho_\theta(\gamma)} \leq 1+||\mu-\mu^*||_\gamma \pi_1$$

Therefore

$$\mathrm{tr}(\bar{Q}^{-1})^{(s)} = \mathrm{tr}[(F'\bar{Q}F)^{-1}]^{(s)}$$

$$= \sum_{i_1<\ldots<i_s} \prod_{j=1}^{s} (1+||\mu-\mu^*||_\gamma \pi_{i_j})^{-1}$$

$$\leq \binom{m}{s} \left[\frac{\rho_\theta(\gamma)}{\rho_\theta(\gamma)-||\mu-\mu^*||_\gamma}\right]^s$$

Setting this into Eq.(29) we obtain

$$\mathrm{tr}[D(e)Q^{-1}]^{(s)} \leq \binom{m}{s} [2/\rho_\theta(\gamma)]^s. \square$$

PROPERTIES OF $Q(\hat{\gamma},\bar{\gamma})$

In the particular case of $\theta(\gamma)=G\gamma$; ($\gamma\in\Gamma$) for some matrix G, we have $Q(\hat{\gamma},\bar{\gamma})=M(\hat{\gamma})$=the estimate of the Fisher information matrix. In the general (nonlinear) case the matrix $M(\hat{\gamma})$ depends only on the first order derivatives of $\mu(\gamma)$ resp. of $\theta(\gamma)$, but $Q(\hat{\gamma},\bar{\gamma})=M(\hat{\gamma})+$ a "correcting term" depending also on the second-order derivatives and taking better into account that the family is curved.

If we set an estimate of γ, which is different from $\hat{\gamma}$, instead of $\bar{\gamma}$ into $Q(\hat{\gamma},\bar{\gamma})$, we obtain an estimate of the Fisher information matrix.

We present some properties of $Q(\hat{\gamma},\bar{\gamma})$:

i) The matrix $Q(\hat{\gamma},\bar{\gamma})$ is symmetric,as follows from Eq. (20).

ii) Lemma 3. The matrix $Q(\hat{\gamma},\bar{\gamma})$ is p.d. if

$$\|\mu(\hat{\gamma})-\mu^*(\hat{\gamma})\| < \varrho_\theta(\hat{\gamma})$$

Proof. Let $z(u)$ be a θ-geodesics such that $z(0)=\hat{\gamma}$. Using Eq. (13) we obtain that

$$\frac{dz'(0)}{du}\, Q\, \frac{dz(0)}{du} = \frac{d\theta'[z(0)]}{du}\Sigma_{\hat{\gamma}}\frac{d\theta[z(0)]}{du} + (\hat{\mu}-\mu^*)'\frac{d^2\theta[z(0)]}{du^2}$$

Hence Q is p.d. iff for every θ-geodesics z the inequality

$$(\hat{\mu}-\mu^*)'\Sigma_{\hat{\gamma}}^{-1/2}\,\Sigma_{\hat{\gamma}}^{1/2}\,\frac{d^2\theta[z(0)]}{du^2} \leq \frac{d\theta'[z(0)]}{du}\,\Sigma_{\gamma}\,\frac{d\theta[z(0)]}{du}$$

holds. Applying the Schwarz inequality on the left-hand side of this inequality and using Eqs. (9) and (10) we obtain the required statement. □

iii) Lemma 4. The matrix $M(\hat{\gamma})-Q(\hat{\gamma},\bar{\gamma})$ is p.d. iff

$$[\mu^*(\hat{\gamma})-\mu(\hat{\gamma})]'\,\frac{d^2\theta[z(0)]}{du^2} > 0$$

for every θ-geodesics $z(u)$ such that $z(0)=\gamma$.

Proof. From Eqs. (20) and (21) it follows that

$$\frac{dz'(0)}{du}\,[M-Q]\,\frac{dz(0)}{du} = (\mu^*-\hat{\mu})'\,\frac{d^2\theta[z(0)]}{du^2}\,. \quad \square$$

The vector $d^2\theta[z(0)]/du^2$ points towards the centre of the curvature of the geodesics z. Hence the nonlinear experiment is worse than a linear experiment with the same M.L. estimate of the information matrix $M(\gamma)$, if $\mu^*(\hat{\gamma})$ is on the same side of the tangent plane to $\textcircled{\scriptsize 4}_r$ as the centres of curvature of all θ-geodesics through the point $\hat{\gamma}$.

REFERENCES

Amari S. (1985): Differential-Geometrical Methods in Statistics. Lecture Notes in Statistics No 28. Springer-Verlag, Berlin-Heidelberg.

Barndorff-Nielsen O.E. (1979): Information and Exponential Families in Statistical Theory. Wiley, Chichester.

(1983): On a formula for the distribution of the maximum likelihood estimator. Biometrika 70, 343-365.

Efron B. (1978): The geometry of exponential families. Ann.Statist. 6, 362-376.

(1980): A distance theorem for exponential families. Prob.and Math.Statist. 1, 95-98.

Fiedler M. (1981): Special Matrices and Their Use in Numerical Mathematics. (Czech.) SNTL, Prague.

Field C. (1982): Small sample asymptotic expansions for multivariate M-estimates. Ann.Statist. 10, 672-689.

Gantmacher F.R. (1966): Matrix Theory (Russian). Nauka, Moscow.

Hougaard P. (1985): Saddle-point aproximations for curved exponential families. Statistics and Probability Letters 3, 161-166.

Pázman A. (1984): Probability distribution of the multivariate nonlinear least squares estimates. Kybernetika 20, 209-230.

(1986a): On the uniqueness of the M.L. estimates in curved exponential families. Kybernetika 22, 124-132.

(1986b): On formulas for the distribution of nonlinear L.S. estimates. Statistics (to appear).

Skovgaard I.M. (1981): Large deviation approximations for maximum likelihood estimators. Preprint 1981/9, Institute of Mathematical Statistics, University of Copenhagen.

Vajda I. (1982): Information Theory and Statistical Decision (Slovak) ALFA, Bratislava.

A. Pázman

Mathematical Institute
Slovak Academy of Sciences
Obrancov mieru 49
814 73 Bratislava
Czechoslavakia

RECENT RESULTS IN ROBUSTNESS RESEARCH

Dieter Rasch

Dummerstorf-Rostock

1. INTRODUCTION

Robustness in statistics means insensitivity of statistical procedures against the violation of assumptions underlying the theoretical development of these procedures. The term was first used in this connection by Box (1953). Later Huber (1964) initiated a theory of robust estimation (robust against outliers and contamination of distributions).

In this paper robustness research is understood in the original sense used by Box as the investigation of the behaviour of statistical procedures used in applied statistics if one or some of the assumptions are violated. Many procedures are based on the normality assumption, and it is this assumption whose robustness is investigated mainly.

General definitions of robustness can be found in Zielinski (1981), but we use the special definition of the robustness of tests and confidence estimations given by Rasch and Herrendörfer (1981). As an example of the latter we give the

Definition: Let $\underline{d}_\alpha$ [1)] be a confidence estimation with respect of a parameter $\theta \in \Omega$ of a probability measure P_θ. Let $\underline{d}_\alpha$ be based on an experimental design V_n and let $1 - \alpha$ be the maximum confidence coefficient of $\underline{d}_\alpha$ as long as $P_\theta \in G$.
Let $H \supset G$ be a class of distributions containing G and let $1 - \alpha(V_n, P)$ be the actual confidence coefficient of $\underline{d}_\alpha$ for any $P \in H$. Then $\underline{d}_\alpha$ is called ε-robust in H, if

1) random variables are underlined

(1) $$\sup_{P \in H} |\alpha(V_n, P) - \alpha)| \le \varepsilon$$

The robustness of a test in H with an actual first kind risk α in G can be defined analoguously. The deviation from normality is measured by the skewness

(2) $$\gamma_1 = \frac{\mu_3}{\sigma^3}$$

and the kurtosis

(3) $$\gamma_2 = \frac{\mu_4}{\sigma^4} - 3$$

For distributions with existing first four moments we have

(4) $$\gamma_2 \ge \gamma_1^2 - 2$$

This and sharper inequalities for unimodal distributions are proved by Guiard (1981). We conjecture that the robustness behaviour of statistical methods for continuous unimodal distributions with fixed first four moments also not vary considerably otherwise all the robustness results are valid only for the distributions for which they were obtained and can not be generalized.

2. METHODS USED IN OUR ROBUSTNESS RESEARCH

The aim of this paper is to present results of research program in robustness which was undertaken by the department of biomathematics at the Research Centre of Animal Production in Dummerstorf-Rostock (GDR). We used three kinds of methods investigating the robustness of standard tests, confidence estimations and selection procedures.

2.1 ANALYTICAL METHODS

An analytical method leads to exact robustness results by mathematical derivations.

Unfortunately such results are very rare. For instance the robustness of the t-test was investigated by such methods only for sample sizes n = 2 and n = 3 but only moderate n-values are of practical importance. In some cases we found analytical methods useful, e.g. in the judgement of the robustness of the formula for calculating selection difference (Rasch and Pierer, 1985) and the robustness of selection procedures for Bechhofer's indifference zone approach (Domröse 1985b, Domröse and Rasch 1987).

Using series expansions to compute the robustness behaviour of tests approximately led to uncontrollable approximation errors which as a rule were larger than the sampling errors of simulation experiments. Therefore we avoided such investigations.

2.2 COMBINATORIAL METHODS

The combinatorial method was developed by Herrendörfer and Rasch (1980) and used to calculate the robustness of the one-sample u- and t-test. This method can be used, if the alternative distribution is discrete with finite support i. e. if it is a k-point--distribution of the form

$$(5) \qquad P_k = \binom{y_1,\dots,y_k}{p_1,\dots,p_k}, \qquad \sum_{i=1}^{k} p_i = 1, \qquad p_i \geq 0$$

Samples of size n of such a distribution have

$$M = \binom{n+k-1}{k-1}$$

possible realizations (for different y_i). For each realization it is possible to calculate the value of a test statistic, a confidence interval or the result of a selection rule and the probability of its occurence. Let for instance $t_1,\dots,t_M$ be the possibly, but not necessarily, different values of the confidence intervals of the mean μ of a distribution and let $\pi_1,\dots,\pi_M$ be the probabilities of their occurence. Then the actual confidence coefficient and the power of the one-sample t-test for the distribution P_k can be calculated.

We used the method for the most extreme interesting case k = 3 (the two-point distributions are found at the parabola $\gamma_2 = \gamma_1^2 - 2$ and are of no practical interest) and demonstrate the method for this case in more detail.

In

$$(6) \qquad P_3 = \binom{y_1 \; y_2 \; y_3}{p_1 \; p_2 \; p_3}, \quad p_1 + p_2 + p_3 = 1, \qquad p_i \geq 0$$

five parameters can be chosen under the side condition (4). Put without loss of generality $E(\underline{y}) = 0$, $V(\underline{y}) = 1$. Further, we fix $\gamma_1(\underline{y}) = \gamma_1$, $\gamma_2(\underline{y}) = \gamma_2$ with fixed values (γ_1,γ_2) fulfilling (4). Then four parameters of P_3 are fixed and let us say y_3 is further at our disposal. Herrendörfer and Feige (1984, 1985) used the method to find the smallest integer n_0 so that for $n > n_0$

the t-test is ε = 0,2 α-robust against P_3. The result in its dependence on y_3 is shown for α = 0,05 and some (γ_1, γ_2)-pairs in table 1.

TABLE 1

Values of n_0 in dependence on y so that the t-test is 0.01-robust for α=0.05 if $n > n_0$

y_3	$\gamma_1=0$ $\gamma_2=7$	$\gamma_1=0$ $\gamma_2=-1.9$	$\gamma_1=1$ $\gamma_2=0$	$\gamma_1=1$ $\gamma_2=.0.5$
2	38	38	32	35
3	66	66	41	54
6	41	41	131	136
10	65	65	211	152
50	94	94	155	185
100	-	-	185	170

2.3 SIMULATION METHOD

Using the simulation method the risk of a statistical procedure under the violation of an assumption are not calculated exactly as by the methods in 2.1 or 2.2, but are estimated by a simulation experiment of planned size N. N random samples or r-tuples of samples (depending on the problem) are generated using random number generators and the statistical procedure under investigation is applied to each single sample (r-tuple of samples). The relative frequency of incorrect decisions is used as an estimate of the relative risk.

Our research program in robustness began with a thorough investigation of pseudo-random number generators (see Feige & al 1984, 1985) and with the determination of the number N of runs needed for a simulation experiment to estimate a risk (Rasch and Schimke 1984, 1985).

We found that N = 10000 repetitions are necessary.

The greatest part of our investigations is connected with the robustness of statistical tests against non-normality.

Non-normal random variables, $\underline{y}$, were generated from N(0,1)-distributed random variables $\underline{x}$ by the power transformation (Fleishman 1978)

(7) $$\underline{y} = a + b\underline{x} + c\underline{x} + d\underline{x}$$

where the coefficients a, b, c are uniquely determined by the side conditions

$$E(\underline{y}) = 0, \quad V(\underline{y}) = 1, \quad \gamma_1(\underline{y}) = \gamma_1, \quad \gamma_2(\underline{y}) = \gamma_2$$

or by two-sided truncation at the points u, v, which are functions of γ_1 and γ_2. (Rasch and Teuscher 1982)

3. ROBUSTNESS RESULTS

What follows is a short overview of the results obtained by our research group. Those who are interested in more details may read the original papers.
A test is suggested as useful for a nominal first kind risk α_{nom} if the estimated (method 2.3) or calculated (method 2.1 or 2.2) actual first kind risk α is between $\alpha - 0.2\alpha$ and $\alpha + 0.2\alpha$ or at least below $\alpha + 0.2\alpha$.

3.1 TESTING HYPOTHESIS FOR A MEAN OF A DISTRIBUTION

3.1.1 EXPERIMENTS WITH FIXED SAMPLE SIZE

We want to test

$$H_0: \mu = \mu_0$$

against

$$H_A: \mu \neq \mu_0$$

Suggestions:

- The t-test may be used for all distributions if the sample size $n \geq 200$ (Herrendörfer and Feige 1984, 1985)
- For continuous distributions and $n \geq 50$ the modified t-test by Johnson (1978) is preferable. It is based on the test statistic (g_1 is the usual estimate of γ_1)

(8) $$t_1 = \frac{\sqrt{n}}{s}\left[(\bar{y} - \mu_0) + \frac{g_1}{6s^2 n} + \frac{g_1}{3s^4\sqrt{n}}(\bar{y} - \mu_0)^2\right]$$

using the usual t-quantiles. (Teuscher 1984, 1985).

3.1.2 SEQUENTIAL EXPERIMENTS (Rasch 1984, 1985)

We suggest for continuous distributions the sequential test based on the sequence

$$(9) \qquad T_2(n) = \frac{n-1}{2}\left[\ln \frac{n-1}{n} s_n^2 - \ln \frac{n-1}{n} s_n^2 + (\bar{y}_n - d)^2\right],$$

$(\mu - \mu_0)^2 = \sigma^2 d^2$ where $\{\bar{y}_n\}$ and $\{s_n^2\}$ are the sequences of sample means and variances.

H_0 is accepted, if

$$T_2(n) < \ln\beta - \ln(1-\alpha),$$

and rejected, if

$$T_2(n) > \ln(1-\beta) - \ln\alpha;$$

otherwise a further element is observed and the process is repeated. Here $\alpha_{real} < \alpha_{nom} - 0.2\alpha_{nom}$. This leads to a poor power, but no better test could be found.

3.2 COMPARING MEANS OF TWO CONTINUOUS DISTRIBUTIONS

3.2.1 EXPERIMENTS WITH FIXED SAMPLE SIZE

Here two kinds of assumptions were violated in the simulation experiments both in using non-normal distributions and unequal variances in both distributions. Figure 1 shows the recommendations found by Tuchscherer and Pierer (1986) (see also Rasch and Guiard (1986) - R IV).

Tables 2 and 3 show the robustness against distributions of the Fleishman-system and truncated normal distributions respectively.

3.2.2 SEQUENTIAL EXPERIMENTS

Frick (1984, 1985) investigated the robustness of sequential two sample tests, and we recommend the Welch-modification of the two sample t-test it inequal variances can occur, otherwise the usual Wald-test can be used.

3.3 MULTIPLE COMPARISONS WITH A CONTROL

Rudolph (1985) investigated the behaviour of some many-one-tests and recommends for sample sizes below 15 the Dunnett-procedure and otherwise Steels (1959) generalization of the two-sample

Recommendation for the two-sample problem for testing means

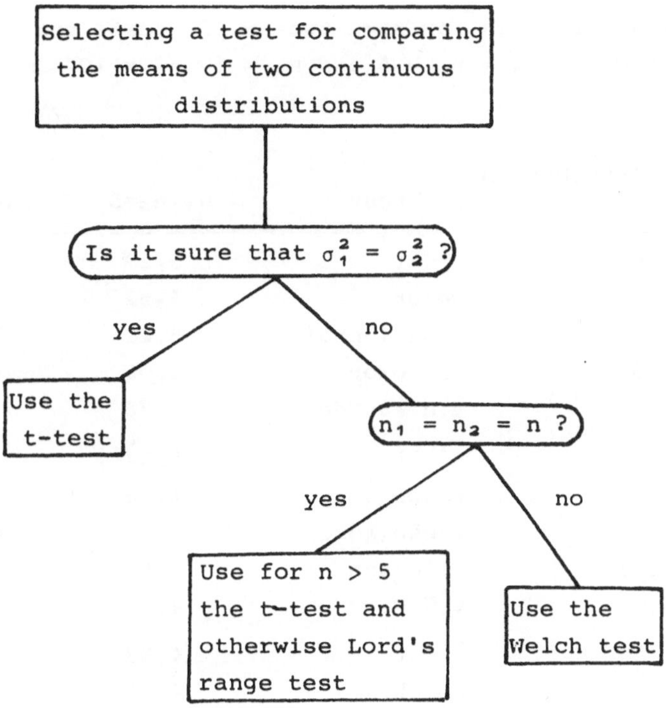

Fig. 1

TABLE 2

Relative frequencies $100\hat{\alpha}$ of rejecting $H_0 : \mu_1 = \mu_2$
if H_0 is valid for some two-sample tests for
$\alpha = 0.05$, $n_1 = n_2 = 5$ and $n_1 = 5$, $n_2 = 10$

N(0,1)-distribution truncated at	test	$n_1=n_2=5$	$n_1=5,\ n_2=10$
	t	5,22	5,17
-1,645	Welch	4,52	5,30
($\gamma_1 \approx 0,34$,	range(lord)	5,29	5,50
$\gamma_2 \approx -0,23$)	Wilcoxon	3,08	4,05
	V.d.Waerden	3,76	5,15
	t	5,34	4,82
-1,282	Welch	4,68	5,49
($\gamma_1 \approx 0,48$	range(Lord)	5,43	5,35
$\gamma_2 \approx -0,14$)	Wilcoxon	3,50	3,84
	V.d.Waerden	4,24	4,87
	t	4,93	4,61
-0,674	Welch	4,13	5,27
($\gamma_1 \approx 0,73$	range(Lord)	5,35	5,35
$\gamma_2 \approx 0,23$)	V.d.Waerden	4,08	4,80
	t	4,96	4,58
0	Welch	3,91	4,96
($\gamma_1 \approx 1$	range(Lord)	5,66	5,38
$\gamma_2 \approx 0,87$)	Wilcoxon	3,40	3,61
	V.d.Waerden	4,17	4,76
	t	4,55	4,46
0,674	Welch	3,49	4,87
($\gamma_1 \approx 1,22$	range(Lord)	5,42	5,62
$\gamma_2 \approx 1,63$)	Wilcoxon	2,90	4,06
	V.d.Waerden	3,72	4,96

TABLE 3

Relative frequencies $100\hat{\alpha}$ of rejecting $H_0 : \mu_1 = \mu_2$ if H_0 is valid for the two-sampled test and $\alpha = 0.005$.

γ_1	γ_2	δ_2^2/δ_1^2	$n_1=n_2=5$	$n_1=5,\ n_2=10$
0	0	1/2	5,39	8,44
		1	5,11	4,99
		2	-	2,85
0	1,5	1/2	5,00	7,35
		1	4,78	4,78
		2	-	2,57
0	3,75	1/2	4,33	7,59
		1	4,38	4,62
		2	-	2,71
0	7	1/2	4,10	6,79
		1	3,99	4,79
		2	-	3,36
1	1,5	1/2	5,31	7,57
		1	4,98	5,27
		2	-	3,36
1,5	3,75	1/2	5,32	7,66
		1	4,24	4,19
		2	-	3,41
2	7	1/2	5,25	7,28
		1	4,08	4,36
		2	-	3,38

Wilcoxon test. First results for multiple test procedures with ordered alternatives are published by Hothorn (1985).

3.4 COMPARISON OF TWO AND MORE VARIANCES

The robustness of tests for comparing two variances was investigated by Nürnberg (1982). Results for the comparison of three variances which seem to be valid also for more than three variances can be found in Nürnberg (1986). In these papers it was shown that all classical tests like Bartlett's test, Cochran's test and the F-test are extremely non-robust against deviations from normality. The investigations were based on samples of equal size $n_1 = n_2 = n = 6.18$ and 42 respectively. Amongst the robust tests the Box-Anderson-test and Levene's s-test can be recommended because they showed a uniformely higher power than other robust (in respect of the first kind risk α) tests.
The Box-Anderson test compares the usual F-statistic with the $(1-\alpha)$-percentile of an F-distribution with

$$f_1 = f_2 = f = \frac{n-1}{1 + \frac{1}{2}g}$$

d. f. Here g_2 is the usual estimate of the kurtosis γ_2.
Levene's s-test is based on the usual one-way-ANOVA F-test performed with $|y_{ij} - \bar{y}_{i.}|$ (i = 1,2, i = 1,2,3 respectively, j = 1,.. ...,n). If this test is significant, the null hypothesis of equal variances is rejected.

3.5 TEST AND CONFIDENCE ESTIMATIONS IN NON-LINEAR REGRESSION

Rasch and Schimke (1983, 1984, 1985) proposed the use of the asymptotic covariance matrix of the least squares estimator $\hat{\underline{\theta}}$ to test the parameters θ of nonlinear regression functions $f(x_i, \theta)$ in a regression model

(10) $\underline{y}_i = y(x_i) = f(x_i, \theta) + \underline{e}_i \qquad (i = 1,\dots,n > p)$

if the $\underline{e}_i$ are i.i.d. $N(0,\sigma^2)$-distributed where $\theta' = (\theta_1,\dots,\theta_p)$ is a vector with p components and f is nonlinear in at least one θ_j $(j = 1,\dots,p)$.
Under mild regularity conditions

$$n(\hat{\underline{\theta}} - \theta)$$

is asymptotically $N(O_p,\Sigma)$-distributed with

(11) $$\Sigma = \lim_{n\to\infty} n\sigma^2(F'F)^{-1}$$

and

(12) $$F = F(\theta) = \frac{\partial f(x_i,\theta)}{\partial \theta j}$$

Let $\hat{F} = F(\hat{\theta})$ be the estimate of F then Rasch and Schimke proposed that

$$H_{jo}: \theta_j = \theta_{jo}$$

be tested against $H_{jo}: \theta_j \neq \theta_{jo}$

using a test statistic

$$t_j = \frac{\hat{\theta}_j - \theta_{jo}}{s\sqrt{c_{jj}}}$$

where

(14) $$\underline{s}^2 = \frac{1}{n-p} \sum_{i=1}^{n} [\underline{y}_i - f(x_i,\hat{\theta})]^2$$

is the estimator of σ^2 and $(\hat{F}'\hat{F})^{-1} = (c_{ij})$. It was shown that even for small n the test based on the test statistic (13) and defined by

(15) $$k_j(\underline{y}) = \begin{cases} 1, & \text{if } |t_j| > t(n-p \mid 1 - \frac{\alpha}{2}) \\ 0, & \text{otherwise} \end{cases}$$

is a α^*-test with $\alpha - 0.2\alpha < \alpha^* < \alpha + 0.2\alpha$ for $\alpha = 0.01$, $\alpha = 0.05$ and $\alpha = 0.10$ $[\underline{y}' = (y_1,\ldots,y_n)]$. The number n_o was determined so that for all $n > n_o$ such a property could be found. In our robustness research we used non-normal $\underline{e}_i$ generated by power transformation (7) and for the (γ_1,γ_2)-pairs (0,1.5), (0,3.75), (11.5), (1.5,3.75) and (2,7). We obtained the following n_o-values for $\alpha = 0.05$ for some special functions over a wide range of θ-values.

No.	Function	p	n_o for normal a_i	n_o for non-normal e_i
1	$\theta_1+\theta_2 e^{\theta_3 x}$	3	4	4
2	$\theta_1+\theta_2 \arctan\theta_3(x+\theta\ )$	4	10	18
3	$\theta_1+\theta_2 \tanh\theta_3(x+\theta_4)$	4	11	15

4	$\frac{1}{1+\theta_2 e^{\theta_3 x}}$	3	6	6
5	$\theta_1 e^{\theta_2 e^{\theta_3 x}}$	3	4	4

TABLE 4

Relative frequency of acceptance of $H_{oj}: \theta_j = \theta_{jo}$ for the arctan-function for a nominal first find risk $\alpha = 0.05$, different n-values and $\theta_1 = \theta_2 = 50$ and normal distribution

(θ_3, θ_4)-values		n					
Distribution	H_0 for	6	8	10	12	14	20
$\theta_3 = -0.015$	θ_1	91,50	93,05	94,27	94,36	94,78	94,71
$\theta_4 = 1$	θ_2	91,45	92,96	94,10	94,55	94,55	94,67
	θ_3	91,57	93,55	94,35	94,62	94,43	94,88
	θ_4	91,39	93,42	94,47	94,42	94,37	94,89
$\theta_3 = -0.045$	θ_1	91,19	93,76	94,58	94,72	94,73	94,75
$\theta_4 = 0.15$	θ_2	91,25	93,79	94,57	94,46	94,55	94,89
	θ_3	91,68	93,86	94,28	94,46	95,08	94,66
	θ_4	91,67	93,76	94,30	94,49	95,00	94,82

Table 4 and 5 demonstrates for the function No 2 that the t-test works fairly well for θ_2, θ_3 and θ_4 and, if the distributions are not too extreme, also for θ_1 as long as the number of measurements is not smaller than 15. For more details see Rasch and Schimke (1983, 1984, 1985), Lippert (1986), Davis (1986) and Gretzebach (1986).

TABLE 5

Robustness Results (Relative frequencies of acceptance of H_0) for the t-test for the parameters of the arctan-function, $n = 15$, $\alpha = 0.05$, $\theta_1 = \theta_2 = 50$, $\theta_3 = -0.03$ and $\theta_4 = 1$

H_0 for:	Parameters of the distribution					
	$\gamma_1=0$ $\gamma_2=1.50$	$\gamma_1=0$ $\gamma_2=3.75$	$\gamma_1=0$ $\gamma_2=7$	$\gamma_1=1$ $\gamma_2=1.5$	$\gamma_1=1.5$ $\gamma_2=3.75$	$\gamma_1=2$ $\gamma_2=7$
θ_1	94,65	95,04	95,46	94,27	93,66	92,88
θ_2	95,07	95,30	95,52	94,95	94,79	95,21
θ_3	95,29	95,04	95,20	94,92	95,03	95,33
θ_4	95,50	95,02	95,06	94,88	94,96	95,58

3.6 SELECTION PROCEDURES

Selection has two meanings. The first kind of selection is identical with natural or artificial selection in plant or animal breeding. Robustness in this sense was investigated by Rasch and Pierer (1985). The second kind of selection is a competitor of multiple test procedures. Instead of formulating a problem as the aim of comparing the expectations of the means of a distributions we can consider the problem of selecting the best out of a distributions if the best distribution is that with the largest mean. Domröse (1984, 1985 a,b) and Domröse and Rasch (1987) found that, if no a priori information regarding the distributions is available , Bechhofer's selection rule (selecting the distribution with the largest sample mean) in to be preferred.

4. REFERENCES

The following books contain a great number of special references. They are cited as RI to VI and P 1/2.

R I Herrendörfer, G. (Federf.) Robustheit I
Probleme der angewandten Statistik, 4,
FZT Dummerstorf-Rostock (1980)

R II Guiard, V. (Federf.) Robustheit II
Probleme der angewandten Statistik, 5,
FZT Dummerstorf-Rostock (1981)

R III Rasch, D. und Herrendörfer, G. (Federf.) Robustheit III
Probleme der angewandten Statistik, 7,
FZT Dummerstorf-Rostock (1982)

R IV Rasch, D. (Federf.) Robustheit IV
Probleme der angewandten Statistik, 13,
FZT Dummerstorf-Rostock (1985)

R V Rudolph, P. E. (Federf.) Robustheit V
Probleme der angewandten Statistik, 15,
FZT Dummerstorf-Rostock (1985)

R VI Rasch, D. and Guiard, V. (Federf.)
Robuste statistische Verfahren
Probleme der angewandten Statistik, 20,
FZT Dummerstorf-Rostock (1986)

P 1 Rasch, D. and Tiku, M. L. (ed)
Robustness of Statistical Methods and Nonparametric Statistics
VEB Deutscher Verlag der Wissenschaften Berlin (1984)

P 2 Rasch, D. and Tiku, M. L. (ed)
Robustness of Statistical Methods and Nonparametric Statistics
Reidel Publ. Co Dortrecht, Boston, Lancaster, Tokyo (1985)

Box, G. E. P. (1953): Non-normality and tests on variances Biometrica 40, 1-34.

Davis, J. (1986): Simulationsuntersuchungen zum t-Test für die Parameter der Wachstumsfunktion $\alpha+\beta \arctan \gamma(\lambda+\delta)$ Jahresarbeit, Sektion Mathematik, Wilhelm Pieck-Universität Rostock.

Domröse, H. (1984, 1985): The robustness of selection procedures investigated by a combinatorial method P 1/2 (1984, 1985a) 26-29.

Domröse, H. (1985b): Zur Robustheit von Auswahlverfahren Dissertation, Wilhelm Pieck Universität Rostock.

Domröse, H. and Rasch, D. (1987): Biometrical Journal 29, in press.

Feige, K. D., Guiard, V., Herrendörfer, G., Hoffmann, J., Neumann, P., Peters H., Rasch D. and Vettermann Th. (1984, 1985): Results of comparisons between different random number generators P 1/2, 30-34.

Fleishman, A. I. (1978): A method for simulating non-normal distributions. Psychometrika 43, 521-532.

Frick, D. (1984, 1985): Robustness of the two-sample sequential t-test P 1/2, 35-36.

Gretzebach, L. (1986): Simulationsuntersuchungen zum t-Test für die Parameter der Funktion $\alpha+\beta\tanh(\gamma+\delta x)$ und optimale Versuchsplanung, Jahresarbeit, Sektion Mathematik, Wilhelm--Pieck-Universität Rostock.

Guiard, V. (1981): Schiefe und Exzess R II, 40-53.

Herrendörfer, G. and Feige, K. D. (1984, 1985): A combinatorial method in robustness research and two applications P 1/2, 53-57.

Herrendörfer, G. and Rasch, D. (198o): Methoden zur Untersuchung der Robustheit - exakte Methode. In: Herrendörfer, G. (Federf.) Robustheit I, Probleme der angewandten Statistik, Heft 4, 79-105.

Hothorn, L. (1985): Simulationsuntersuchungen zur Robustheit von k-Stichprobenlokationstests mit geordneter Alternativhypothese (Teil 1) R V, 116-134.

Huber, P. I. (1964): Robust estimation of a location parameter. Ann-Math. Statist. 35, 73-101.

Johnson, N. J. (1978): Modified t-tests and confidence intervals for asymetrical populations. Jour. Amer. Statist. Assoc. 73, 536-44.

Lippert, S. (1986): Untersuchung des t-Tests bei Nichtlinearität und Nichtnormalität am Beispiel der logistischen Funktion,

Jahresarbeit, Sektion Mathematik, Wilhelm-Pieck-Universität Rostock.

Nürnberg, G. (1982): Beiträge zur Versuchsplannung für die Schätzung von Varianzkomponenten und Robustheitsunter-suchungen zum Vergleich zweier Varianzen, Probleme der angewandten Statistik, 6, FZT Dummerstorf-Rostock

Nürnberg, G. (1986): Robuste Tests zum Vergleich von mehr als zwei Varianzen in R VI.

Rasch, D. (1984, 1985): Robustness of three sequential one-sample tests against non-normality P 1/2, 100-103.

Rasch, D. and Herrendörfer, G. (1981): Robustheit statistischer Methoden Rostock Math. Kolloq. 17, 87-104.

Rasch, D. and Pierer, H. (1985): Die Robustheitseigenschaften der ublichen Formeln zur Berechnung des Selektionserfolges und des neuen Verfahrens ROSE. R IV, 1-39.

Rasch, D., and Schimke, E. (1983): Distribution of estimators in exponential regression. Skandin. Jour. of Statist. 10, 293-300.

Rasch, D. and Schimke, E. (1984, 1985): A test for exponential regression and its robustness P 1/2, 104-112.

Rasch, D. and Teuscher, F. (1982): Das System der gestutzten Normalverteilungen, 7. Sitzungsbericht der IG Mathem. Statistik, MG DDR, Berlin, 1982, 51-60.

Rudolph, P. E. (1985): Die Robustheit von Prozeduren zum Vergleich von Behandlungen mit einer Kontrolle bei gleichen und ungleichen Varianzen R V, 81-115.

Teuscher, F. (1984, 1985): Simulation studies on robustness of the t- and u-test against truncation of the normal distribution P 1/2, 145-151.

Tuchscherer, A. and Pierer, H. (1985): Simulationsuntersuchungen zur Robustheit verschiedener Verfahren zum Mittelwertvergleich im Zweistichprobenproblem (Simulationsergebnisse) R V, 1-42.

Zielinsky, R. (1981): Robust statistical procedures: A general approach. Inst. Math. Polish Acad. Sci, Preprint No 254.

D. Rasch

17

Research Centre of Animal Production
Dummerstorf-Rostock
of the Academy of Agricultural Sciences
of the GDR

ANALYTICAL METHODS IN PROBABILITY THEORY

Kazimierz Urbanik

Wrocław

Key words: convolutions, characteristic functions, integral transforms.

ABSTRACT

An analogue of the method of characteristic functions for generalized convolutions is discussed. Moreover, the set of weak characteristic functions is described in terms of stable distributions. Finally, some applications to analysis are mentioned.

The method of characteristic functions is a useful analytic tool especially for proving limit theorems for sums of independent random variables. Indeed, the distribution of a sum of independent random variables is the convolution of the distribution functions of the individual terms; the calculation of this convolution is in most cases rather complicated. The main property of the characteristic functions allows a very simple calculation of the characteristic function of a sum of independent random variables from the characteristic functions of its terms, as it is just their product which exhibits a successful applicability to probability theory. This method can be extended to a wide class of operations on probability measures. In this paper we restrict ourselves to the simplest case of non-negative random variables.

We denote by P the set of all probability measures defined on Borel subsets of the positive half-line $R_+ = [0,\infty)$. The set P is endowed with the topology of weak convergence. For $\mu \in P$ and $a > 0$ we define the map T_a by setting $(T_a\mu)(E) = \mu(a^{-1}E)$. By δ_c we denote the probability measure concentrated at the point c. For a subset A of R_+, I_A denotes its indicator. By ω_s $(s > 0)$ we denote the probability measure with the density function $sx^{-s-1}I_{[1,\infty)}(x)$. Further, for any pair μ,ν from P by $\mu\nu$ we denote the probability distribution of the product XY of two independent random variables with probability distributions μ and ν respectively. It is clear that $T_a\mu = \delta_a\mu$.

A continuous commutative and associative P-valued binary operation $\circ$ on P is called a *generalized convolution* if it is distributive with respect to convex combinations and maps T_a $(a > 0)$ with δ_0 as the unit element. Moreover, the key axiom postulates the existence of positive norming constants c_n and a measure $\gamma \in P$ other than δ_0 such that $T_{c_n}\delta_1^{\circ n} \to \gamma$ where $\delta_1^{\circ n}$ is the n-th power of δ_1 under $\circ$. For basic properties of generalized convolutions we refer to Urbanik (1964), (1973), (1984) and (1986).

Mamy examples of generalized convolutions are to be found in various branches of probability theory. Now we shall quote some of them. It is clear that each generalized convolution $\circ$ is uniquely determined by the expressions $\delta_a \circ \delta_b$ with $a,b > 0$.

α-convolutions $*_\alpha (\alpha > 0)$: $\delta_a *_\alpha \delta_b = \delta_{(a^\alpha+b^\alpha)^{1/\alpha}}$. These convolutions correspond to the operations $(X^\alpha + Y^\alpha)^{1/\alpha}$ on independent random variables X and Y. For $\alpha = 1$ we get the ordinary convolution, i.e. $*_1 = *$.

Max-convolution $*_\infty$: $\delta_a *_\infty \delta_b = \delta_{\max(a,b)}$. This convolution is induced by the operation $\max(X,Y)$ on independent random variables X and Y.

Symmetric convolution $*_{1,1}$: $\delta_a *_{1,1} \delta_b = \frac{1}{2}\delta_{|a-b|} + \frac{1}{2}\delta_{a+b}$. This convolution is induced on R_+ by the ordinary convolution of symmetric probability measures on the whole real line.

Kingman convolutions $*_{1,\beta}$ $(\beta > 1)$: $\delta_a *_{1,\beta} \delta_b$ is the probability measure with the density function

$$4^{-1}a^{-3}b^{-3}B(\tfrac{1}{2},\tfrac{\beta}{2})^{-1}(x^2(a^2+b^2) - (a^2-b^2)^2 - x^4)^{(\beta-3)/2}I_{[|a-b|,a+b]}(x)$$

where B is the Beta function. These convolutions have been introduced in Kingman (1962) for the study of spherically symmetric random walk in Euclidean spaces.

Kendall convolution Δ : $\delta_a \Delta \delta_b = (1 - \frac{a}{b})\delta_b + \frac{a}{b}T_b\omega_2$ if $0 < a \le b$. This convolution is induced by Kendall operation on random sets (Kendall (1974)).

Kucharczak convolutions $\circ_\alpha$ $(0 < \alpha < 1)$: $\delta_a \circ_\alpha \delta_b$ is the probability measure with the density function

$$\pi^{-1}\cdot\sin\pi\alpha\cdot a^\alpha b^\alpha x^{-\alpha}(x-a-b)^{-\alpha}(x-a)^{-1}(x-b)^{-1}(2x-a-b)I_{[(a^\alpha+b^\alpha)^{1/\alpha},\infty)}(x).$$

This convolutions have been introduced in Kucharczak-Urbanik (1987).

Vol'kovich convolutions $\circ_{1,\beta}$ $(0 < \beta < \frac{1}{2})$: $\delta_a \circ_{1,\beta} \delta_b$ is the probability measure with the density function

$$2B(\beta,\tfrac{1}{2}-\beta)^{-1}a^{2\beta}b^{2\beta}x(x^2-(a-b)^2)(x^2-(a+b)^2)^{-\beta-\frac{1}{2}}I_{[0,a+b]}(x)$$

These convolutions have been introduced in Vol'kovich (1984) in connection with the study of multidimensional weakly stable distributions.

The next example of a generalized convolution ∇_α $(0 < \alpha < 1)$ given in Kucharczak-Urbanik (1987) is also connected with weakly stable distributions. Given $0 < \alpha < 1$, we put

$$\mu = (2-2^{-\alpha}) \sum_{n=0}^{\infty} 2^{-1-n(1+\alpha)} T_{2^n}\omega_\alpha .$$

One can prove that for every pair $a,b \geq 0$ there exists a unique probability measure $\rho(a,b)$ fulfilling the equation

$$T_a\mu *_\infty T_b\mu = \mu\rho(a,b) ,$$

where $*_\infty$ is the max-convolution. Setting $\delta_a \nabla_\alpha \delta_b = \rho(a,b)$ we get a generalized convolution.

To prepare the way for the extension of the method of characteristic functions we introduce auxiliary notions. Let m_0 be the sum of δ_0 and the Lebesgue measure on R_+. By P_0 we denote the subset of P consisting of all absolutely continuous with respect to m_0 measures. We say that the generalized convolution $\circ$ admits a *weak characteristic function* if there exists a one-to-one correspondence $\mu \leftrightarrow \hat{\mu}$ between measures μ from P and real-valued functions $\hat{\mu}$ from $L_\infty(R_+,m_0)$ commuting with convex combinations and scale changes, i.e. $(T_a\mu)^\wedge(t) = \hat{\mu}(at)$. Further, the key condition postulates $(\mu\circ\nu)^\wedge = \hat{\mu}\hat{\nu}$. Moreover, the functions $\hat{\lambda}$ are continuous for $\lambda \in P_0$ and the convergence $\mu_n \to \mu$ is equivalent to the convergence $\hat{\mu}_n \to \hat{\mu}$ in the $L_1(R_+,m_0)$-topology of $L_\infty(R_+,m_0)$. It is clear that every scale change $\hat{\mu}(at)$ $(a > 0)$ of a weak characteristic function $\hat{\mu}(t)$ is also a weak characteristic function. The main result in this respect has been proved in Urbanik (1986), Th. 4.1.

THEOREM 1. *Each generalized convolution admits a weak characteristic function being uniquely determined up to a scale change. Moreover, this function is an integral transform*

$$\hat{\mu}(t) = \int_0^\infty \Omega(tx)\mu(dx)$$

with kernel Ω *from* $L_\infty(R_+, m_0)$.

Many limit theorems involving generalized convolutions can be derived by means of the weak characteristic function. For detailed information we refer to Bingham (1971) and (1984), Jajte (1976) and (1977), Jurek (1981) and (1985), Kłosowska (1977), Urbanik (1964), (1973) and (1985). Now we shall quote some examples of kernels of weak characteristic functions.

α-convolution: $\Omega(t) = \exp(-t^\alpha)$.

Max-convolution: $\Omega(t) = I_{[0,1]}(t)$.

Symmetric convolution: $\Omega(t) = \cos t$.

Kingman convolution: $\Omega(t) = \Gamma(\frac{\beta}{2})(\frac{2}{t})^{\beta/2-1}J_{\beta/2-1}(t)$, where J_γ is the Bessel function.

Kendall convolution: $\Omega(t) = (1-t)I_{[0,1]}(t)$.

Kucharczak convolution: $\Omega(t) = \Gamma(\alpha)^{-1}\Gamma(\alpha,t)$, where $\Gamma(\alpha,t)$ is the incomplete Γ-function.

Vol'kovich convolution: $\Omega(t) = 2^{1-\beta}t^\beta\Gamma(\beta)^{-1}K_\beta(t)$, where K_β is the Macdonald function.

The convolution $_\alpha$:

$$\Omega(t) = (1-2^{(1+\alpha)[\log_2 t]} - (2-2^{-\alpha})(1-2^{[\log_2 t]})t^\alpha)I_{[0,1]}(t),$$

where the square brackets denote the integer part.

If the kernel Ω is equal to a continuous function m_0-almost every where on R_+, then we say that $\mu \to \hat{\mu}$ is the *characteristic function* of the convolution in question. Taking the continuous version of $\hat{\mu}$ we infer that the convergence $\mu_n \to \mu$ is then equivalent to the uniform convergence $\hat{\mu}_n \to \hat{\mu}$ on every compact subset of R_+, i.e. the situation is the same as for ordinary convolution. We have seen that the indicator $I_{[0,1]}$ is the kernel of the weak characteristic function for the max-convolution. Consequently, the max-convolution is the simplest example of a generalized convolution which does not admit a characteristic function.

The uniqueness up to a scale change of the weak characteristic function enables us to associate with every generalized convolution t the subset $C(\circ)$ of $L_\infty(R_+, m_0)$ defined as follows: $f \in C(\circ)$ if and only if $f = \hat{\mu}$ m_0-almost every where for some $\mu \in P$. For the ordinary convolution $*$ making use of Bernstein Theorem we find that the set $C(*)$ consists of all functions equal m_0-almost everywhere to continuous on R_+, completely monotone on the open half--line $(0,\infty)$ and assuming the value 1 at the origin functions. Further, for the symmetric convolution $*_{1,1}$, by Bochner's Theorem, the set $C(*_{1,1})$ consists of all functions equal m_0-almost everywhere to continuous positive definite functions on R_+ taking the value 1 at the origin. Our aim is to give a description of the set $C(\circ)$ for arbitrary generalized convolution $\circ$. To prepare the way for such description we consider first a simple limit problem. For the remainder of this paper we fix the weak characteristic function for a generalized convolution in question.

A measure λ from P is said to be $\circ$-*stable* if $\lambda \neq \delta_0$ and $T_{a_n}\mu^{\circ n} \to \lambda$ for a measure $\mu \in P$ and a norming sequence a_n of positive numbers. In Urbanik (1986) the following result has been established.

THEOREM 2. *For every generalized convolution* $\circ$ *there exists a constant* $\kappa(\circ)$ $(0 < \kappa(\circ) \le \infty)$ *such that for every* $p \in (0, \kappa(\circ)]$ *there exists a measure* $\sigma_p \in P$ *with* $\hat{\sigma}_p(t) = e^{-t^p}$ *if* $p < \infty$ *and* $\hat{\sigma}_\infty(t) = I_{[0,1]}(t)$. *Moreover, the set of* $\circ$*-stable measures consists of all measures of the form* $T_a\sigma_p$ $(a > 0,\ 0 < p \le \kappa(\circ))$.

The measure σ_p is called the standard $\circ$-stable measure with exponent p. Of course, the constant $\kappa(\circ)$ does not depent upon the choice of the weak characteristic function and is called the *characteristic exponent* for $\circ$. The measure σ_κ where $\kappa = \kappa(\circ)$ is called the *characteristic measure* for $\circ$. A generalized convolution is completely described by its characteristic exponent and measur More precisely, it has been proved in Urbanik (1986), Th. 4.3 that *if* $\kappa(\circ) = \kappa(\circ')$ *and the characteristic measures of* $\circ$ *and* $\circ'$ *are identical, then* $\circ = \circ'$. *Moreover, the* $\circ$*-stable measures* σ_p *with* $p < \kappa(\circ)$ *are equivalent to the Lebesgue measure on* R_+. We denote by g_p $(p < \kappa(\circ))$ their density function and adopt this notation in the case $p = \kappa(\circ)$ provided the characteristic measure is absolutely continuous with respect to the Lebesgue measure. We conclude this section with some examples of characteristic exponents and $\circ$-stable measures.

Ordinary convolution $*$: $\kappa(*) = 1$, $\sigma_1 = \delta_1$. When we in the sequel are dealing with a $*$-stable measure σ_p $(0 < p < 1)$ we alway use the notation φ_p for its density function. The density function $\varphi_{\frac{1}{2}}$ can be written down in terms of elementary functions. Namely

$$\varphi_{\frac{1}{2}}(x) = 2^{-1}\pi^{-\frac{1}{2}}x^{-\frac{3}{2}}\exp(-4^{-1}x^{-1}) .$$

It has been shown in Zolotarev (1983), that

$$\varphi_{\frac{1}{3}}(x) = 3^{-1}\pi^{-1}x^{-\frac{3}{2}}K_{\frac{1}{3}}(3^{-\frac{3}{2}}\cdot 2x^{-\frac{1}{2}})$$

where $K_{\frac{1}{3}}$ is the Macdonald function and

$$\varphi_{\frac{2}{3}}(x) = (3\pi)^{-\frac{1}{2}}x^{-1}W_{\frac{1}{2},\frac{1}{6}}(27^{-1}\cdot 4x^{-2})\exp(-27^{-1}\cdot 2x^{-2})$$

where $W_{p,q}$ is the Whittaker function.

Max-convolution $*_\infty$: $\kappa(*_\infty) = \infty$, $\sigma_\infty = \delta_1$ and $g_p(x) = px^{-p-1}\exp(-x^{-p})$ $(0 < p < \infty)$ (the Weibull-Gnedenko distributions). It has been proved in Urbanik (1986), Prop. 2.3 and Lemma 2.1, that $\kappa(\circ) = \infty$ if and only if $\circ = *_\infty$.

Symmetric convolution $*_{1,1}$: $\kappa(*_{1,1}) = 2$. The density functions g_1 and g_2 can be expressed by means of elementary functions. Namely,

$$g_1(x) = 2\pi^{-1}(1+x^2)^{-1}, \quad g_2(x) = \pi^{-\frac{1}{2}}\exp(-4^{-1}x^2).$$

Moreover, it has been shown in Zolotarev (1983) that

$$g_{\frac{2}{3}}(x) = (3\pi)^{-\frac{1}{2}}x^{-1}W_{-\frac{1}{2},\frac{1}{6}}(27^{-1}\cdot 4x^{-2})\exp(27^{-1}\cdot 2x^{-2}) .$$

Kingman convolution $*_{1,\beta}$ $(1 < \beta)$: $\kappa(*_{1,\beta}) = 2$,

$$g_2(x) = 4^{1-\beta}\Gamma(\beta - \tfrac{1}{2})^{-1}x^{2\beta-2}\exp(-4^{-1}x^2)$$

(the Maxwell distribution),

$$g_1(x) = 2B(\tfrac{1}{2},\beta)^{-1}x^{2\beta-2}(1+x^2)^{-\beta}$$

and for $0 < p < 2$

$$g_p(x) = 4^{\beta-1}\Gamma(\beta - \tfrac{1}{2})^{-1}x^{2\beta-2}\int_0^\infty y^{\frac{1}{2}-\beta}\varphi_{\frac{p}{2}}(y)\exp(-4^{-1}y^{-1}x^2)\,dy .$$

Kendall convolution $\triangle$: $\kappa(\triangle) = 1$ and

$$g_p(x) = (p(1-p) + p^2x^{-p})x^{-p-1}\exp(-x^{-p}) \qquad (0 < p \le 1)$$

Kucharczak convolution $\circ_\alpha$ $(0 < \alpha < 1)$: $\kappa(\circ_\alpha) = \alpha$ and

$$\sigma_p([0,x)) = x^{1-\alpha}\int_0^x (x-y)^{\alpha-1}\varphi_p(y)dy \quad (0 < p \le \alpha).$$

Vol'kovich convolution $\bullet_{1,\beta}$ $(0 < \beta < \frac{1}{2})$: $\kappa(\bullet_{1,\beta}) = 2\beta$

and for $0 < p \le 2\beta$

$$g_p(x) = \beta^{-1}B(\tfrac{1}{2},\beta+\tfrac{1}{2})^{-1}(\varphi_p(x)+(1-2\beta)\int_0^1(1-y^2)^{\beta-\frac{3}{2}}y(\varphi_p(x)-\varphi_p(xy))dy).$$

Convolution ∇_α $(0 < \alpha < 1)$: $\kappa(\nabla_\alpha) = \alpha$ and for $0 < p \le \alpha$

$$\sigma_p([0,x)) = 2^{1+\alpha}(2^{1+\alpha}-1)^{-1}(1+p\alpha^{-1}x^{-p})\exp(-x^{-p}) -$$
$$- (2^{1+\alpha}-1)^{-1}(1+p2^p\alpha^{-1}x^{-p})\exp(-2^px^{-p}).$$

We now come to a description of the set $C(\circ)$. The following result has been established in Urbanik (1986), Th. 4.5.

THEOREM 3. *Let* $0 < p < \kappa(\circ)$. *A function* f *from* $L_\infty(R_+,m_0)$ *belongs to* $C(\circ)$ *if and only if*

$$\lim_{t\to 0} t^{-1}\int_0^t f(u)du = f(0) = 1$$

and the function $\int_0^\infty f(t^{\frac{1}{p}}x)g_p(x)dx$ *is completely monotone on* $(0,\infty)$. *For continuous functions* f *this result can be extended to the case* $p = \kappa(\circ)$.

The above Theorem enables us to manufacture some rather unexpected applications to analysis starting from concrete generalized convolutions. For the sake of simplicity we restrict ourselves to continuous functions. Let D denote the set of all real-valued bounded and continuous functions on R_+ taking the value 1 at the origin. For ordinary convolution $*$ and $p = \frac{1}{3},\frac{1}{2}$ and $\frac{2}{3}$ we get, by Theorem 3, the following statement.

COROLLARY 1. *Let* $f \in D$. *Then the functions*

$$f(t), \quad \int_0^\infty f(t^{\frac{3}{2}}x)x^{-1}W_{\frac{1}{2},\frac{1}{6}}(x^{-2})\exp(-\tfrac{1}{2}x^{-2})dx,$$

$$\int_0^\infty f(t^2x)x^{-\frac{3}{2}}\exp(-x^{-1})dx \quad \textit{and} \quad \int_0^\infty f(t^3x)x^{-\frac{3}{2}}K_{\frac{1}{3}}(x^{-\frac{1}{2}})dx$$

are or are not simultaneously completely monotone on $(0,\infty)$.

Similarly, for symmetric convolution $*_{1,1}$ and $p = \frac{2}{3}, 1$ and 2 we obtain a relationship between the positive definiteness and the complete monotonicity.

COROLLARY 2. *Let* $f \in D$. *Then the following properties are equivalent.*

(i) f *is positive definite,*

(ii) $\int_0^\infty f(t^{\frac{1}{2}}x)\exp(-x^2)dx$ *is completely monotone on* $(0,\infty)$,

(iii) $\int_0^\infty f(tx)(1+x^2)^{-1}dx$ *is completely monotone on* $(0,\infty)$,

(iv) $\int_0^\infty f(t^{\frac{3}{2}}x)x^{-1}W_{-\frac{1}{2},\frac{1}{6}}(2x^{-2})\exp(x^{-2})dx$ *is completely monotone on* $(0,\infty)$.

Taking Kingman convolution $*_{1,\beta}$ $(\beta > 1)$ and $p = 2$ we get the following statement.

COROLLARY 3. *A function* f *from* D *has the representation*

$$f(t) = \Gamma(\tfrac{\beta}{2})2^{\beta/2-1}t^{1-\beta/2}\int_0^\infty J_{\beta/2-1}(tx)x^{1-\beta/2}\mu(dx)$$

for some $\mu \in P$ *if and only if the function*

$$\int_0^\infty f(t^{\frac{1}{2}}x)x^{2\beta-2}\exp(-x^2)dx$$

is completely monotone on $(0,\infty)$.

Further, for Kendall convolution $\triangle$ we get the following result.

COROLLARY 4. *Let* $0 < p \leq 1$. *A function* f *from* D *has the representation*

$$f(t) = \int_0^{t^{-1}} (1-tx)\mu(dx)$$

for some $\mu \in P$ *if and only if the function*

$$\int_0^\infty f(t^{\frac{1}{p}}x)(1-p+px^{-p})x^{-p-1}\exp(-x^{-p})dx$$

is completely monotone on $(0,\infty)$.

Starting from Kucharczak convolution $\circ_\alpha$ and $p = \frac{1}{2}$ it is easy to establish, by a routine computation, the following fact.

COROLLARY 5. *Let* $\frac{1}{2} \leq \alpha < 1$. *A function* f *from* D *has the representation*

$$f(t) = \Gamma(\alpha)^{-1}\int_0^\infty \Gamma(\alpha,tx)\mu(dx)$$

for some $\mu \in P$ *if and only if the function*

$$\int_0^\infty\int_1^\infty f(t^2x^2y)(x^2-1)^{\alpha-1}x^{1-2\alpha}(2y^{-\frac{5}{2}}-y^{-\frac{3}{2}})\exp(-y^{-1})dxdy$$

is completely monotone on $(0,\infty)$.

Similarly, starting from Vol'kovich convolution $\circ_{1,\beta}$ and $p = \frac{1}{2}$ we obtain the following statement.

COROLLARY 6. *Let* $\frac{1}{4} \leq \beta < \frac{1}{2}$, $f \in D$ *and*

$$F(t) = \int_0^\infty f(t^2x)x^{-\frac{3}{2}}\exp(-x^{-1})dx .$$

Then f *has the representation*

$$f(t) = 2^{1-\beta}\Gamma(\beta)^{-1}t^\beta\int_0^\infty K_\beta(tx)x^\beta\mu(dx)$$

for some $\mu \in P$ *if and only if the function*

$$F(t) + (2-4\beta)\int_1^\infty (x^4-1)^{\beta-\frac{3}{2}} x^{1-4\beta}(F(t) - x^2F(tx))dx$$

is completely monotone on $(0,\infty)$.

REFERENCES

Bingham N.H. (1971): Factorization theory and domains of attraction for generalized convolution algebras. In : Proc. London Math. Soc. 23, 16-30.

(1984): On a theorem of Kłosowska about generalized convolutions. In : Colloq. Math. 48, 117-125.

Jajte R. (1976): Quasi-stable measures in generalized convolution algebras. In : Bull. Acad. Polon. Sci. Sér. Sci. Math. Astronom. Phys. 24, 505-511.

(1977): Quasi-stable measures in generalized convolution algebras II. In : Bull. Acad. Polon. Sci. Sér. Sci. Math. Astronom. Phys. 25, 67-72.

Jurek Z.J. (1981): Some characterizations of the class L in generalized convolution algebras. In : Bull. Acad. Polon. Sci. Sér. Sci. Math. Astronom. Phys. 29, 409-415.

(1985): Limit distributions in generalized convolution algebras. In : Probab. Math. Statistics. 5, 113-135.

Kendall D.G. (1974): Foundations of a theory of random sets. In : Stochastic geometry (ed. E.F. Harding and D.G. Kendall), Wiley, 322-376.

Kingman J.F. (1962): Random walks with spherical symmetry. In : Colloq. Comb. Methods in Prob. Theory, Aarhus, 40-46.

Kłosowska M. (1977): On the domain of attraction for generalized convolution algebras. In : Rev. Roum. Math. Pures et Appl. 22, 669-677.

Kucharczak J. and Urbanik K. (1987): Transformations preserving weak stability. In: Bull. Pol. Acad: Math. (in print).

Urbanik K. (1964): Generalized convolutions. In : Studia Math. 23, 217-245.

(1973): Generalized convolutions II. In : Studia Math. 45, 57-70.

(1984): Generalized convolutions III. In : Studia Math. 80, 167-189.

(1985): Limit behaviour of medians. In : Bull. Pol. Acad: Math. 33, 413-419.

(1986): Generalized convolutions IV. In : Studia Math. 83, 57-95.

Vol'kovich V.E. (1984): Multidimensional B-stable distributions and realizations of generalized convolutions. In : Problems of Stability of Stochastic Models, 40-54 (in Russian).

Zolotarev V.M. (1983): One - dimensional stable distributions. Nauka, (in Russian).

Wrocław University, Institute of Mathematics
Pl. Grunwaldzki 2/4, 50-384 Wrocław
Poland

COMMUNICATION

CONFIDENCE INTERVALS FOR VARIANCE COMPONENTS IN BALANCED RANDOM MODELS

R. Ahmad and S. M. Mostafa

Glasgow, Scotland

Key words: Balanced random models, variance components, confidence limits

ABSTRACT

For general balanced random models, confidence limits for the variance components are constructed by using three different approaches. To illustrate and constrast the procedures, for a particular model some simulation results are provided when the errors are independent as well as when they are related through an autoregressive series.

INTRODUCTION

The problem of obtaining confidence intervals for variance components is not discussed as widely in the literature as that of getting point estimators for variance components. Some of the investigators who have been in the past concerned with this problem are: Satterthwaite (1946), Bross (1950), Tukey (1951), Moriguti (1954), Bulmer (1957),Healy (1961), Williams (1962), Healy (1963), Boardman (1974) and Jeyaratnam and Graybill (1980). A comprehensive bibliography and review on this area is given by Sahai (1979), Khuri and Sahai (1985) and Sahai et al (1985). In essence, there are three approaches towards this problem. These are: (i) Moriguti-Bulmer (MB), (ii) Tukey-Williams (TW), and (iii) Bross-Healy (BH) procedures; the last one depending on Fisher (1935) arguments on fiducial intervals.

In this note, we extend all the three techniques to cover the general balanced random models - which include the balanced nested and cross classification designs without interaction. To illustrate the development, a Monte Carlo study is carried out to compare and contrast the MB, TW and BH procedures for the two-way nested model. Our investigation is carried over to two cases: (a) When the errors are assumed independent, and (b) when the errors are dependent through an autoregressive series.

CONFIDENCE INTERVALS IN GENERAL BALANCED RANDOM MODELS

Consider a general balanced random model given by

$$Y = aI + \sum_{i=1}^{p} U_i b_i + e \qquad (I' = (1,1,\ldots,1)). \tag{1}$$

where Y is an observable random vector, a an unknown constant, U_i is the respective design matrix for random effects b_i, b_i is an unobservable random effect vector from $N(0, \sigma_i^2)$ and e is a random error vector from $N(0, \sigma_o^2)$. The b_i's may consist of crossed or nested random effects, and are assumed to be independent of e.

Let S_i^2 and v_i be the sum of squares and degrees of freedom associated with the random effects b_i $(i=1,2,\ldots,p)$. The pair (S_o^2, v_o) is similarly associated with error e. If T_i denotes the expected mean squares (EMS) for S_i^2, then we know that $S_i^2 \sim T_i\ \chi^2_{v_i}$. Let $\chi^2_{\alpha U v_i}$ and $\chi^2_{\alpha L v_i}$ be the upper and lower χ^2 points which enclose $(1-\alpha)$ of the underlying distribution with v_i d.f. Similarly, we define $F_{\alpha U v_i, v_j}$ and $F_{\alpha L v_i, v_j}$ for the F-distribution.

THE BALANCED NESTED RANDOM MODEL CASE:- In this model, the EMS take the form: $T_o = \sigma_o^2$ and $T_j = T_{j-1} + k_j\sigma_j^2$ for $j=1,2,\ldots,p$, where k_j's are constants depending on U_j's. By using the chi-square distribution property for S_i^2, the relationships among T_j $(j=0,1,\ldots,p)$, and extending Tukey-Williams arguments in conjunction with a result of Graybill (1976; p.624, Thm. 15.3.5), we get TW-type confidence limits for variance components as

$$P\left[\frac{S_j^2(1-F_{\alpha U v_j,\, v_{j-1}}/F_j)}{k_j\,\chi^2_{\alpha U v_j}} \leqslant \sigma_j^2 \leqslant \frac{S_j^2(1-F_{\alpha L v_j, v_{j-1}}/F_j)}{k_j\,\chi^2_{\alpha L v_j}}\right] \geqslant 1-2\alpha \tag{2}$$

for $j=1,2,\ldots,p$, where $F_j = (v_{j-1}\ S_j^2)(v_j\ S_{j-1}^2)^{-1}$. Note that the inequalities above have been justified by using Bonferroni's inequality, see Miller (1981).

On the other hand, now we use a slightly different approach of Moriguti (1954) and Bulmer (1957) to get MB-type confidence interval for variances components. Since we can obtain the variance components $k_j\ \sigma_j^2$ from differences $T_j - T_{j-1}$, we get

$$P\left[\frac{S_{j-1}^2}{k_j\, v_{j-1}}\left\{\frac{F_j}{F_{\alpha U v_j,\,\infty}} - 1 + \frac{F_{\alpha U v_j,\, v_{j-1}}}{F_j}\left(1 - \frac{F_{\alpha U v_j,\, v_{j-1}}}{F_{\alpha U v_j,\,\infty}}\right)\right\} \leq\right.$$

$$\left.\leq \sigma_j^2 \leq \frac{S_{j-1}^2}{k_j\, v_{j-1}}\left\{\frac{F_j}{F_{\alpha L v_j,\,\infty}} - 1 + \frac{F_{\alpha L v_j,\, v_{j-1}}}{F_j}\left(1 - \frac{F_{\alpha L v_j,\, v_{j-1}}}{F_{\alpha L v_j,\,\infty}}\right)\right\}\right] = \tag{3}$$

$$= 1-\alpha \quad (j=1,2,\ldots,p).$$

The above expression gives Moriguti-Bulmer type confidence interval in contrast to that of Tukey-Williams type.

A GENERAL CROSS CLASSIFICATION RANDOM MODEL WITHOUT INTERACTION:- In this case the EMS take the form: $T_o = \sigma_o^2$, $T_j = \sigma_o^2 + k_j\,\sigma_j^2$ $(j=1,2,\ldots,p)$. By using a similar reasoning as above and replacing F_j in expression (2) by $F_j^* = v_o S_j^2 / v_j S_o^2$, a general TW-type confidence interval statement for this case takes the form

$$P\left[\frac{S_j^2(1 - F_{\alpha U v_j,\, v_o}/F_j^*)}{k_j\, \chi^2_{\alpha U v_j}} \leq \sigma_j^2 \leq \frac{S_j^2(1 - F_{\alpha L v_j,\, v_o}/F_j^*)}{k_j\, \chi^2_{\alpha L v_j}}\right] \geq 1-2\alpha \tag{4}$$

Next, from relations of the expected mean squares in this model, one can obtain the variance components, $k_j\,\sigma_j^2$, from the differences $T_j - T_o$ for $j=1,2,\ldots,p$. Again, by using Moriguti-Bulmer arguments in extended form, we can get a MB-type confidence interval for variance components as

$$P\left[\frac{S_o^2}{k_j\, v_o}\left\{\frac{F_j^*}{F_{\alpha U v_j,\,\infty}} - 1 + \frac{F_{\alpha U v_j,\, v_o}}{F_j^*}\left(1 - \frac{F_{\alpha U v_j,\, v_o}}{F_{\alpha U v_j,\,\infty}}\right)\right\} \leq\right.$$

$$\left.\leq \sigma_j^2 \leq \frac{S_o^2}{k_j\, v_o}\left\{\frac{F_j^*}{F_{\alpha L v_j,\,\infty}} - 1 + \frac{F_{\alpha L v_j,\, v_o}}{F_j^*}\left(1 - \frac{F_{\alpha L v_j,\, v_o}}{F_{\alpha L v_j,\,\infty}}\right)\right\}\right] = \tag{5}$$

$$= 1-\alpha \quad (j=1,2,\ldots,p).$$

THE FIDUCIAL LIMITS FOR THE BALANCED NESTED AND CROSSED RANDOM MODELS:-

Bross (1950) and Healy (1963), respectively, derived the approximate and exact fiducial intervals of a variance component in the one-way random model. The reasoning behind this approach was that of fiducial arguments of Fisher (1935). Healy stated that the approximate limits which were derived by Bross coincide with the exact limits when any one of a certain pair of degrees of freedom or the associated ratio of mean square variates become large.

For the two models under consideration, by using the extended version of Bross-Healy arguments and employing the structural relationships among T_j's, we get the following two expressions which give confidence interval statements for variance components in the balanced nested and crossed classification models, respectively,

$$P\left[\frac{S^2_{j-1}(F_j-1)(F_j-F_{\alpha U v_j,\, v_{j-1}})}{k_j\, v_{j-1}\,(F_j\, F_{\alpha U v_j,\, \infty} - F_{\alpha U v,\, v_{j-1}})} \leqslant \sigma^2_j \leqslant \right. \tag{6}$$

$$\left. \leqslant \frac{S^2_{j-1}\,(F_j-1)(F_j-F_{\alpha L v_j,\, v_{j-1}})}{k_j\, v_{j-1}(F_j\, F_{\alpha L v_j,\, \infty} - F_{\alpha L v_j,\, v_{j-1}})}\right] = 1-\alpha$$

and

$$P\left[\frac{S^2_0\,(F^*_j-1)(F^*_j-F_{\alpha U v_j,\, v_0})}{k_j\, v_0\,(F^*_j\, F_{\alpha U v_j,\, \infty} - F_{\alpha U v_j,\, v_0})} \leqslant \sigma^2_j \leqslant \right. \tag{7}$$

$$\left. \leqslant \frac{S^2_0\,(F^*_j-1)(F^*_j-F_{\alpha L v_j,\, v_0})}{k_j\, v_0(F^*_j\, F_{\alpha L v_j,\, \infty} - F_{\alpha L v_j,\, v_0})}\right] = 1-\alpha$$

for $j=1,2,\ldots,p$.

MONTE CARLO STUDIES

To illustrate the previous development, here we provide some simulation studies to compare and contrast the performance of TW, MB and BH type procedures for constructing confidence limits for variance components. For this purpose, in this note we consider the two-way nested design model given by $Y_{ijk} = u+a_i+b_{ij}+e_{ijk}$ (i;j;k=1,2,..., I;J;K) where $a_i \sim N(0, \sigma_2^2)$, $b_{ij} \sim N(0, \sigma_1^2)$, $e_{ijk} \sim N(0,\sigma_0^2)$ and a_i, b_{ij} and e_{ijk} are statistically independent.

The simulation study includes: (a) the percentage coverage of the true value of variance components in replicated simulation design, and (b) the expected width of the confidence interval (EWCI). The sampling experiment was repeated 500 times with $\alpha = \cdot 05$. Other models were also considered, which gave similar results, and will be reported elsewhere. The first table contains the simulation study results for the model when the errors are assumed independent. The second table shows the simulation study results under the first order autoregressive scheme for the errors with different values for ρ, the autocorrelation. Notice that in this later case the errors are related by the relationship: $e_{ijk} = e^*_{ijk} + \rho e_{ij(k-1)}$ with $-1 < \rho < 1$, where $e^*_{ijk} \sim N(0, \sigma_0^2)$. Furthermore, note that in this case through the error structure the autoregressive model reduces to the independent case - but now with fewer observations.

CONCLUSIONS AND REMARKS

All the procedures give coverage at the nominal rate of 93% to 98% with chosen $\alpha = \cdot 05$. The expected confidence interval widths are quite small when the ratios σ_i^2/σ_0^2 are small. In general, we notice that the MB and BH expected widths are slightly smaller than those for TW procedure. However, taking a broad overview, we find that in almost all cases under study there are no appreciable differences among the three procedures and that all three are easy in computation. Finally, a confidence interval may take negative values, so when the lower limit is negative it should be put equal to zero; and when both the limits are negative then σ_i^2 should be taken as zero.

From the table II it can be seen that under an autoregressive series of the random errors the similarity among the three procedures is still there. However, we notice that the EWCI's are smaller when the autoregressive parameter, ρ, is negative as compared with when it is positive or zero.

The above procedures can be extended to other models, for example to crossed classification models with interaction and several other models if treated separately.

TABLE I

Comparison between TW, MB and BH procedures for a confidence interval on σ_i^2, in the two-way nested random model with I,J,K = 5,5,6

(a) Ratio σ_2^2/σ_0^2	TW Procedure		MB Procedure		BH Procedure	
	% Coverage	EWCI	% Coverage	EWCI	% Coverage	EWCI
0.5	0.972	3.940	0.972	3.936	0.952	3.877
0.1	0.972	7.806	0.972	7.799	0.948	7.686
1.5	0.944	12.179	0.94	12.001	0.936	12.084
2.0	0.956	15.138	0.956	15.125	0.928	15.254
2.5	0.952	20.131	0.944	19.556	0.956	19.996
3.0	0.956	23.216	0.956	23.196	0.944	23.019
3.5	0.96	26.127	0.956	26.090	0.951	25.905
4.0	0.94	30.163	0.936	29.698	0.924	30.318
4.5	0.968	36.366	0.968	36.336	0.964	36.004
5.0	0.986	37.880	0.984	37.846	0.952	37.364

(b) Ratio σ_1^2/σ_0^2	% Coverage	EWCI	% Coverage	EWCI	% Coverage	EWCI
0.5	0.948	1.173	0.948	1.169	0.944	1.158
1.0	0.96	2.114	0.96	2.110	0.96	2.102
1.5	0.956	3.148	0.948	3.115	0.952	3.131
2.0	0.956	4.078	0.956	4.073	0.956	4.067
2.5	0.928	4.780	0.928	4.775	0.928	4.769
3.0	0.94	6.009	0.94	6.004	0.94	5.998
3.5	0.96	6.991	0.956	6.972	0.956	6.966
4.0	0.948	7.864	0.948	7.859	0.948	7.853
4.5	0.96	9.073	0.96	9.068	0.96	9.062
5.0	0.952	10.030	0.948	9.952	0.948	9.946

TABLE II

Comparison of TW, MB and BH procedures for a confidence interval on σ_i^2, in the two-way nested random model with I,J,K = 5,6,7

	The value ρ	TW Procedure		MB Procedure		BH Procedure	
(a)	$\sigma_2^2/\sigma_o^2 = 0.5$	% Coverage	EWCI	% Coverage	EWCI	% Coverage	EWCI
	0	0.96	4.225	0.956	4.160	0.952	4.197
	0.25	0.96	4.314	0.96	4.310	0.96	4.283
	0.50	0.98	4.761	0.976	4.689	0.972	4.691
	0.75	0.964	7.120	0.964	7.107	0.92	7.112
	-0.25	0.972	4.223	0.964	4.216	0.972	4.192
	-0.50	0.968	4.280	0.968	4.277	0.96	4.250
	-0.75	0.932	3.948	0.932	3.946	0.936	3.925
(b)	$\sigma_1^2/\sigma_o^2 = 0.5$	% Coverage	EWCI	% Coverage	EWCI	% Coverage	EWCI
	0	0.952	1.034	0.952	1.034	0.952	1.034
	0.25	0.952	1.176	0.948	1.160	0.932	1.137
	0.50	0.94	1.476	0.94	1.463	0.936	1.410
	0.75	0.976	3.242	0.968	3.196	0.948	3.024
	-0.25	0.94	0.974	0.936	0.964	0.94	0.957
	-0.50	0.94	0.969	0.938	0.965	0.932	0.961
	-0.75	0.948	0.964	0.948	0.962	0.948	0.960

REFERENCES

Boardman, T.J. (1974): Confidence interval for variance components - A Comparative Monte Carlo Study. Biometrics, 30, 251-262.

Bross, I. (1950), Fiducial interval for variance components. Biometrics, 6, 136-144

Bulmer, M.G. (1957): Approximate confidence limits for components of variance. Biometrika, 44, 159-167.

Fisher, R.A. (1935): The fiducial argument in statistical inference. Ann. Eugen. Lond., 6, 391-398.

Gaylor, D.W., Lucas, H.L., Anderson, R.L. (1970): Calculation of expected mean squares by the abbreviated Doolittle and square root methods. Biometrics, 26, 641-655.

Graybill, F.A. (1976): Theory and Applications of Linear Models. Duxbury Press, N. Scituate, Mass., USA.

-/-

Healy, M.J.R. (1963): Fiducial limits for a variance component. J. Roy. Statist. Soc. Ser. B, 25, 128-130.

Healy, W.C. (1961): Limits for a variance component with an exact confidence coefficient. Ann. Math. Statist., 32, 466-476.

Howe, W.G. (1974): Approximate confidence limits on the means of X plus Y where X and Y are two table independent random variables. J. Amer. Statist. Assoc., 69, 789-794.

Jeyaratnam, S. and Graybill, F.A. (1980): Confidence intervals on variance components in 3-factor cross-classification models. Technometrics, 22, 375-380.

Khuri, A.I. and Sahai, H. (1985): Variance components analysis: a selective literature survey. Inter. Statist. Rev., 53, 279-300.

Miller, J.G. Jr. (1981): Simultaneous Statistical Inference, 2nd Edn., Springer-Verlag, New York.

Moriguti, S. (1954): Confidence limits for a variance component. Rep. Stat. Appl. Res. JUSE, 3, 7-19.

Sahai, H. (1979): A bibliography on variance components. Int. Statist. Rev., 47, 177-222.

Sahai, H., Khuri, A.I. and Kapadia, C.H. (1985): A second bibliography on variance components. Comm. Statist., A14, 63-115.

Satterthwaite, F.E. (1941): Synthesis of variance. Psychometrika, 6, 309-316.

Satterthwaite, F.E. (1946): An approximate distribution of estimates of variance components. Biometrics Bulletin, 2, 110-114.

Tukey, J.W. (1951): Components in regression. Biometrics, 7, 33-69.

Welch, B.L. (1956): On linear combinations of several variances. J. Amer. Statis. Assoc., 51, 132-148.

Williams, J.S. (1962), A confidence interval for variance components. Biometrika, 49, 278-281.

R. Ahmad S. M. Mostafa,
Department of Mathematics,
University of Strathclyde,
Livingstone Tower,
26 Richmond Street,
GLASGOW. G1 1XH.
SCOTLAND.

ON INDISCERNIBLE ESTIMATORS OF STATIONARY PROCESSES

Vladimír Albrecht

Prague

Key words: Gaussian stationary processes, Bayesian discrimination, prediction, signal estimation

ABSTRACT

Prediction and signal estimation are considered in the class of Gaussian non-deterministic stationary processes. In this class the traditional Minimum Mean Square Error (MMSE) criterion yields the estimato whose distribution is singular with respect to the distribution of the estimated process. Processes with singular probability distributions are discernible by the Bayes test. Different class of estimators is obtained when instead of the MMSE method the criterion of indiscernibility is employed.

INTRODUCTION

Let $\{Y(t), X(t)\}$, $t = 0, \pm 1, \pm 2, \ldots$ be a bivariate jointly stationary process. Consider the problem of estimation of $\{Y(t)\}$ based on observation of $\{X(t)\}$. Particularly, if

(1) $$Y(t) = X(t + s), \quad s > 0,$$

then we have the problem of prediction of $X(t)$. When

(2) $$X(t) = Y(t) + N(t),$$

where $\{N(t)\}$ is a stationary process uncorrelated with $\{Y(t)\}$, then, i this connection, $\{Y(t)\}$ is interpreted as a signal and we speak about signal extraction problem. The best prediction or generally the optimu estimation of $\{Y(t)\}$ traditionally means to construct the estimator according to the MMSE criterion, i.e. each random variable $Y(t)$ is estimated by $\hat{Y}(t)$ that is chosen in order to minimize the quantity

(3) $$E\,|Y(t) - \hat{Y}(t)|^2.$$

Effective solution of this minimization problem may be complicated, however, it is usually attainable when seeking for the estimator in the class of linear filters of the observed process, i.e. assuming that

(4) $$\hat{Y}(t) = \sum_u g(u)\, X(t-u).$$

Then the problem consists in finding sequence $\{g(u)\}$ that minimizes (3). If $\{Y(t), X(t)\}$ is moreover Gaussian then the MMSE estimator lies in the class of linear filters, hence consideration of (4) is not a restriction (see Bhansali, Karavellas 1983).

If $u = 0,1,2,\ldots$ in (4) then the estimator is referred to as a "physically realizable" or "non-anticipative" but also "one-sided" filter, if $u = 0, \pm1, \pm2,\ldots$ in (4) then we shall speak about two-sided filter. Typically, s-step predictor of $X(t)$ is given only by an one--sided filter, i.e.

(5) $$\hat{X}(t+s) = \sum_{u=0}^{\infty} g_s(u)\, X(t-u),$$

whereas the signal extractor may be either of one-sided or two-sided form. In particular, assume that $\{Y(t)\}$ and $\{X(t)\}$ have spectral densities $f_Y(\omega)$ and $f_X(\omega)$, respectively, $f_X(\omega) > 0$ all ω. Then the two-sided signal extractor is of the form

(6) $$\hat{Y}(t) = \int_{-\pi}^{\pi} h(\omega)\, e^{it\omega}\, dZ_X(\omega),$$

where

(7) $$h(\omega) = f_Y(\omega) / f_X(\omega)$$

and $Z_X(\omega)$ is from the spectral representation of $X(t)$, i.e.

(8) $$X(t) = \int_{-\pi}^{\pi} e^{it\omega}\, dZ_X(\omega),$$

(cf. Pristley 1981, p.774). Hence,

(9) $$\hat{Y}(t) = \sum_{u=-\infty}^{\infty} g(u)\, X(t-u)$$

with $\{g(u)\}$ being the Fourier coefficients of $h(\omega)$ from (7).

Note that both formulas (5) and (9) define new random processes whose spectral densities differ from spectral densities of the estimated processes. Particularly, (6) is a stationary process with spectral density

(10) $$f_{\hat{Y}}(\omega) = \frac{f_Y(\omega)}{f_X(\omega)}\, f_Y(\omega).$$

This formula indicates that the MMSE signal extractor will underestimate the frequency components with pure signal to noise ratio. Generally, the MMSE estimator has different "dynamical" properties than the estimated processes. These differences may be considerable when a comparatively long trajectory of the signal is recovered by the MMSE estimator.

Differences between dynamical properties of the process and its MMSE estimator follow from the evident fact that these two processes have different probability distributions. Although it is not convenient to require that the estimator should have "similar" probability distribution as the estimated process, we must realize that in many classes of stochastic processes such a requirement is rather substantial than negligible. This requirement should be taken into account in those classes, where the MMSE method yields the estimator whose probability distribution is extremely dissimilar from that of the estimated process. Typically, the class of probability distributions of Gaussian stationary non-deterministic process consists only of mutually singular distributions. Hence, if both $\{Y(t)\}$ and its MMSE estimator $\{\hat{Y}(t)\}$ belong to this class, then these two processes generate trajectories from different subsets of the trajectory space, more exactly, if P_Y and $P_{\hat{Y}}$ are probability distributions corresponding to $\{Y(t)\}$ and $\{\hat{Y}(t)\}$, respectively, then there exists a measurable subset A of the infinite-product-Borel-line space R^∞ so that

$$P_Y(A) = 0, \; P_{\hat{Y}}(A) = 1$$

and

$$P_Y(R^\infty - A) = \;\;\; , \; P_{\hat{Y}}(R^\infty - A) = 0,$$

i.e. the MMSE estimator produces trajectories that are not generated by the estimated process and vice versa.

In practice only finite trajectory segments are recovered. However, since singularity of distributions of random processes is closely connected with their statistical discernibility, the above statement is relevant also when estimating finite number of random variables of an unknown process.

Discernibility of stationary processes is usually based on application of the Bayes test (see e.g. Grenander 1974). The result summarized in the next paragraph imply that processes discernible by this test have mutually singular probability distributions. Hence, constructing estimator $\{\tilde{Y}(t)\}$ that is not discernible in the Bayesian manner from $\{Y(t)\}$ we may eliminate the paradoxal singularity of distributions of $\{Y(t)\}$ and its MMSE estimator.

BAYESIAN DISCERNIBILITY OF GAUSSIAN STATIONARY PROCESSES AND INDISCERNIBLE ESTIMATORS

Assume that $\{Y(t)\}$ is a zero mean Gaussian stationary process

with spectral density $f_k(\omega)$ under the respective hypothesis H
k = 1, 2. Suppose that

(11) $$\int_{-\pi}^{\pi} \ln f_k(\omega)d\omega > -\infty.$$

It is well known that this inequality is statified iff

(12) $$Y(t) = \sum_{u=0}^{\infty} g(u)\ \varepsilon(t-u),$$

where $\{\varepsilon(t)\}$ is a sequence of uncorrelated random variables with zero mean and variance σ^2. The right hand side of (12) indicates that $Y(t+1)$ can't be predicted from previous variables without error and due to this property $\{Y(t)\}$ is referred to as the non-deterministic process. We restrict ourselves to the class of Gaussian non-deterministic processes that is further denoted by G.

Recall that probability distribution of any Gaussian stationary process is determined by its spectral density, thus let P_k denote the distribution corresponding to $f_k(\omega)$, k = 1, 2, P_k^n denotes restriction of P_k on the n-dimensional space and finally, let p_k^n be the corresponding Gaussian probability density.

Let $y^n = [y(1),\ldots, y(n)]$ be a given observation of $Y^n = [Y(1), \ldots, Y(n)]$. Assume that H_1 and H_2 are apriori equiprobable. Then the Bayes test of H_1 against H_2 is based on the likelihood

$$r(y^n) = p_1^n(y^n)/p_2^n(y^n).$$

The critical region for H_1 is

(13) $$W_n = \{y^n: r(y^n) < 1\}$$

(cf. Grenander 1974). The probability of erroneouss decision between $H_1 : P_1$ and $H_2 : P_2$ is then given by the quantity

$$e_n(P_1, P_2) = \frac{1}{2}\ [P_1^n(W^n) + P_2^n(\overline{W}^n)].$$

If

(14) $$\lim_{n\to\infty} e_n(P_1, P_2) = 0$$

then distributions P_1 and P_2 will be called *discernible*. If P_1 and P_2 are discernible then evidently

$$\lim_{n\to\infty} P_1(W^n) = 0 \quad \text{and} \quad \lim_{n\to\infty} P_2(W^n) = 1$$

and this implies that discernible distributions are always mutually singular.

It follows from Markov inequality that

$$(15) \qquad e_n(P_1, P_2) \le \min_{0\le\alpha\le 1} H_\alpha(P_1^n, P_2^n), \quad \text{all } n,$$

where $H_\alpha(P_1^n, P_2^n)$ is the α-entropy of P_1^n and P_2^n, i.e.

$$H_\alpha(P_1^n, P_2^n) = \int (p_1^n/p_2^n)^\alpha dP_2^n.$$

It is shown in previous paper (Albrecht 1984) that

$$(16) \qquad \lim_{n\to\infty} \frac{1}{2} \ln H_{\frac{1}{2}} (P_1^n, P_2^n) = 0 \quad \text{iff } f_1(\omega) = f_2(\omega) \text{ a.e. .}$$

Combining (15) and (16) implies that equation (14) is satisfied iff

$$\int_{-\pi}^{\pi} |f_1(\omega) - f_2(\omega)| \, d\omega > 0.$$

Hence, processes with different probability distributions (= processes with different spectral densities) are discernible. Hence, probability distributions corresponding to different spectral densities (simply: different probability distributions) are discernible.

We shall say that two processes from G are discernible, if their probability distributions are discernible by the Bayes test. Recalling the stochastic equivalence of two processes (see e.g. Gichman and Skorochod 1971, p. 61) we see that processes from G are indiscernible iff they are stochastically equivalent.

Now, let $\tilde{Y} = \{\tilde{Y}(t)\}$ be an estimator of $Y = \{Y(t)\}$. Assume that $Y \in G$ and $\tilde{Y} \in G$. We want to recover a finite part of trajectory of $\{Y(t)\}$, say $Y(1),\ldots,Y(n)$, by $\tilde{Y}(1),\ldots,\tilde{Y}(n)$. Let $\tilde{y}(1),\ldots,\tilde{y}(n)$ be a realization of $\tilde{Y}(1),\ldots,\tilde{Y}(n)$. The above statements imply that unless Y and $\tilde{Y}$ are stochastically equivalent then the probability that $\tilde{y}(1), \ldots,\tilde{y}(n)$ will be classified as a realization of $Y(1),\ldots, Y(n)$ asymptotically vanishes. This is the case of MMSE estimates. The possibility of avoiding this paradoxal feature of the MMSE estimators is to construct the estimator that is not discernible from the estimated process. This is the *indiscernible estimator* and we know that it must be stochastically equivalent with the estimated process.

THE INDISCERNIBLE PREDICTION

Let $\{Y(t)\}$ be a time series from G. Given $Y(t-1)$, $Y(t-2)$, $Y(t-3)$, ... we want to construct the indiscernible estimator of $Y(t+1)$. We know that for any $s = 1, 2, \ldots$ we have

$$(17) \qquad Y(t-s) = \sum_{k=0}^{\infty} c(k)\, \varepsilon(t-s-k).$$

Let now γ be $N(0,\sigma^2)$ variable that is independent from $\varepsilon(t+1-u)$, $u = 1, 2,\ldots$ Then

(18) $$\tilde{Y}(t+1) = \sum_{k=1}^{\infty} c(k)\ \varepsilon(t+1-k) + c(0)\gamma$$

has clearly the same structure as (17), thus (17) is the one-step predictor of $Y(t)$ given $Y(t-1), Y(t-2),\ldots$.

Note that the MMSE predictor of $Y(t+1)$ is

(19) $$\hat{Y}(t+1) = \sum_{k=1}^{\infty} c(k)\ \varepsilon(t+1-k),$$

(cf. e.g. Priestley 1981, p. 740). Hence, we can write

(20) $$\tilde{Y}(t+1) = \hat{Y}(t+1) + c(0)\gamma.$$

Whereas $\hat{Y}(t+1)$ has the mean quadratic error $c^2(0)\sigma^2$, the error of the indiscernible predictor is

(21) $$E\ |Y(t+1) - \tilde{Y}(t+1)|^2 = 2\ c^2(0)\sigma^2,$$

i.e. it is two times greater than the error of the MMSE predictor.

Generally, the indiscernible m-step predictor of $Y(t)$ given $\{Y(t-u)\}_{u=1}^{\infty}$ may be written as

(22) $$\tilde{Y}(t+m) = \hat{Y}(t+m) + \sum_{k=0}^{m-1} c(k)\gamma(k),$$

where $\gamma(k)$, $k = 0,\ldots,m-1$ are independent $N(0,\sigma^2)$ variables that are independent from $\varepsilon\{(t+m-u)\}_{u=0}^{\infty}$.

In practice, the indiscernible m-step prediction requires generation of m variables with $N(0,\sigma^2)$ distribution. Thus, in difference to the MMSE method the numerical value of $\tilde{Y}(t+m)$ is not determined uniquely by the sequence $\{Y(t-u)\}_{u=0}^{\infty}$.

Remark. In our numerical study with Gaussian AR processes random variables $\{\gamma(k)\}$ were generated according to the Box-Müller (1958) method. It should be noted that when processing non-stationary AR models then the indiscernible predictor exhibited surprising robustness: its mean quadratic error was, as a rule, considerably smaller than the error of the MMSE predictor.

THE INDISCERNIBLE SIGNAL EXTRACTION

Assume that $\{Y(t)\}$ and $\{N(t)\}$ from (2) are two uncorrelated stationary Gaussian processes, $\{Y(t)\} \in G$. Then apparently $\{X(t)\} \in G$. Let $f_X(\omega)$ and $f_Y(\omega)$ be spectral densities of $\{X(t)\}$ and $\{Y(t)\}$, respectively. Let $R_k(\tau)$ denote the covariance function corresponding to $f_k(\omega)$, i.e.

(23) $$f_k(\omega) = \frac{1}{2\pi} \sum_{\tau=-\infty}^{\infty} R_k(\tau)\, e^{-i\tau\omega}, \quad k = X, Y$$

and consider this expansion for komplex z. Assume that $\ln f_k(z)$ is analytic in an annulus $A = \{z: \rho < z < 1/\rho,\ \rho < 1\}$. Then (see Pristley (1981, pp. 733-735)

(24) $$f_k(\omega) = \exp(c_k)\ |G_k(e^{-i\omega})|^2, \quad k = X, Y$$

with $G_k(z)$ being of the form

(25) $$G_k(z) = \exp \sum_{u=1}^{\infty} c_k(u)\, z^u, \qquad k = X, Y$$

for $z \in A$,

Let

(26) $$\tilde{Y}(t) = \int_{-\pi}^{\pi} h(\omega)\, e^{it\omega}\, dZ_X(\omega),$$

where $Z_X(\omega)$ is from (8) and

(27) $$h(\omega) = \exp \frac{c_Y - c_X}{2} \exp \sum_{u=1}^{\infty} (c_Y(u) - c_X(u))\, e^{-iu\omega} .$$

This form of $h(\omega)$ implies that (26) can be written as

(28) $$\tilde{Y}(t) = \sum_{u=0}^{\infty} a(u)\, X(t-u),$$

where $\{a(u)\}$ are Fourier coefficients of $h(\omega)$. Simultaneously, $\{\tilde{Y}(t)\}$ is also Gaussian and its spectral density is evidently equal to $f_Y(\omega)$. Hence formula (26) gives the one-sided indiscernible signal extractor given $X(t), X(t-1), \ldots$.

Example. Let

$$f_X(\omega) = \frac{|1 - \beta e^{-i\omega}|}{2\pi |1 - \alpha e^{-i\omega}|^2}, \quad f_Y(\omega) = \frac{k}{2\pi |1 - \alpha e^{-i\omega}|^2}$$

with $0 < |\alpha| < 1$, $0 < |\beta| < 1$, $\alpha\beta > 0$, $\alpha \neq \beta$, and $k = (\alpha-\beta)(1-\alpha\beta)/\alpha$. Then the indiscernible signal estimator given $X(t), X(t-1), \ldots$ is given by

(29) $$\tilde{Y}(t) = k \sum_{u=0}^{\infty} \beta^u\, X(t-u).$$

It can be shown that this estimator differs from the MMSE extractor only by a multiplicative constant (cf. Priestley 1981, pp. 778-779).

REFERENCES

Albrecht, V. (1984) On the convergence rate of probability of error in Bayesian discrimination between two Gaussian processes. In: Proc. of the Third Prague Symposium on Asymptotic Statistics,

1983, Elsevier Sci. Publ. B.V., Amsterdam, 165-175.

Albrecht, V. (1985): Estimation of evoked EEG activity by maximum--entropy signal estimator. In: Proc. of IFIP-IMIA Conf. on Medical Decision Making, 1985, North Holland, Amsterdam, 173-176.

Bhansali, R. J., Karavellas, D. (1983): Wiener filtering (with emphasis on frequency-domain approaches). In: Handbook of Statistics 3 - Time series in the frequency domain, 1983, North Holland, Amsterdam, 1-19.

Box, G. E. P., Müller, M. A. (1958): A note on the generation of random normal deviates. Ann. Math. Statist. 29, 610-613.

Gichman, I. I., Skorochod, A. V. (1971): Theory of stochastic processes, Nauka, Moscow (in Russian).

Grenander, U. (1974): Large sample discrimination between two Gaussian processes with different spectra. Ann. Statist. 2, 347-352.

Perez, A. (1973): Asymptotic discernibility of random processes. In: Proc. of the First Prague Symp. on Asymptotic Statistics, 1972, Charles Univ. Press, Prague, 311-322.

Priestley, M. B. (1981): Spectral analysis and time series. Academic Press, London.

Dr. Vladimír Albrecht
Department of Biomathematics
Institute of Physiology
Czechoslovak Academy of Sciences
Vídeňská 1083
142 20 Prague 4
Czechoslovakia

ON THE VARIANCE OF FIRST PASSAGE TIMES IN THE EXPONENTIAL CASE

Gerold Alsmeyer

Kiel

Key words: Variance of first passage times, asymptotic expansions, non-linear renewal theory

ABSTRACT

For i.i.d. random variables $x_1, x_2, \dots$ with positive mean, finite variance and exponential right tail distribution, asymptotic expansions up to vanishing terms will be derived for the variance of first passage times of the form $T = T(b) = \inf\{n \geq 1: s_n > nf(b/n)\}$, $b \geq 0$, where $s_n = x_1 + \dots + x_n$ and f is a strictly increasing, positive and three times continuously differentiable function on $(0,\infty)$. In particular, it will be shown that the excess over the boundary $s_T - Tf(b/T)$ is exponentially distributed and independent of T extending a result which is known when $f(x) = x$.

1. INTRODUCTION

Consider a sequence of i.i.d. random variables $x_1, x_2, \dots$ with positive mean $\tilde{\mu}$, finite variance $\tilde{\sigma}^2$ and distribution function F. Assume that x_1 has an exponential right tail distribution, i.e. for some $\lambda, C > 0$ and for all $t > 0$

(1.1) $\quad 1 - F(t) = C \exp(-\lambda t)$.

Let $s_n = x_1 + \dots + x_n$ for $n \geq 1$ and for $b \geq 0$

(1.2) $\quad \tau = \tau(b) = \inf\{n \geq 1: s_n > b\}$.

Then one may easily show, cf. Woodroofe(1982), p.19, that $s_\tau - b$, the so-called excess over the boundary, is exponentially distributed with parameter λ and independent of τ f.a. $b \geq 0$. Therefore this special case, henceforth called exponential case, admits an exact computation of $E\tau$ by using Wald's identity, namely

(1.3) $E\tau = \tilde{\mu}^{-1}(b + \lambda^{-1})$.

From theorem 5 of Lai & Siegmund(1979) one may further conclude

(1.4) $\mathrm{Var}\tau = \tilde{\sigma}^2\tilde{\mu}^{-3}(b + \lambda^{-1}) - \tilde{\mu}^{-2}\lambda^{-2}$,

utilizing that $\mathrm{Cov}(s_\tau - b,\tau) = 0$ here.

For arbitrary i.i.d. sequences $x_1, x_2, \ldots$, however, only the asymptotic distribution of $s_\tau - b$, as $b \to \infty$, is available leading to approximations for $E\tau$ and $\mathrm{Var}\tau$, cf. Woodroofe(1982), ch.2, and Lai & Siegmund(1979), theorem 5.

Switching to time-dependent boundaries h_b, i.e.

(1.5) $T = T(b) = \inf\{n \geq 1: s_n > h_b(n)\}$,

approximations for ET and VarT are even more difficult to obtain and there have been many contributions to these problems. Lai & Siegmund(1977,1979) developed an approach, called nonlinear renewal theory, which requires a transformation of T into the form

(1.6) $T = \inf\{n \geq 1: S_n + \xi_n > a\}$,

where $a = a(b) \to \infty$ as $b \to \infty$. Here $(S_n)_{n\geq1}$ is the sum process of a new i.i.d. sequence $X_1, X_2, \ldots$ with positive mean μ and $(\xi_n)_{n\geq1}$ constitutes a sequence with slowly changing paths in a certain sense. Moreover, ξ_n has to be independent of $X_{n+1}, X_{n+2}, \ldots$ Provided that X_1 has a nonarithmetic distribution and under suitable conditions on $(\xi_n)_{n\geq1}$, Lai & Siegmund determined the asymptotic distribution of the "new" excess $S_T + \xi_T - a$ and derived an expansion for ET up to vanishing terms as $b \to \infty$. For a broad class of differentiable boundaries h_b, Alsmeyer(1985) developed a different approach towards an expansion for ET by involving a certain reverse stopping time closely related with T and using the ideas of Lai & Siegmund in a generalized way. In the context of nonlinear renewal theory, Alsmeyer & Irle(1985) have derived an expansion for $\mathrm{Var}T - \mathrm{Cov}(S_T + \xi_T - a, T)$ up to vanishing terms as $b \to \infty$, but have not been able to produce a similar result for VarT. In a so far unpublished thesis, Zhang(1984) has obtained such an expansion for VarT under similar conditions on h_b as Alsmeyer(1985) when x_1 has a directly Riemann integrable density. However, his approach suffers from the drawback of being very complicated and technical since it requires deep Fourier analytical tools. In this paper, we will present a much simplified and purely probabilistic alternative in the exponential case for certain differentiable boundaries.

To be precise, let f be a strictly increasing, positive and three times continuously differentiable function on $(0,\infty)$ and suppose that $f(0) = \lim_{x\downarrow 0} f(x)$ exists. Suppose further that $f(0) < \tilde{\mu}$ and that $xf(b/x)$ does not decrease in x for all $b \geq 0$. Define T as in (1.5) with $h_b(x) = xf(b/x)$ f.a. $x > 0$. Clearly, $T < \infty$ a.s. f.a. $b \geq 0$, since $n^{-1}s_n \to \tilde{\mu}$ a.s. by SLLN and $f(b/n) \to f(0) < \tilde{\mu}$ as $n \to \infty$. In the context of nonlinear renewal theory, T may be rewritten as

(1.7) $\quad T = \inf\{n \geq 1: ng(s_n/n) > b\}$,

where $g(x) = f^{-1}(x^+)$, $x^+ = x1_{(0,\infty)}(x)$. A Taylor expansion of $g(s_n/n)$ about $\tilde{\mu}$ yields that T has the form given in (1.6) with

(1.8) $\quad X_n = g(\tilde{\mu}) + g'(\tilde{\mu})(x_n - \tilde{\mu})$, i.e. $S_n = ng(\tilde{\mu}) + g'(\tilde{\mu})(s_n - n\tilde{\mu})$ and

$$\xi_n = \begin{cases} g''(\tilde{\mu})s_n^{*2}/2 + g^{(3)}(\zeta_n)n^{-1/2}s_n^{*3}/6 , & \text{if } s_n > 0 \\ -S_n , & \text{if } s_n \leq 0 \end{cases},$$

where $|\zeta_n - \tilde{\mu}| \leq |n^{-1}s_n - \tilde{\mu}|$ and $s_n^* = n^{-1/2}(s_n - n\tilde{\mu})$.

Note that ζ_n is a random variable measurable with respect to $x_1, \dots, x_n$, so that ξ_n is indeed independent of $X_{n+1}, X_{n+2}, \dots$
Clearly, τ as defined in (1.2) has the above form with $f(x) = x$. Another class of boundaries which frequently occurs in the literature is given by $h_b(x) = b^{1-\rho}x^{\rho}$, $0 < \rho < 1$. The associated f's in this case are easily seen to be $f(x) = x^{1-\rho}$, $0 < \rho < 1$. The next result for T as in (1.7) in the general case is stated for reference.

Theorem A

Provided that X_1 is nonarithmetic, $E|x_1|^p < \infty$ for some $p > 2$ and

(A.1) $\quad P\{T \leq \delta b\} = o(b^{-1})$, as $b \to \infty$,

for some $\delta > 0$, the following assertions hold:

(A.2) $\quad b^{-1}T \to g(\tilde{\mu})^{-1}$ a.s. ;

(A.3) $\quad b^{-1/2}(T - g(\tilde{\mu})^{-1}b) \xrightarrow{d} N(0, g(\tilde{\mu})^{-3}g'(\tilde{\mu})^2\tilde{\sigma}^2)$

and

(A.4) $\quad g(\tilde{\mu})ET = b - g''(\tilde{\mu})\tilde{\sigma}^2/2 + \Delta_1 + o(1)$,

as $b \to \infty$, where Δ_1 denotes the first moment of the limiting distribution of the "new" excess $Tg(s_T/T) - b$.

(A.2) follows from lemma 4.1. of Woodroofe(1982). In a more general context, (A.3) and (A.4) have been proved by Lai & Siegmund (1977,1979).
Concerning the the variance of T, the next result is due to Alsmeyer & Irle(1985):

<u>Theorem B</u>
In the situation of theorem A let $E|x_1|^p < \infty$ for some $p > 4$ and

(B.1) $P\{T \leq \delta b\} = o(b^{-2})$, as $b \to \infty$,

for some $\delta > 0$. Then there are events $E = E_b$, $b \geq 0$, such that $P(E^c) = o(b^{-2})$ and

(B.2) $$\mathrm{Var}T - 2g(\tilde{\mu})^{-1}\mathrm{Cov}(1_E(S_T+\xi_T-b),T) = g(\tilde{\mu})^{-3}g'(\tilde{\mu})^2\tilde{\sigma}^2(b - \tfrac{1}{2}g''(\tilde{\mu})\tilde{\sigma}^2) + g(\tilde{\mu})^{-2}(\Delta_1^2 - \Delta_2 + g'(\tilde{\mu})g''(\tilde{\mu})\tilde{\gamma} + g'(\tilde{\mu})g^{(3)}(\tilde{\mu})\tilde{\sigma}^4 + \tfrac{1}{2}g''(\tilde{\mu})^2\tilde{\sigma}^4) + o(1) ,$$

as $b \to \infty$, where $\tilde{\gamma} = E(x_1 - \tilde{\mu})^3$ and Δ_1, Δ_2 are the first and second moment, resp., of the limiting distribution of $S_T + \xi_T - b$. Furthermore

(B.3) $\mathrm{Cov}(1_E(S_T+\xi_T-b),T) = o(b^{1/2})$, as $b \to \infty$.

The theorem shows that it remains to consider $\mathrm{Cov}(1_E(S_T+\xi_T-b),T)$ towards an expansion for VarT. In section 3, we will derive an expansion up to vanishing terms for this covariance in the exponential case. For this purpose the exact distribution of the original excess $s_T - Tf(b/T)$ will be computed in section 2 as well as its independence of T been proved. The desired result then follows by using the ideas of nonlinear renewal theory and a relation between original excess and "new" excess $S_T + \xi_T - b = Tg(s_T/T) - b$. In order to assess the accuracy of our approximation for VarT, we will finally present some numerical values from a Monte Carlo study in section 4.

2. THE EXCESS OVER THE BOUNDARY

Our first theorem states that in the exponential case the excess distribution for general stopping times T with nondecreasing boundaries is the same as for $\tau(b)$ given in (1.2). Furthermore independence of excess and stopping time is preserved.

<u>2.1. Theorem</u>
Let $x_1, x_2, \ldots$ be as stated in the introduction and let $a_1, a_2, \ldots$

be a sequence of nondecreasing nonnegative numbers. Define

(2.1) $T = \inf\{n \geq 1: s_n > a_n\}$.

Then for all $t > 0$

(2.2) $P\{s_T - a_T > t, T < \infty\} = P\{T < \infty\}\exp(-\lambda t)$.

If $T < \infty$ a.s., then $s_T - a_T$ is independent of T.

Proof:

Let $t > 0$. Since $\tau(a_n) \geq (=) T$ on $\{T \geq (=) n\}$ f.a. $n \geq 1$, we obtain

$$(2.3) \quad P\{T = n, s_T - a_T > t\} = P\{T = n, s_{\tau(a_n)} - a_n > t\}$$

$$= P\{T > n-1, s_{\tau(a_n)} - a_n > t\} - P\{T > n, s_{\tau(a_n)} - a_n > t\}$$

$$= \int_{(-\infty, a_{n-1}]} P\{R(a_n - x) > t\}\, Q_{n-1}(dx)$$

$$- \int_{(-\infty, a_n} P\{R(a_n - x)\, Q_n(dx) ,$$

where $R(y) = s_{\tau(y)} - y$ f.a. $y \geq 0$ and $Q_k(dx) = P\{T > n, s_n \equiv dx\}$. Now (2.2) follows from the fact that $R(y)$ is exponentially distributed with parameter λ f.a. $y \geq 0$. Moreover, if $T < \infty$ a.s., then the independence of $s_T - a_T$ and is also implied.

Now let us return to T as defined in (1.5) with $h_b(x) = xf(b/x)$ and f as introduced in the previous section. Since $nf(b/n)$ is nondecreasing in n f.a. $b \geq 0$ by assumption, we infer from theorem 2.1. that the "original" excess

(2.4) $R_b = s_T - tf(b/T)$

is exponentially distributed with parameter λ and independent of T f.a. $b \geq 0$. In order to examine the "new" excess

(2.5) $R_b' = Tg(s_T/T) - b = S_T + \xi_T - b$,

S_n, ξ_n as defined in (1.8), we have to provide a representation of R_b' in terms of R_b. Write

$$R_b' = T\big(g(s_T/T) - b/T\big) = T\big(g(s_T/T) - g(f(b/T))\big) .$$

By making a Taylor expansion of $g(s_T/T)$ about $f(b/T)$ $[s_T, f(b/T) > 0\ !]$, we obtain

(2.6) $R_b' = g'(f(b/T))R_b + g''(\eta_T)R_b^2/2T$,

where η_T is an intermediate point between s_T/T and $f(b/T)$. It is easily seen now that in the situation of theorem A

(2.7) $R_b' \xrightarrow{d} g'(\tilde{\mu})W$, as $b \to \infty$,

where W is an exponentially distributed random variable with parameter λ, because from (A.2) with ζ_T as given in (1.8)

(2.8) $f(b/T) \to f(g(\tilde{\mu})) = \tilde{\mu}$ a.s. and $\zeta_T \to \tilde{\mu}$ a.s., as $b \to \infty$.

3. AN EXPANSION FOR VAR(T)

Let us henceforth assume the exponential case and that T is defined as in (1.5) with $h_b(x) = xf(b/x)$, f as given in the introduction. Furthermore, the following notations will be used:

(3.1) $\mu = EX_1 = g(\tilde{\mu})$, $\sigma^2 = \mathrm{Var}X_1 = g'(\tilde{\mu})^2\tilde{\sigma}^2$, $\gamma = E(X_1 - \mu)^3 = g'(\tilde{\mu})\,\tilde{\gamma}$,

$n^* = b^{-1/2}(n - ET)$, $s_n^* = n^{-1/2}(s_n - n\tilde{\mu})$, $S_n^* = n^{-1/2}(S_n - n\mu) = g'(\tilde{\mu})s_n^*$.

The next theorem is the main result of this paper:

<u>3.1. Theorem</u>.

Assume the situation of theorem B. Then, as $b \to \infty$,

(3.2) $$\mathrm{Var}T = g(\tilde{\mu})^{-3}g'(\tilde{\mu})^2\tilde{\sigma}^2\left(b - \tfrac{1}{2}g''(\tilde{\mu})\tilde{\sigma}^2 + g'(\tilde{\mu})\lambda^{-1}\right) - g(\tilde{\mu})^{-2}\left(g'(\tilde{\mu})^2\lambda^{-2} + 2g'(\tilde{\mu})g''(\tilde{\mu})\tilde{\sigma}^2\lambda^{-1} - \tfrac{1}{2}g''(\tilde{\mu})^2\tilde{\sigma}^4 - g'(\tilde{\mu})g''(\tilde{\mu}) - g'(\tilde{\mu})g^3(\tilde{\mu})\tilde{\sigma}^4\right) + o(1) ,$$

and, in particular,

(3.3) $$\mathrm{Cov}(1_E R_b', T) = -g(\tilde{\mu})^{-1}g'(\tilde{\mu})g''(\tilde{\mu})\tilde{\sigma}^2\lambda^{-1} + o(1) ,$$

where $E = E_b$, $b \geq 0$, are as stated in theorem B.

Proof:

Since x_1 has an exponential right tail, clearly, $X_1 = g(\tilde{\mu}) + g'(\tilde{\mu})(x_1 - \tilde{\mu})$ is nonarithmetic. Furthermore, it follows from (2.7) that $\Delta_1 = g'(\tilde{\mu})\lambda^{-1}$ and $\Delta_2 = 2g'(\tilde{\mu})^2\lambda^{-2}$. The following properties of T and the events E have been provided by Alsmeyer & Irle(1985):

(C.1) $T \leq \max\{L, 2b/\mu\}$ on E for sufficiently large b, where L is an integer-valued r.v. with finite third moment;

(C.2) $b(1-\varepsilon(b)) < \mu T < b(1+\varepsilon(b))$ on E for positive numbers $\varepsilon(b)$ such that $\varepsilon(b) \downarrow 0$ as $b \to \infty$;

(C.3) $1_E T^{*2}$, $b \geq b_0$, are uniformly integrable for some $b_0 > 0$;

(C.4) $1_E S_T^{*4}$, $b \geq b_1$, are uniformly integrable for some $b_1 > 0$;

(C.5) $E(1_{E^c}(T - ET)^2) = o(1)$ and $E(1_{E^c}(S_T - \mu T)^2) = o(1)$ as $b \to \infty$;

(C.6) $|s_T - \tilde{\mu}T| \leq \varepsilon_T T$ for suff. large b on E, where $\varepsilon_n \downarrow 0$ as $n \to \infty$.

Using (B.2) with Δ_1, Δ_2 as stated above, it obviously remains to prove (3.2). We obtain from (2.6) that

(3.3) $\mathrm{Cov}(1_E R_b', T) = \mathrm{Cov}(1_E g'(f(b/T))R_b, T) + \frac{1}{2}\mathrm{Cov}(1_E g''(\eta_T)R_b^2/T, T)$

where $|\eta_T - f(b/T)| \le |s_T/T - f(b/T)| = TR_b$.

Set $n_1 = n_1(b) = b(1-\varepsilon(b))$, $n_2 = n_2(b) = b(1+\varepsilon(b))$ and $E' = E_b' = \{n_1 < \mu T < n_2\}$. Then $E \subset E'$ for suff. large b by (C.2). Moreover, since $g' \circ f$ is continuous and b/T uniformly bounded on E', we infer for some constant C and suff. large b

$$\mathrm{Cov}(1_{E'-E} g'(f(b/T))R_b, T) \le C(ER_b^2)^{1/2}(E(1_{E^c}(T - ET)^2))^{1/2} = o(1) ,$$

as $b \to \infty$, by (C.5). Hence,

$$\mathrm{Cov}(1_E g'(f(b/T))R_b, T) = \mathrm{Cov}(1_{E'} g'(f(b/T))R_b, T) + o(1) .$$

By theorem 2.1., R_b and $1_{E'}T$ are independent. Therefore, as $b \to \infty$

(3.4) $\mathrm{Cov}(1_{E'} g'(f(b/T))R_b, T) = ER_b \int_{E'} g'(f(b/T))(T - ET)\, dP$

and $ER_b = \lambda^{-1}$ f.a. $b \ge 0$.

By making a Taylor expansion of $g'(f(b/T))$ about $\mu^{-1} = g(\tilde{\mu})^{-1}$ and recalling that $g = f^{-1}$ on $(0,\infty)$, we obtain further, as $b \to \infty$,

(3.5) $\int_{E'} g'(f(b/T))(T - ET)\, dP = g'(\tilde{\mu}) \int_{E'} (T - ET)\, dP$

$$- \int_{E'} g''(f(\theta_T^{-1}))f'(\theta_T^{-1})\theta_T^{-2}(T - ET)(T - b/\mu)/b\, dP =: I_1 - I_2 ,$$

where $|\theta_T - \mu^{-1}| \le |T - b/\mu|/b$. One may easily conclude from (C.5) that

(3.6) $I_1 = g'(\tilde{\mu}) \int_{E'^{\circ}} (T - ET)\, dP = o(1)$, as $b \to \infty$.

Since $ET = \mu^{-1}b + O(1)$, as $b \to \infty$, by theorem A,

(3.7) $I_2 = \int_{E'} g''(f(\theta_T^{-1}))f'(\theta_T^{-1})\theta_T^{-2}(T - ET)^2/b\, dP$

$$+ O(1) \int_{E'} g''(f(\theta_T^{-1}))f'(\theta_T^{-1})\theta_T^{-2}(T - ET)/b\, dP =: I_{21} + I_{22}.$$

From $P(E'^c) = o(b^{-2})$, $1_{E'}\theta_T \overset{P}{\to} \mu^{-1}$, continuity of g'', f and f' and from (A.3) it is easily seen that

(3.8) $V_{1,T} = g''(f(\theta_T^{-1}))f'(\theta_T^{-1})\theta_T^{-2}T^* 1_{E'} \overset{d}{\to} N(0, g(\tilde{\mu})g''(\tilde{\mu})^2\tilde{\sigma}^2)$,

as $b \to \infty$, whence by (C.3)

(3.9) $I_{22} = b^{-1/2}\, O(1)\, EV_{1,T} = o(b^{-1/2})$, as $b \to \infty$.

Similar arguments yield

(3.10) $V_{2,T} = g''(f(\theta_T^{-1}))f'(\theta_T^{-1})\theta_T^{-2}T^{*2} 1_{E'} \overset{d}{\to} g(\tilde{\mu})^{-1}g'(\tilde{\mu})g''(\tilde{\mu})\tilde{\sigma}^2\chi_1^2$

and again, by (C.3),

(3.11) $I_{21} = EV_{2,T} = g(\tilde{\mu})^{-1}g'(\tilde{\mu})g''(\tilde{\mu})\tilde{\sigma}^2 + o(1)$, as $b \to \infty$.

Now, (3.3) - (3.11) yield (3.2), if we still ensure that the second term on the right hand side of (3.3) tends to 0 as $b \to \infty$. However, $g''(\eta_T)$ is uniformly bounded on E by a suitable constant M f.a. suff. large b, so that

$$\mathrm{Cov}(1_E g''(\eta_T)R_b^2/T,T) \lesssim M\, ER_b^2\, E(1_{E'}|T-ET|/T) ,$$

where we again have used the independence of R_b and $1_{E'}|T-ET|/T$. Moreover, $|T-ET|/T \le M'\varepsilon'(b)$ for suitable M' and $\varepsilon'(b)$ such that $\varepsilon'(b) \downarrow 0$ as $b \to \infty$. This is a consequence of the definition of E'. Finally, since $ER_b^2 = 2\lambda^{-2}$, the assertion follows completing the proof of the theorem.

<u>3.2. Remarks</u>

(a) Theorem 3.1. remains true when all assumptions imposed on f only hold in a neighborhood of $\tilde{\mu}$ and provided that (B.1) is still true. However, the independence of R_b and T is lost, and we have to define a suitable stopping time T_0 with stopping boundary f_0 such that $T = T_0$ on a sufficiently large subevent and f_0 is as required in the previous theorem. We will not detail this more general case, because the involved calculations are messy and do not require any new idea.

(b) Similar but more tedious calculations as in theorem 3.1. also yield an expansion for ET up to terms $o(b^{-1})$ as $b \to \infty$. Although we will not state such an expansion, we should note that in contrast to the case where $f(x) = x$ many additional terms arise in it even in this very nice situation where excess and stopping time are independent.

(c) Defining the two-sided version of T, namely

(3.12) $\overline{T} = \inf\{n \ge 1: |s_n| > nf(b/n)\}$,

one may easily verify that theorem A, theorem B and theorem 3.1. remain true for $\overline{T}$. Again, we only have to show that $T = \overline{T}$ on a sufficiently large subevent. We omit the details.

4. MONTE CARLO RESULTS

Let us finally compare in an example the obtained approximations for ET(b), VarT(b) and $\mathrm{Cov}(R_b', T(b))$ with the corresponding "true" values computed by a Monte Carlo study. For all pairs $(b,\rho) \in \{3, 4, 5, 8, 10\} \times \{0.1, 0.2, 0.3, 0.4\}$ and standard exponentials $x_1, x_2, \ldots$ we have generated 10000 replicas of the stopping time

$$T_\rho(b) = \inf\{n \geq 1 : s_n > bn^\rho\} .$$

The results are given in Table 1 ($ET_\rho(b)$), Table 2 ($VarT_\rho(b)$) and Table 3 ($Cov(R_b', T_\rho(b))$ where for each pair (b,ρ) the first entry denotes the simulated value and the second entry the approximated value of the respective quantity. Note that here

$$R_b' = T_\rho(b)\left(s_{T_\rho(b)}/T_\rho(b)\right)^\lambda - b^\lambda , \quad \lambda = 1/(1-\rho) .$$

The simulations were done on a Siemens 7760.

TABLE 1

Simulated and approximated value for $ET_\rho(b)$

ρ \ b	3	4	5	8	10
.1	4.46 4.43	5.73 5.71	7.07 7.02	11.17 11.13	14.01 13.96
.2	5.06 5.04	6.77 6.75	8.61 8.57	14.56 14.55	18.94 18.88
.3	5.98 5.93	8.38 8.37	11.14 11.09	20.68 20.63	28.06 27.94
.4	7.47 7.35	11.24 11.19	15.78 15.73	33.27 33.11	47.74 47.53

TABLE 2

Simulated and approximated value for $VarT_\rho(b)$

ρ \ b	3	4	5	8	10
.1	4.09 4.13	5.66 5.70	7.37 7.32	12.42 12.39	16.01 15.89
.2	5.92 6.07	8.63 8.74	11.55 11.58	21.10 20.92	27.86 27.69
.3	9.26 9.74	14.38 14.72	20.28 20.28	39.93 39.74	54.63 54.69
.4	16.06 17.64	27.28 28.31	41.41 40.92	89.47 89.20	129.06 129.24

TABLE 3

Simulated and approximated values for $Cov(R_b', T_\rho(b))$

ρ \ b	3	4	5	8	10
.1	- .103 - .137	- .108 - .137	- .117 - .137	- .140 - .137	- .103 - .137
.2	- .348 - .390	- .392 - .390	- .431 - .390	- .403 - .390	- .321 - .390
.3	- .822 - .875	-1.061 - .875	- .963 - .875	- .988 - .875	-1.008 - .875
.4	-2.287 -1.852	-2.662 -1.852	-2.536 -1.852	-2.141 -1.852	-2.444 -1.852

Table 3 indicates that the convergence of $\mathrm{Cov}(R_b', T_\rho(b))$ to the limiting value given in (3.3) is not very fast. Moreover, the simulations have shown a rather large standard error for this quantity.

REFERENCES

Alsmeyer, G. (1985): Extended renewal theory for curved differentiable boundaries. Submitted.

Alsmeyer, G. & Irle, A. (1985): Asymptotic expansions for the variance of stopping times in nonlinear renewal theory. To appear in Stoch. Proc. and Appl.

Lai, T.L. & Siegmund, D. (1977): A nonlinear renewal theory with applications in sequential analysis I. Ann. Stat. 5, pp. 946-954.

Lai, T.L. & Siegmund, D. (1979): A nonlinear renewal theory with applications to sequential analysis II. Ann. Stat. 7, pp. 60-76.

Woodroofe, M. (1982): Nonlinear renewal theory in sequential analysis. CBMS Regional Conf. Series in Appl. Math. 39, Soc. for Industr. and Appl. Math., Philadelphia.

Zhang, C. (1984): Random walk and renewal theory. Ph. D. thesis, Columbia University.

Mathematisches Seminar
Olhausenstrasse 40
2300 Kiel
WEST GERMANY

ON STATIONARY DISTRIBUTIONS OF SOME TIME SERIES MODELS

Jiří Anděl, Manuel Garrido

Prague, Madrid

Key words: autoregressive process, bilinear model, moment problem, random parameter, stationary distribution, threshold model

ABSTRACT

If a time series model and its stationary distribution are given, then the hard problem is to find the corresponding distribution of a strict white noise used in the model. We solve a version of this problem when some moments of the stationary distribution are given.

1. INTRODUCTION

One of the simplest time series models is the stationary autoregression of the first order defined by

$$X_t = a X_{t-1} + Y_t \tag{1.1}$$

where $a \in (-1,1)$ and Y_t are i.i.d. random variables. If the distribution of Y_t is normal, then X_t have also normal distribution. There are several papers devoted to the problem how to find a distribution of Y_t if a stationary distribution of X_t is given. A method introduced by Gaver and Lewis (1980) can be briefly summarized as follows. Consider the characteristic functions $\omega(u) = E\exp(iuX_t)$ and $\psi(u) = E\exp(iuY_t)$. From (1.1) we get the condition

$$\omega(u) = \omega(au)\psi(u) . \qquad (1.2)$$

Since $\omega(u)$ is supposed to be known, (1.2) yields immediately the function $\psi(u)$. A survey of results obtained by this procedure can be found in Anděl (1983). This approach has some disadvantages:

(i) In many cases the problem has no solution which is equivalent to the fact that $\psi(u)$ computed from (1.2) is not a characteristic function. It can be rather difficult to prove that a function $\psi(u)$ is not a characteristic function.

(ii) If $\psi(u)$ is a characteristic function, it is generally not easy to find the corresponding distribution function explicitly. Even in the simple case that X_t has exponential distribution the distribution of Y_t has a continous and a discrete component.

(iii) Only very rarely such a method can be used also for other models.

Consider for example, so called threshold autoregressive process

$$X_t = \begin{cases} \alpha X_{t-1} + Y_t & \text{for} \quad X_{t-1} \leq 0 , \\ -\alpha X_{t-1} + Y_t & \text{for} \quad X_{t-1} > 0 , \end{cases}$$

where $0 < \alpha < 1$. If $Y_t \sim N(0,1)$, then it is proved that the stationary density of X_t is

$$f(x) = [2(1-\alpha^2)/\pi]^{1/2} \exp\{-(1-\alpha^2)x^2/2\}\Phi(-\alpha x) ,$$

see Anděl et al. (1984). If Y_t has the Cauchy distribution $C(0,1)$, then X_t has the stationary distribution

$$g(x) = (2A/\pi^2)\{-[4A^2x^2+(1-A^2+x^2)^2]^{-1}[x \ln A^{-2}(1+x^2)$$
$$+ A^2-1+x^2) \operatorname{arctg} x] + (2A)^{-1}[(1+A)^2+x^2]^{-1}(1+A)\pi\},$$

where $A = \alpha/(1-\alpha)$, see Anděl and Bartoň (1986). From here it is clear that even in very simple models the correspondence between the distribution of X_t and Y_t is so complicated that only exceptionally we can expect explicit results.

2. A METHOD BASED ON MOMENTS

In many practical cases a distribution is not given completely and only a finite number of its moments is known. Denote

$$m_k = E\,X_t^k\,, \qquad s_k = E\,Y_t^k$$

for a stationary process X_t with a strict white noise Y_t. If moments $m_1,\dots,m_n$ are given, then in some models it is easy to calculate $s_1,\dots,s_n$. For example, it follows from (1.1) that

$$m_k = \sum_{i=0}^{k} \binom{k}{i} a^i m_i \; s_{k-i}\,, \qquad k = 0,1,\dots$$

and from here we get a recurrent formula

$$s_k = (1 - a^k)\, m_k - \sum_{i=1}^{k-1} \binom{k}{i} a^i m_i \; s_{k-i}\,. \tag{2.1}$$

Of course, $s_o = m_o = 1$. If remains to solve two following problems.

(a) If numbers $s_1,\dots,s_n$ are calculated, decide whether they are moments or not.

(b) If it is proved that $s_1,\dots,s_n$ are moments, find at least one distribution such that $s_1,\dots,s_n$ are its moments.

Here we restrict ourselves to the case that $n = 2r$ and that no other restrictions on the distribution of Y_t are given. The details about problems of this kind and the proofs of the following two theorems can be sound in Krejn and Nudelman (1973), p.246--247.

<u>Theorem 2.1.</u> A sequence of numbers $\{s_k\}_o^{2r}$ is a system of moments on $(-\infty,\infty)$ if and only it the matrix $A = (s_{i+j})_{i,j=o}^{r}$

is positive semidefinite.

If $n = 4$, we have several possibilities how to find a distribution with moments $s_1, \dots, s_4$. For example, we can use a distribution from Pearson's or from Johnson's system. Another modern and very popular family of distribution is generalized Tukey's lambda system. An information about it and tables facilitating the computations can be found in Ramberg et al. (1979). However, such systems are not in our disposal if $n > 4$. In such case we can use only the following classical result.

Theorem 2.2 Let $s_1, \dots, s_{2r}$ be such numbers that the matrix $A = (s_{i+j})_{i,j=0}^{r}$ is positive semidefinite. Let ξ be an arbitrary real number. Then the polynomial

$$Q(z) = \begin{vmatrix} s_0 & s_1 & \dots & s_{r-1} & 1 & 1 \\ s_1 & s_2 & \dots & s_r & \xi & z \\ \dots & \dots & \dots & \dots & \dots & \dots \\ s_{r+1} & s_{r+2} & \dots & s_{2r} & \xi^{r+1} & z^{r+1} \end{vmatrix}$$

has $r + 1$ different real roots $z_1, \dots, z_{r+1}$. (One of them is ξ.) The system of linear equations

$$z_1^k p_1 + \dots + z_{r+1}^k p_{r+1} = s_k \qquad (2.2)$$

for $k = 0, \dots, r$ has a unique solution $p_1, \dots, p_{r+1}$. This solution is non-negative and satisfies (2.2) also for $k = r+1, \dots, 2r$.

3. SOME SPECIAL CASES

If we are able to calculate $s_1, \dots, s_n$ from $m_1, \dots, m_n$, then the results given above lead either to a construction of a distribution function with the moments s_k or to the conclusion that the problem has no solution because s_k do not satisfy the moment condition given in Theorem 2.1.

Consider first the model (1.1). Using (2.1) we get

$$s_1 = (1 - a)\, m_1 ,$$

$$s_2 = (1 - a^2)\, m_2 - 2\, a\, (1 - a)\, m_1^2 ,$$

$$s_3 = (1 - a^3)\, m_3 - 3a(1 + 2a)(1 - a)\, m_1 m_2 + 6a^2(1 - a) m_1^3 ,$$

$$s_4 = (1 - a^4)\, m_4 - 4a(1 - a)(1 + a + 2a^2)\, m_1 m_3 +$$
$$+ 12a^2 (1 - a)(1 + 3a)\, m_1^2 m_2 - 6a^2 (1 - a^2) m_2^2 -$$
$$- 24a^3 (1 - a)\, m_1^4 .$$

For example, let $m_1 = 0$, $m_2 = 1/3$, $m_3 = 0$, $m_4 = 1/5$ be the first four moments of the rectangular distribution on $[-1,1]$. Consider only the case $a > 0$. Then the matrix A is positive semidefinite for $a \in [0, \frac{1}{2}]$. If $\xi = 0$, then $Q(z)$ has the roots

$$z_{13} = \mp [(3 - 7a^2)/5]^{1/2} , \quad z_2 = 0$$

and from the system (2.2) we get

$$p_1 = p_3 = (5/6)(1 - a^2)/(3 - 7a^2), \qquad p_2 = (4/3)(1 - 4a^2)/(3 - 7a^2).$$

Numerical results and other special cases concerning the model (1.1) can be found in Anděl (1986a).

Now, let

$$X_t = b_t X_{t-1} + Y_t \tag{3.1}$$

where b_t are i.i.d. random variables independent of $\{Y_t\}$. Let $v_k = E\, b_t^k$ It is known that $\{X_t\}$ is stationary if and only if $|v_1| < 1$ and $v_2 < 1$ hold. (In this special case this condition can be reduced to $v_2 < 1$ only.) Process $\{X_t\}$ is called autoregressive process of the order 1 with random parameter. The stationarity conditions were derived in Anděl (1976). . The class of AR models with random parameters is investigated in Nicholls and Quinn (1982) . We get from (3.1)

$$s_k = (1 - v_k) m_k - \sum_{i=1}^{k-1} \binom{k}{i} v_i m_i s_{k-i} . \tag{3.2}$$

Especially,

$$s_1 = (1 - v_1)m_1 ,$$

$$s_2 = (1 - v_2)\, m_2 - 2v_1\,(1 - v_1)\, m_1^2 ,$$

$$s_3 = (1 - v_3)m_3 - 3(v_1 - 2v_1v_2 + v_2)m_1m_2 + 6v_1^2(1 - v_1)m_1^3 ,$$

$$s_4 = (1 - v_4)m_4 - 4(v_1 + v_3 - 2v_1v_3)m_3m_1 - 6v_2(1 - v_2)\cdot$$
$$\cdot\, m_2^2 + 12v_1(v_1 - 3v_1v_2 + 2v_2)m_1^2m_2 - 24v_1^3(1 - v_1)m_1^4 .$$

It is clear that (2.1) is the special case of (3.2) when $v_i = a^i$, since it corresponds to $b_t = a$ with probability 1.

<u>Theorem 3.1.</u> Let $\{X_t\}$ be a stationary process given by (3.1) where $EY_t = 0$ and $EY_t^2 > 0$. If $P(|b_t| > 1) > 0$, then there exists no distribution of Y_t such that X_t has a normal distribution.

<u>Proof</u>. We have from (3.2)

$$m_{2r} = (1 - v_{2r})^{-1}\left[s_{2r} + \sum_{i=1}^{2r-1} \binom{2r}{i} v_i\, m_i\, s_{2r-i} \right].$$

Because $EY_t = 0$, (3.1) yields $m_1 = 0$. If there exists a distribution of Y_t such that X_t are normal, it would be $m_{2i-1} = 0$ for $i = 1,2,\ldots,$ and thus we would have

$$m_{2r} = (1 - v_{2r})^{-1}\left[s_{2r} + \sum_{i=1}^{r-1} \binom{2r}{2i} v_{2i}\, m_{2i}\, s_{2(r-i)} \right].$$

Let $P(|b_t|>1) > 0$. Then there exist $\varepsilon > 0$ and $\delta > 0$ such that $P(|b_t| > 1 + \varepsilon) \geqq \delta$ and

$$v_{2r} = \int b_t^{2r}\, dP = \int_{|b_t| \leqq 1+\varepsilon} b_t^{2r}\, dP + \int_{|b_t| > 1+\varepsilon} b_t^{2r}\, dP \geqq$$
$$\geqq (1 + \varepsilon)^{2r}\, \delta > 1 \quad \text{for some } r = r_0 .$$

For such $r = r_0$ one would have $m_{2r} < 0$.

Consider the simplest superdiagonal bilinear model

(3.3) $$X_t = b\, X_{t-2}\, Y_{t-1} + Y_t ,$$

where Y_t are i.i.d. variables. Since

$$m_k = \sum_{i=0}^{k} \binom{k}{i} b^i m_i s_i s_{k-i} ,$$

we get from here the fomulas

$$(3.4) \qquad m_k = (1 - b^k s_k)^{-1} \sum_{i=0}^{k-1} \binom{k}{i} b^i m_i s_i s_{k-i} ,$$

$$(3.5) \qquad s_k = (1 + b^k m_k)^{-1} \left[m_k - \sum_{i=1}^{k-1} \binom{k}{i} b^i m_i s_i s_{k-i} \right].$$

From (3.4) we can calculate $m_1, \ldots, m_n$ when $s_1, \ldots, s_n$ are given and (3.5) can be used for computation s_k from m_k. For example, let $Y_t \sim N(0,1)$. Because $s_1 = 0$, $s_2 = 1$, $s_3 = 0$, $s_4 = 3$, (3.4) yields $m_1 = 0$, $m_2 = (1 - b^2)^{-1}$, $m_3 = 0$, $m_4 = 3(1 + b^2)/[(1-b^2) . (1 - 3b^4)]$. We have

$$A = \begin{pmatrix} 1 & 0 & (1 - b^2)^{-1} \\ 0 & (1 - b^2)^{-1} & 0 \\ (1-b^2)^{-1} & 0 & 3(1+b^2)/[(1-b^2)(1-3b^4)] \end{pmatrix},$$

$$|A| = 2(1 - b^2)^{-3}(1 - 3b^4)^{-1} .$$

The matrix A is positive semidefinite if and only if $|b| \leq 3^{-1/4} \doteq 0.7598$. (It implies, that for $|b| > 3^{-1/4}$ no stationary distribution with finite moments to the fourth order exists.) On the other side, if a stationary distribution exists, then its coefficient of skewness is 0 and its coefficient of excess must be $m_4 m_2^{-2} - 3 = 6 b^4/(1 - 3b^4)$.

If we want to get a stationary distribution of X_t with the moments $m_1 = 0$, $m_2 = 1$, $m_3 = 0$, $m_4 = 3$, i.e. with the first four moments of $N(0,1)$, then we get from (3.5) $s_1 = 0$, $s_2 = (1 + b^2)^{-1}$, $s_3 = 0$, $s_4 = 3(1 + b^4)/[(1 + b^2)^2(1 + 3b^4)]$. In this case get for $A = (s_{i+j})^2_{i,j=0}$ that $|A| = 2(1 + b^2)^{-3}(1 + 3b^4)^{-1}$ and, surprisingly, this matrix A is positive definite for every real b. From s_3 and s_4 we calculate that the coefficient of skewness is 0 and the coefficient of excess is $s_4 s_2^{-2} - 3 = -6b^4/(1+3b^4)$. If we choose a distribution with zero mean, variance $(1 + b^2)^{-1}$, skewness 0 and the above excess for input variables Y_t, the output variables X_t will have the moments $m_1 = m_3 = 0$, $m_2 = 1$, $m_4 = 3$.

Our last example concerns a threshold model

$$X_t = \begin{cases} a X_{t-1} + Y_t & \text{if} \quad X_{t-1} \leq u, \\ b X_{t-1} + Y_t & \text{if} \quad X_{t-1} > u. \end{cases}$$

If we wish to get a stationary distribution function F of X_t, how to choose the distribution function G of Y_t ?
Denote

$$p = P(X_t \leq u), \qquad q = P(X_t > u),$$

$$w_k = E(X_t^k \mid X_t \leq u), \qquad w_k^* = E(X_t^k \mid X_t > u).$$

Then we come to the following condition for the moments:

$$m_k = \sum_{i=0}^{k} \binom{k}{i} (a^i w_i p + b^i w_i^* q)\, s_{k-i}.$$

For the special case that we wish $X_t \sim N(0,1)$ and $u = 0$, we have

$$p = q = 0.5, \quad m_{2r-1} = 0, \quad m_{2r} = 2^{-r}(2r)!/r!,$$

$$w_r^* = 2^{r/2}\pi^{-1/2}\,\Gamma\left(\frac{r+1}{2}\right), \qquad w_r = (-1)^r w_r^*$$

for $r = 1,2,\ldots$ If we restrict ourselves to the first four moment we have

$$0 = s_1 + (b-a)(2\pi)^{-1/2} ,$$

$$1 = s_2 + (b-a)(2/\pi)^{1/2} s_1 + (b^2+a^2)/2 ,$$

$$0 = s_3 + 3(b-a)(2\pi)^{-1/2} s_2 + \tfrac{3}{2}(b^2+a^2)s_1 + (b^3-a^3)(2/\pi)^{1/2},$$

$$3 = s_4 + 2(b-a)(2/\pi)^{1/2} s_3 + 3(b^2+a^2) s_2 +$$

$$+ 4(b^3-a^3)(2/\pi)^{1/2} s_1 + 3(b^4 + a^4)/2 .$$

The solution is

$$s_1 = (2\pi)^{-1/2} (a - b) ,$$

$$s_2 = 1 + \pi^{-1}(a - b)^2 - (a^2 + b^2)/2 ,$$

$$s_3 = (2\pi)^{-1/2} (a-b) [3 + (3 - \pi)(a - b)^2/\pi] ,$$

$$s_4 = 3 [1 + 2(a - b)^2/\pi - a^2 - b^2] + 3 a^2b^2 +$$

$$+ (a - b)^2(a^2 + 4ab + b^2)/\pi + 2(3 - \pi)(a - b)^4/\pi^2 .$$

This procedure can be easily extended to the threshold autoregressive model of the first order with several thresholds.

An extension of this method to general linear processes is described in Anděl (1986 b).

REFERENCES

Anděl J. (1976): Autoregressive series with random parameters. Math. Operationsforsch. Statist. 7, 735-741.

(1983): Marginal distributions of autoregressive processes. In: Trans. of 9 th Prague Conf. Inf. Th. etc., 127-135, Academia, Prague.

(1986 a): On simulating AR (1) process with given moments of its marginal distribution. Sent for publication.

(1986 b): On linear processes with given moments. Sent for publication.

Anděl J., Netuka I., Zvára K. (1984): On threshold autoregressive processes. Kybernetika 20, 89-106.

Anděl J., Bartoň T. (1986): A note on a threshold AR(1) model with Cauchy innovations. J. Time Series Anal. 7, 1-5.

Gaver D.P., Lewis P-A.W. (1980): First - order autoregressive gamma sequences and point processes. Adv. Appl. Prob. 12, 727-745.

Krejn M.G., Nudelman A.A. (1973): Problema momentov Markova i ekstremalnyje zadači. Nauka, Moskva

Nicholls D.F., Quinn B.G. (1982): Random Coefficient Autoregressive Models: An Introduction. Lecture Notes in Statistics 11, Springer, New York-Heidelberg--Berlin.

Ramberg J.S., Dudewicz E.J., Tadikamalla P.R., Mykytka E.F. (1979): A probability distribution and its uses in fitting data. Technometrics 21, 201-214.

Department of Statistics
Charles University
Sokolovská 83
186 00 Prague 8
Czechoslovakia

Faculty of Informatics
Polytechnical University
Carretera Valencia km. 7
Madrid 31
Spain

BRAINWARE FOR SEARCHAL PSEUDOBOOLEAN OPTIMIZATION

Alexander N. Antamoshkin

Kemerovo

Key words: searchal pseudoboolean optimization, random search, regular search, identification, brainware, software

ABSTRACT

An approach to the solving of applied pseudoboolean optimization problems is discussed. This approach presupposes the construction of a set of random search algorithms, which are adaptable "for the problem" parametrically and structurally, for a suboptimal (rough) solving of the problem and a gathering of an a posteriori information about characteristics of the problem. It also presupposes the construction of a set of regular algorithms, which are optimal by the labour-charge on the classes of problems, for exact solving of the problem on the basis of the indicated a posteriori information. The sets of algorithms have been realized on a computer.

Many applied problems are reduced to the optimization problem of pseudoboolean functions:

$$f: B_{2^n} \rightarrow R^1,$$

where $B_{2^n} = \{X \in R^n: x_j = 0 \vee 1\}$

(see e. g. Hammer and Rudeanu (1968), Antamoshkin (1983, 1986)).

The efficient brainware for pseudoboolean optimization for the case when objective function (and the condition functions) of the pseudoboolean optimization problem has an analytical representation has been considered by Hammer and Rudeanu (1968). However for the applied optimization problems absence of an analytical representation of objective function is characteric. As a rule in practice the objective function and condition functions of pseudoboolean optimization problems are given by an algorithm (the set of rules for definition of the function meanings). This case (the searchal optimization) will be considered.

For a suboptimal (rough) solving of a pseudoboolean optimization problem the method of variable probabilities (MIVER) is proposed (see Antamoshkin and Saraev (1985)). This method produces the set of random search algorithms united by the common scheme. The scheme of method and the detailed description of it are adduced in Antamoshkin (1986). Here we shall point out the following only:

The algorithm of random search with adaptation, the modifications of it, the local and global algorithms of random search with return (Antamoshkin (1981)); the zonal algorithms of indicated algorithms. All enumerated above algorithms with variable size of change probabilities steps (Antamoshkin (1986)) form the basis of MIVER. The procedures of parametric and structural adaptation of the algorithms of MIVER "for the problem" have been developed (Antamoshkin (1981)) and have been realized by programs in the software package "MIVER" (see Antamoshkin and Saraev (1985)). This procedures permit to gain necessary reability of the solving of pseudoboolean optimization problem with the minimum labourcharge of the method. The basic algorithms of MIVER had been realized by some program modules, too (see Antamoshkin and others(1985b,c)).

Besides a certain concrete suboptimal solution the algorithms of MIVER provide an a posteriori information about the realized pseudoboolean optimization problem based on the possibility to identify the objective functional of the problem with one of the classes of pseudoboolean functions. However, this identification requires the beforehand division of the whole set of pseudoboolean functions to the classes and the analysis of characteristics of these classes. The various approaches to the classification of the pseudoboolean function set are considered in Antámoshkin (1985) (the "extensive" classification) and Antamoshkin and Vaingaus (1986) (the "detailed" classification). The principal difference of the approaches is in number of the classes. The principal demand to the classification is in covering of the whole pseudoboolean function set by the defined classes.

The analysis of characteristics of the space B_2n and the pseudoboolean function classes permits to construct the regular (exact) algorithms of search,unimprovable by the labour-charge on the function classes. These algorithms had been described in Antamoshkin and others(1985, 1986). Estimates of the labour-charge of the algorithms have been adduced in Antamoshkin (1985). The all regular algorithms have been realized in program modules - see Antamoshkin and others(1985, 1986).

By means of constructed brainware a number of concrete applied problems of planning and control had been solved (see e.g. Antamoshkin and Saraev (1985), Antamoshkin and Vaingaus (1986), Antamoshkin and Antamoshkina (1985), Antamoshkin and others (1984), Antamoshkin (1983, 1986)).

REFERENCES

Antamoshkin A. (1981): Adaptation of one class of random search algorithms. Avtomatika i vychisliteljnaja technika, No 3, 54.

(1983): The solving of applied problems of optimization of functionals with boolean variables. VSNTO, Kemerovo.

(1985): Regular optimization of pseudoboolean function. In: 12th IFIP Conference on System Modelling and Optimization. Budapest.

(1986): Optimization of functionals with boolean variables. Univer. Press, Tomsk.

Antamoshkin A., Antamoshkina O. (1985): Integer models and algorithms of distributive problems of economy. In: International Conference "Parametric Optimization and Related Topics". Humboldt-Univer., Berlin 5, 6.

Antamoshkin A., Vaingaus A. (1986): Software for reduction of factor space dimension and selection of informative signs. In: COMPSTAT 1986, Rome.

Antamoshkin A., Saraev V. (1985): Software for definition of treatment. In: International Working Conference in Computer-aided Medical Decision Making. Praha.

Antamoshkin A. and others (1984): On formalization of some problems of planning of agrarian-industrial complexes. Avtomatika, No 2, 84-87.

(1985) In: OFAP (Sectoral Fund of Algorithms and Programs), Minvuza SSSR, Mosk. Univers., Moscow, Inv. No M85119, M85021, M85022.

(1986): In: OFAP, Inv. No M86006, M86007, M86009, M86020, M86021.

Hammer (Ivànescu) P. L., Rudeanu S. (1968): Boolean Methods in Operations Research and Related Areas. Springer-Verlag OHG, Berlin.

Kemerovo State University
SSSR, 650043, Kemerovo,
Krasnaj 6

ASYMPTOTIC APPROXIMATIONS FOR THE EXTREME VALUE DISTRIBUTION OF NONSTATIONARY DIFFERENTIABLE NORMAL PROCESSES

Karl Breitung
München

Key words : Normal processes, extreme value distribution, crossings, method of Laplace

ABSTRACT

For the crossing rates of nonstationary normal processes with differentiable sample paths asymptotic approximations are derived using the method of Laplace. These results are used for obtaining bounds for the extreme value distribution of the process.

1 INTRODUCTION

In Cramer/Leadbetter (1967) (chap.13) crossing problems of differentiable nonstationary normal processes are discussed. The results for the mean number of crossings are used to obtain bounds for the extreme value distribution of the process. Berman (1985) gives results for the extreme value distribution of normal processes with nonconstant variance without assuming differentiability.

Here we will consider only the case of a one-dimensional normal process $x(t)$ with differentiable sample paths. The results are obtained by first deriving asymptotic expansions for the mean number of crossings and then using these for bounding the extreme value distribution.

Now let $x(t)$ be a normal process on $[0,1]$ with mean zero and $var(x(t)) = \sigma^2(t) = r(t,t)$, $r(t,t)$ is the covariance function of the process. We assume that $r(t,u)$ is twice continuously differentiable with partial derivatives $r_{ij}(t,t)$ and that $\sigma(t)$ is three times continuously differentiable.

Further we assume that $x(t)$ has continuous sample paths with probability one. Using the notation of Cramer/Leadbetter we write :

(1) $\gamma^2(t) = \mathrm{var}(x'(t)) = r_{11}(t,t)$

(2) $\mu(t) = \dfrac{\mathrm{cov}(x(t),x'(t))}{\gamma(t)\ \sigma(t)} = \dfrac{\sigma'(t)}{\gamma(t)}$

This gives for the conditional distribution of $x'(t)$:

(3) $E(x'(t);\ x(t)=u) = u\cdot\dfrac{\sigma'(t)}{\sigma(t)}$

(4) $\mathrm{var}(x'(t);\ x(t)=u)=\gamma^2(t)(1-\mu^2(t))=\gamma^2(t)-\sigma'^2(t).$

For these functions we assume that for all $t\in[0,1]$ $\sigma(t)>o$ and $|\mu(t)|<1$, i.e. the joint distribution of $x(t)$ and $x'(t)$ is always nonsingular. Writing $p_t(u,z)$ for the bivariate normal density of $x(t)$ and $x'(t)$, we obtain for $E(C_u)$, the mean number of crossings of the level u by $x(t)$ during $[0,1]$, using formula (13.2.5) in Cramer/Leadbetter (The formula is for zero crossings, we obtain u-crossings by setting $m(t)=u$) :

(5) $E(C_u) = \int\limits_o^1 (\int\limits_R |z| p_t(u,z)dz)dt$

This can be rewritten, with $p_t(u)$ the normal density of $x(t)$ at u:

(6) $E(C_u) = \int\limits_o^1 p_t(u)\ (\int\limits_R |z|\ \dfrac{p_t(u,z)}{p_t(u)}\ dz)dt =$

$= \int\limits_o^1 E(|x'(t)|;x(t)=u)p_t(u)dt$

Now, this integral can be interpreted as the first absolute moment of a random variable. Let τ be a random variable uniformly distributed on $[0,1]$, which is independent of the process $x(t)$. Then we find for the probability distribution and the probability density $p(u)$ of $x(\tau)$:

$$P(x(\tau) > u) = \int_0^1 \Phi(-\frac{u}{\sigma(t)})dt$$

$$p(u) = \int_0^1 p_t(u)dt$$

This yields :

$$E(|x'(\tau)|; x(\tau)=u) =$$

$$= \frac{1}{p(u)} \int_0^1 E(|x'(t)|; x(t)=u) p_t(u)dt =$$

(7) $$= \frac{E(C_u)}{p(u)}$$

From $E(C_u)$ the expected number of upcrossings $E(U_u)$ and downcrossings $E(D_u)$ can be derived using the relations (see Cramer/ Leadbetter, p.289) :

(8a) $$E(U_u) + E(D_u) = E(C_u)$$

(8b) $$E(U_u) - E(D_u) = \Phi(\frac{-u}{\sigma(0)}) - \Phi(\frac{-u}{\sigma(1)})$$

For the maximum $M(1) = \max_{t \in [0,1]} x(t)$ an upper bound is given by :

(9) $$P(M(1)>u) \leq P(x(0)>u) + E(U_u) = \Phi(\frac{-u}{\sigma(0)}) + E(U_u)$$

In the following we often use the relations :

$$u^{-1}\varphi(u) \sim \Phi(-u) \text{ for } u \to \infty$$

$$\Phi(\frac{-u}{c}) = o(\Phi(-u)) \quad \text{for } c > 1 \text{ and } u \to \infty$$

2 ASYMPTOTIC EXPANSION OF INTEGRALS BY THE METHOD OF LAPLACE

In Erdelyi (1956), p.37 the following result is given : Let g and h be functions on the interval (α,β) for which

$$(10)\qquad I(u) = \int_{\beta}^{\alpha} g(t)\, \exp(u^2 h(t))dt$$

exists for all sufficiently large positive u, let h be real, continuous at $t=\alpha$, continuously differentiable for $\alpha<t<\alpha+\eta$, $\eta>o$ and such that $h'<o$ for $\alpha<t<\alpha+\eta$, $h(t)\leq h(\alpha)-\varepsilon$, $\varepsilon>o$ for $\alpha+\eta\leq t\leq\beta$; suppose that $h'(t)\sim -a(t-\alpha)^{\nu-1}$ and $g(t)\sim b(t-\alpha)^{\lambda-1}$ as $t\to\alpha$, $\lambda>o$, $\nu>o$ then

$$(11)\qquad I(u) \sim \frac{b}{\nu}\,\Gamma(\frac{\lambda}{\nu})\,(\frac{\nu}{a\,u^2})^{\lambda/\nu}\, \exp(u^2 h(\alpha))\quad (u\to\infty).$$

Similar results can be obtained, if h(t) achieves its global maximum at a point $c\in(\alpha,\beta)$ (for details see Bleistein/Handelsman (1975)). If $h(t)<h(c)$ for all other t and $h''(c)<0$, we get, if g(t) is n times continuously differentiable near c with $g(c)=g'(c)=\ldots=g^{(n-1)}(c)$ and $g^{(n)}(c)\neq o$:

a) if n is odd :

$$(12)\qquad I(u)\sim o\cdot u^{-(n+1)}\, \exp(u^2 h(c))$$

b) if n is even :

$$(13)\qquad I(u) \sim \frac{(n-1)!!}{n!}\sqrt{2\pi}\, g^{(n)}(c)\,(-h''(c))^{-\frac{n+1}{2}}\, u^{-(n+1)} \exp(u^2(c))$$

3 A SINGLE MAXIMUM IN THE INTERIOR

Consider now the case, that the variance function $\sigma^2(t)$ has exactly one global maximum in [0,1] at the point $t_o\in(o,1)$ with $\sigma'(t_o)=o$ and $\sigma''(t_o)<o$. For p(u) we obtain using equation (13) the asymptotic approximation :

(14) $$p(u) \sim \sqrt{\frac{\sigma(t_o)}{-\sigma''(t_o)}}\, u^{-1} \exp\left(\frac{-u^2}{2\sigma^2(t_o)}\right)$$

For the first moment :

$$E(x'(\tau)\ ;\ x(\tau)=u) = \\ = \frac{u}{p(u)} \int_o^1 \frac{\sigma'(t)}{\sigma(t)}\, p_t(u)\,dt$$

Since $\sigma'(t_o)=o$ and $\sigma''(t_o)<0$, with equation (12) :

(15) $$E(x'(\tau);\ x(\tau)=u) \to o \quad \text{for } u \to \infty$$

For the second moment

$$E(x'^2(\tau)\ ;\ x(\tau)=u) = \\ = \frac{1}{p(u)} \int_o^1 (Var(x'(t);x(t)=u)+(E(x'(t);x(t)=u))^2)\,p_t(u)\,dt$$

(16) $$= \frac{1}{p(u)} \int_o^1 ((\gamma^2(t)-\sigma'^2(t)) + (u\cdot\frac{\sigma'(t)}{\sigma(t)})^2)\,p_t(u)\,dt$$

Splitting up in two integrals , we get for the first integral using equation (13) with n=o :

(17) $$\frac{1}{p(u)} \int_o^1 (\gamma^2(t)-\sigma'^2(t))\,p_t(u)\,dt \sim \gamma^2(t_o)$$

and for the second with n=2 and

$$\frac{d}{dt^2} \left(\frac{\sigma'(t_o)}{\sigma(t_o)}\right)^2 = 2\left(\frac{\sigma''(t_o)}{\sigma(t_o)}\right)^2 :$$

(18) $$\frac{u^2}{p(u)} \int_o^1 \left(\frac{\sigma'(t)}{\sigma(t)}\right)^2 p_t(u)\,dt \sim -2\sigma(t_o)\sigma''(t_o)$$

Combining these integrals :

(19) $$E(x'^2(\tau)\ ;\ x(\tau)=u) \to \gamma^2(t_o) - \sigma(t_o)\sigma''(t_o)$$

For all higher moments $E(x'^{l}(\tau);\ x(\tau)=u)$ we can obtain in a similar way, by reducing these moments to conditional moments under the condition $\tau=t$, asymptotic expansions:

$$E(x'^{l}(\tau);\ x(\tau)=u) \rightarrow \begin{cases} (l-1)!!(\gamma^2(t_o)-\sigma(t_o)\sigma''(t_o))^{\frac{l}{2}} & l \text{ even} \\ 0 & l \text{ odd} \end{cases}$$

Therefore due to the moment convergence theorem given in Billingsley (1986),p.402, the conditional distribution of $x'(\tau)$ under the condition $x(\tau)=u$ converges in distribution and quadratic mean towards a normal distribution with mean zero and variance $\gamma^2(t_o)-\sigma(t_o)\sigma''(t_o)$. This gives for the crossings :

$$E(C_u) = p(u)E(|x'(\tau)|;\ x(\tau)=u) \sim$$

$$\sim \sqrt{\frac{\sigma(t_o)}{-\sigma''(t_o)}}\, u^{-1}\exp\left(\frac{-u^2}{2\sigma^2(t_o)}\right) \sqrt{\frac{\gamma^2(t_o)-\sigma(t_o)\sigma''(t_o)}{\pi}}\sqrt{2}$$

$$(21) \quad \sim \sqrt{\frac{-4\gamma^2(t_o)}{\sigma(t_o)\sigma''(t_o)} + 4}\ \Phi\left(\frac{-u}{\sigma(t_o)}\right)$$

Since due to equation (8b) in this case $E(U_u) = E(C_u)/2$, we get an upper bound for the maximum with equation (9) :

$$(22) \quad P(M(1) > u) \leq \Phi\left(\frac{-u}{\sigma(0)}\right) + \sqrt{\frac{-\gamma^2(t_o)}{\sigma(t_o)\sigma''(t_o)} + 1}\ \Phi\left(\frac{-u}{\sigma(t_o)}\right)$$

A lower bound is :

$$(23) \quad P(x(t_o)>u) = \Phi\left(\frac{-u}{\sigma(t_o)}\right) \leq P(M(1)>u)$$

4 MAXIMUM AT ONE ENDPOINT

Consider now the case that the variance function has its global maximum in [0,1] at 0 with $\sigma'(0)<0$ and $\sigma(t)<\sigma(o)$ for all other t. In this case we obtain using eq.(11) with $\nu=1$ and $\lambda=1$:

(24) $$p(u) \sim \frac{1}{\sqrt{2\pi}} \frac{\sigma^2(0)}{\sigma'(0)} u^{-2} \exp\left(\frac{-u^2}{2\sigma^2(0)}\right)$$

For the conditional moments of the random variables $u^{-1}x'(\tau)$ we obtain in the same way for $u \to \infty$:

(25) $$E(u^{-1}x'(\tau); x(\tau)=u) \to \frac{\sigma'(0)}{\sigma(0)}$$

(26) $$\mathrm{var}(u^{-1}x'(\tau);\ x(\tau)=u) \to 0$$

Therefore the conditional distribution of $u^{-1}x'(\tau)$ converges towards a one-point distribution concentrated at $\frac{\sigma'(0)}{\sigma(0)}$ in quadratic mean. This yields for $E(C_u)$:

(27) $$E(C_u) = p(u)\cdot u\cdot E(|u^{-1}x'(\tau)|;\ x(\tau)=u) \sim$$
$$\sim \frac{\sigma(0)}{u}\,\varphi\left(\frac{u}{\sigma(0)}\right) \sim \Phi\left(\frac{-u}{\sigma(0)}\right)$$

From equation (8b) follows :

$E(C_u) \sim E(D_u)$ and $E(U_u) = o(E(C_u)) =$

(28) $$= o\left(\Phi\left(-\frac{u}{\sigma(0)}\right)\right)$$

Since obviously :

$$P(x(o)>u) = \Phi\left(-\frac{u}{\sigma(0)}\right) \leq P(M(1)>u)$$

with equation (9)

(29) $$P(M(1)>u) \sim \Phi\left(\frac{-u}{\sigma(0)}\right)$$

If the maximum is at 1, we obtain in a similar way equation (29) with $\sigma(o)$ replaced by $\sigma(1)$.

5 SEVERAL MAXIMUM POINTS

Consider now that the variance function $\sigma^2(t)$ achieves its global maximum σ_o^2 in [0,1] at exactly k points $t_1,\ldots,t_k$.
We obtain using the corollary 4.2.4. of the normal comparison Lemma in Leadbetter et al. (1983) at page 84 :

$$(30) \quad |P(\max_{i=1,k} x(t_i)<u) - \Phi^k(\frac{u}{\sigma_o})| \leq C_o \exp(\frac{-u^2}{\sigma_o^2(1+c)})$$

with C_o a constant and $c=\max_{i\neq j}|r(t_i,t_j)/(\sigma(t_i)\sigma(t_j)|$
Therefore if $c<1$, i.e. there is no total dependence between some of the $x(t_i)$:

$$(31) \quad P(\max_{i=1,..,k} x(t_i)<u)= \Phi^k(\frac{u}{\sigma_o})+ o(\Phi(\frac{-u}{\sigma_o}))$$

$$(32) \quad P(\max_{i=1,..,k} x(t_i)\geq u)=k\cdot\Phi(\frac{-u}{\sigma_o}) + o(\Phi(\frac{-u}{\sigma_o}))$$

This yields a lower bound :

$$(33) \quad P(M(1)>u)\geq k\Phi(\frac{-u}{\sigma_o}) + o(\Phi(\frac{-u}{\sigma_o}))$$

An upper bound can be derived by splitting up [0,1] in k disjoint intervals I_i, each containing one of the t_i's and then using the result of paragraph 3 and 4 for computing an asymptotic approximation for the expected number of upcrossings of level u in this interval.

REFERENCES

Berman S. (1985): The maximum of a Gaussian process with nonconstant variance,Vol.21,No.2, 383-391

Billingsley P. (1986): Probability and Measure,Wiley,New York

Bleistein N.and Handelsman R.A. (1975): Asymptotic Expansions of Integrals. Holt,Rinehard and Winston,New York

Cramer H. and Leadbetter M.R. (1967): Stationary and Related Stochastic Processes.Wiley,New York

Erdelyi A.(1956): Asymptotic Expansions.Dover, New York

Leadbetter M.R.,Lindgren G. and Rootzen H. (1983) : Extremes and Related Properties of Random Sequences and Processes,Springer,New York

Seminar für Angewandte Stochastik
der Universität München
Akademiestr.1/IV
D-8000 München 40
FRG

SUFFICIENCY AND STANDARD CLASSES OF STATISTICAL PROBLEMS

Tadeusz Bromek

Warsaw

Key words: reduction of statistical problem, standard class, sufficiency

ABSTRACT

Takeuchi and Akahira (1975) considered prediction problems in which decision rules were evaluated by means of so called performance characteristics. i.e., a family of joint distributions of this rule and of the predicted random element. A similar notion was considered by Barra (1971) in estimation and hypotheses testing problems. Thus, the general idea is that the solutions are determined by the families of joint distributions of the respective decisions and the considered goal of investigations. In this paper the notion of standard classes of such problems are defined. For any problem in a standard class and any statistic there is defined a reduced problem. The reduced problem is said to be equivalent to the primary one if for any solution of the former, there exists a solution of the latter, such that both solution are determined by identical families of joint distributions. The implications for sufficient and prediction sufficient statistics are indicated.

1. INTRODUCTION

A statistical problem consists of the following elements:

i. Statistical space $M = (\Omega, \mathcal{A}, \mathcal{P})$, where $(\Omega, \mathcal{A})$ is a sample space and $\mathcal{P}$ is a family of distribution on $\mathcal{A}$.

ii. A measurable space of observables $(\mathcal{T}, \mathcal{F})$ and a measurable function $T : (\Omega, \mathcal{A}) \to (\mathcal{T}, \mathcal{F})$, i.e., observable statistics.

iii. A family of measurable functions $I_P : (\Omega, \mathcal{A}) \to (\mathcal{I}, \mathcal{J})$ called a goal of investigation.

iv. A decision space $(\mathbf{D}, \mathcal{D})$ under consideration.

v. A specific non-empty set Δ of decision rules - markov kernels from $(\mathcal{T}, \mathcal{F})$ to $(\mathbf{D}, \mathcal{D})$, i.e., for $\delta \in \Delta$, $\delta_B(\cdot) = \delta(B, .)$ is for fixed $B \in \mathcal{D}$ a Borel measurable function, and $\delta_t(.) = \delta(., t)$ is for fixed $t \in \mathcal{T}$ a distribution on $\mathcal{D}$.

vi. A set of solutions $\Delta^* \subset \Delta$.

Comments:

Ad ii. A detailed discussion of observability is given in Dąbrowska, Pleszczyńska (1980). Let us mention that in the case of full observability we may take $(\mathcal{T}, \mathcal{F}) = (\Omega, \mathcal{A})$ and the identity function as T.

Ad iii. Typical cases for the goal of the investigation are estimation of distribution parameters, testing hypotheses, prediction problems and so on. For estimation $(\mathcal{I}, \mathcal{J})$ can be the real line with Borel algebra. and $I_P(\omega) = \varkappa(P)$. for every $P \in \mathcal{P}$, and $\omega \in \Omega$, where $\varkappa(P)$ is a parameter to be estimated. For prediction problems $\mathcal{I}$ is a set of values of a predicted random variable, I, endowed with an algebra and $I_P(\omega) = I(\omega)$ for every $P \in \mathcal{P}$ and $\omega \in \Omega$. However, we can deal with more intricate problems, e.g., if we are looking for a value of the regression function of the random variable X on the random variable Y in a point y, then $I_P(\omega) = E_P(X/Y=y)$, $\omega = (x,y) \in R^2$, $P \in \mathcal{P}$.

Ad iv. In the most common cases, the decision space is equal to the goal space. However, in the case of confidence bounds for a distribution parameter, the goal of investigation is the parameter itself, while the decision space consists of intervals. Another difference between the sets $\mathcal{I}$ and $\mathbf{D}$ is if we add to $\mathbf{D}$ the suspended decision.

Ad v and vi. At the very beginning of stating a statistical problem, we can restrict the set of all decision rules to some subset, e.g., non-randomized rules. Usually the set of solutions Δ^* is a set of decision rules which are solutions of some optimization problem.

2. STANDARD CLASSES

For a given statistical problem the joint distribution $Q_{P,\delta}$ of decisions taken according to a decision rule δ and the goal I_P is defined as follows:

$$Q_{P,\delta}(B\times C) = \int_{I_P^{-1}(C)} \delta_B(T(\omega))dP, \quad B\in\mathcal{D}, \; C\in\mathcal{J}, \; P\in\mathcal{P}, \; \delta\in\Delta$$

Therefore, the elements of a statistical problem determine an indexing function q from $\mathcal{P}\times\Delta$ to the family of all distributions on $\mathcal{D}\times\mathcal{J}$, given by $q(P,\delta) = Q_{P,\delta}$.

Let us notice that in a case of inference for the parameter of distributions when $I_P(\omega) \equiv \varkappa(P)$. $Q_{P,\delta}$ is a degenerate distribution concentrated on $D\times\{\varkappa(P)\}$, $P\in\mathcal{P}$.

Since we want to evaluate decision rules on the basis of distributions $Q_{P,\delta}$ only, it is natural to introduce the following equivalency in Δ:

for $\delta,\delta'\in\Delta$, $\delta\approx\delta'$ iff $\forall_{P\in\mathcal{P}}\; Q_{P,\delta} = Q_{P,\delta'}$.

Also we often postulate that Δ is closed with respect to this equivalency:

(2.1) $$(\delta\in\Delta, \delta\approx\delta') \Longrightarrow \delta'\in\Delta.$$

Let

$S(\delta^*,\mathcal{P},\Delta,q,P_1,\ldots,P_k,\delta_1,\ldots,\delta_m)$, where k,m are non-negative integers be a formula satisfying the following conditions:

i) symbols $P_1,\ldots,P_K$ follow quantifiers $\forall_{P_i\in\mathcal{P}}, \exists_{P_i\in\mathcal{P}}$,

ii) symbols $\delta_1,\ldots,\delta_m$ follow quantifiers $\forall_{\delta_i\in\Delta}, \exists_{\delta_i\in\Delta}$,

iii) symbols q, $P_1,\ldots,P_K$, $\delta_1,\ldots,\delta_m,\delta^*$ occur only in terms $q(P_i,\delta_j)$, $q(P_i,\delta^*)$,

iv) symbols $\mathcal{P}$, Δ occur only as the domains of quantifiers.

For a formula of the above form and a statistical problem, we substitute into the formula the function q determined by the problem in place of terms, and the elements of the problem in place of corresponding symbols. Any decision rule δ^* which together with the remaining substitutions satisfies the formula is considered as a solution defined by the formula.

Definition (2.1)

For a given formula S, a family of statistical problems in which the set Δ^* consists of rules δ^*, which together with the triple $(\mathcal{P},\Delta,q)$ determined by the problem, satisfying the formula S is called a <u>standard class of statistical problem defined by the formula S</u>.

Examples

Consider a problem of uniformly minimizing a risk function under some restriction. The risk function is given by:

$$R(P,\delta) = \int_{\Omega} \left(\int_{D} L(I_P(\omega),d)\, d\delta_{T(\omega)} \right) dP, \quad \text{where}$$

$L : \mathcal{I} \times \mathcal{D} \to R$ is the loss function.
The risk function is determined by the loss function L and the distribution $Q_{P,\delta}$ through

$$R(P,\delta) = \int_{\mathcal{I}\times\mathcal{D}} L(i,d)\, dQ_{P,\delta}.$$

We are looking for a rule $\delta \in \Delta$ minimizing $R(P,\delta)$ uniformly with respect to $P \in \mathcal{P}$ and satisfying the following restriction:

$$\int_{\Omega}\left(\int_{D} u(I_P(\omega),d)\, d\delta_{T(\omega)}\right) dP = 0, \quad \text{where } u : \mathcal{I}\times\mathcal{D} \to R \text{ is a given function.}$$

The restriction is determined by the function u and the distribution $Q_{P,\delta}$ through

$$\int_{\mathcal{I}\times\mathcal{D}} u(i,d)\, dQ_{P,\delta} = 0.$$

Therefore, the set of rules Δ^* which are solutions is defined by the following formula $S_{L,u}$

$$S_{L,u} = \forall_{P\in\mathcal{P}} \left[E_{P,\delta^*}(u) = 0 \wedge \forall_{\delta\in\Delta} \left(E_{P,\delta}(u) = 0 \Rightarrow E_{P,\delta}(L) \geq E_{P,\delta^*}(L) \right) \right],$$

where $E_{P,\delta}$ denotes the expectation with respect to $Q_{P,\delta}$.
If $I = D = R$, $I_P(\omega) \equiv \varkappa(P)$, $P \in \mathcal{P}$ $L(i,d) = (i-d)^2$, $u(i,d) = i-d$, the above formula defines a uniformly minimum unbiased estimator of the parameter $\varkappa(P)$ in the family $\mathcal{P}$. If we consider a prediction problem with $I_P(\omega) = I(\omega)$, $P \in \mathcal{P}$, the above formula defines the best mean-square error predictor of the random variable $I(\omega)$.

The second example is a class of problems of finding minimax decision rules. It is given by the following formula S_L:

$$S_L = \forall_{\delta\in\Delta} \forall_{P_1\in\mathcal{P}} \forall_{a>0} \exists_{P_2\in\mathcal{P}} \left(E_{P_1,\delta^*}(L) < E_{P_2,\delta}(L) + a \right).$$

The above may be written more compactly as:

$$S_L = \dot{\forall}_{\delta \in \Delta} \left(\sup_{P \in \mathcal{P}} E_{P,\delta^*}(L) \leq \sup_{P \in \mathcal{P}} E_{P,\delta}(L) \right).$$

3. REDUCTION OF STATISTICAL PROBLEMS

A reduction of a statistical problem consists of reductions of the sample space and set of observables. Obviously reduction should be done in such way that we can define the reduced goal and reduced observable statistic equivalent to primary goal and observable statistic. We also give further conditions on the reduction which guarantee an equivalence of solutions of primary problem and the reduced one.

The sample space is reduced by a measurable function $f : (\Omega, \mathcal{A}) \to (\bar{\Omega}, \bar{\mathcal{A}})$. Therefore $\bar{\mathcal{P}} = \{P^f, P \in \mathcal{P}\}$ where P^f is distribution of f according to P, is the family of induced distributions on the reduced space. The reduction function f has to satisfy the following conditions:

$$(3.1) \quad \forall_{P_1, P_2 \in \mathcal{P}} \quad P_1^f = P_2^f \Rightarrow I_{P_1} = I_{P_2},$$

$$(3.2) \quad \forall_{P \in \mathcal{P}} \forall_{\omega, \omega' \in \Omega} \quad f(\omega) = f(\omega') \Rightarrow I_P(\omega) = I_P(\omega').$$

When (2.1) and (2.2) are satisfied we define the reduced goal $\bar{I}_{\bar{P}} : (\bar{\Omega}, \bar{\mathcal{A}}) \to (\mathcal{I}, \mathcal{J})$ as follows:

$$\bar{I}_{\bar{P}}(\bar{\omega}) = I_P(\omega) \quad \text{for} \quad \bar{P} = P^f \quad \text{and} \quad \bar{\omega} = f(\omega).$$

The set of observables is reduced by a measurable function $g : (\mathcal{T}, \mathcal{F}) \to (\bar{\mathcal{T}}, \bar{\mathcal{F}})$. The reduction functions f and g have to satisfy the following concordance condition:

$$(3.3) \quad \forall_{\omega, \omega' \in \Omega} \quad f(\omega) = f(\omega') \Rightarrow gT(\omega) = gT(\omega').$$

When (3.3) is satisfied we define reduced observable statistics $\bar{T} : (\bar{\Omega}, \bar{\mathcal{A}}) \to (\bar{\mathcal{T}}, \bar{\mathcal{F}})$ as follows:

$$\bar{T}(\bar{\omega}) = gT(\omega), \quad \text{for} \quad \bar{\omega} = f(\omega).$$

A decision rule $\bar{\delta}$ for the reduced observable space, i.e., markov kernel from $(\bar{\mathcal{T}}, \bar{\mathcal{F}})$ to $(D, \mathcal{D})$ defines a decision rule for the primary observable space:

$\delta(t,B) = \bar{\delta}(g(t),B)$, $t \in \mathcal{T}$, $B \in \mathcal{D}$, where we use the notation $\delta = \bar{\delta} \circ g$.

Let Δg denote the subset of Δ consisting of such rules. The last condition imposed upon the reduction function g is:

(3.4) $\Delta g \neq 0$.

The non-empty reduced set of decision rules $\bar{\Delta}$ is defined as

$$\bar{\Delta} = \{ \bar{\delta} : \ \delta = \bar{\delta} \circ g, \ \delta \in \Delta \} .$$

For the joint distribution of the goal and decisions for the reduced and primary problem, if f,g satisfy (3.1), (3.2), (3.3) we have:

(3.5) $Q_{\bar{P},\bar{\delta}} = Q_{P,\delta}$, for $\bar{P} = P^f$, $\delta = \bar{\delta} \circ g$.

For a statistical problem from a standard class given by a formula S and any reductions f and g satisfying (3.1), (3.2), (3.3), (3.4) we define a reduced set of solutions $\bar{\Delta}^*$ as a set of those decision rules which together with $\bar{\mathcal{P}}, \bar{\Delta}$ and indexing function $\bar{q}$ given by (3.5) satisfy the formula S. Therefore, we have defined all elements of the reduced statistical problem. Obviously the primary and reduced problems belong to the same standard class.

Definition (3.1)

In a standard class, the reduced problem is called <u>equivalent to the primary one</u> iff for every $\delta^* \in \Delta^*$ there exists $\bar{\delta}^* \in \bar{\Delta}^*$ such that, for every $P \in \mathcal{P}$, $Q_{P,\delta^*} = Q_{\bar{P},\bar{\delta}^*}$, $\bar{P} = P^f$.

The following property of the reduction together with (3.1), (3.2), (3.3), (3.4) is sufficient for the equivalency of the reduced problem in the standard class:

(3.6) $\forall_{\delta \in \Delta} \ \exists_{\bar{\delta} \in \bar{\Delta}}$ such that $\delta \approx \bar{\delta} \circ g$.

Let us define a notion of sufficiency which is on extention of the classical notions of sufficiency and prediction sufficiency (conf. Takeuchi, Akahira (1975)).

Definition (3.2)

The reduction function g is called <u>sufficient for the statistical problem</u> iff

1° g is sufficient for the family of distributions $\{P^T : P \in \mathcal{P}\}$

2° for every $t \in T$, the conditional distribution of I_P under condition $T(\omega) = t$ is equal to the conditional distribution of I_P under condition $g(T(\omega)) = \bar{t}$. for $\bar{t} = g(t)$.

Let us remark that in the case of parametric inference. when the goal function does not depend on the elements of the sample space, the above definition reduces to the sufficiency for the family $\{P^T, P \in \mathcal{P}\}$. While for the prediction problem, when the goal function does not depend on the distributions, it is the prediction sufficiency definition.

Theorem (3.1)

For every statistical problem in a standard class satisfying (2.1) if the reduction functions f and g satisfy (3.1), (3.2), (3.3), (3.4) and g is sufficient for the problem, then the reduced problem is equivalent to the primary one.

Skech of proof:

Let $P_{\bar{t}}$ be the conditional distribution of T under the condition $g(t) = \bar{t}$. By the sufficiency assumption it is the same for all P^T. Given a decision rule $\delta \in \Delta$ we define a decision rule $\bar{\delta} \in \bar{\Delta}$ as:

$\bar{\delta}(\bar{t},B) = \int_{\mathcal{T}} \delta_B(t)\,dP_{\bar{t}}$. Following the proof of th. 1 in Takeuchi, Akahira (1975), taking into account (3.5), we get $\bar{\delta} \approx \delta_g$ which is the sufficient condition for equivalency of the reduced problem.

Corollary (3.1)

Consider a statistical problem of parametric inference in a standard class satisfying (2.1). If the reduction functions f and g, where $f = gT$, satisfy (3.1), (3.2), (3.3), (3.4) and if g is a sufficient statistic for the family $(P^T, P \in \mathcal{P})$, then the reduced problem is equivalent to the primary one. Moreover, the reduced problem is of full observability.

Remark

In parametric estimation problems usually we assume Δ to be the set of non-randomized decision rules, thus the condition (2.1) is not satisfied. However, we may extend Δ to be the set of all rules and, then apply Corollary (3.1). In the case of convex loss function we get then from the Rao-Blackwell theorem that the solution is again non randomized rule.

REFERENCES

Barra J.R. (1971): Notions fundamentales de statistique mathematique. Dunod, Paris

Dąbrowska D., Pleszczyńska E. (1980): On partial observability in statistical models. Math. Operationsforsch. Statist. Ser. Statistics, vol. 11, no 1, 49-59

Takeuchi K., Akahira M. (1975): Characterization of prediction sufficiency in term of risk functions. Ann. Statist. vol. 3, no 4, 1018-1024

Institute of Computer Science
Polish Academy of Sciences
00-901 Warsaw, P.O. Box 22
Poland

SYNCHRONIZABILITY OF MULTIVALUED ENCODINGS

R.M. Capocelli, L. Gargano, U. Vaccaro

Salerno

ABSTRACT

Multivalued encodings constitute a generalization of ordinary codes in that they allow each source symbol to be encoded by several different codewords. In this paper the problem of synchronizability is considered and a necessary and sufficient condition for a multivalued encoding to be synchronizable is given.

INTRODUCTION

In each communication system information flows through a channel which accepts sequences of symbols from the channel input alphabet and produces sequences of symbols from the channel output alphabet. The channel input usually consists of a sequence of words chosen in a subset of the set containing all finite strings constructed on the input alphabet. Such a subset is referred to as the codeword set or, simply, the code. The main goal of information transmission is to design a code that, for each channel output sequence, allows to reconstruct the transmitted information, that is, to find out the input sequence of codewords that originated the received output sequence. In pratical situations there is some noise on the channel and the occurence of transmission errors is to be aspected. Errors can be classified in three kinds: the change of one symbol into another one (substitution error), the cancellation of one symbol of the transmitted sequence (erasure error) and the insertion of a new symbol in the transmitted sequence (insertion error). If a word w is transmitted and no errors occur during the transmission, then w is produced as output; if errors occur any one of a set of words may occur as output, depending on the errors caused by the noise. Note that if the channel allows not only substitution errors, but also erasure and insertions errors, the output sequences associated to an input sequence may have different lengths. From a theoretical point of view the set of codewords in which a single transmitted codeword can be changed because of the noise might be infinite. In practical situations, however, by ignoring the sequences of errors having very small probability, this set can be considered finite.

There are other situations that can be described as transmission over a channel allowing various kinds of errors. An interesting example is constitued by the information-theoretic approach to problems such as automatic continuous speech recognition and character recognition processes. It has been pointed out by Bahl and Jelinek (1975), Jelinek, Bahl and Mercer (1975), (1983) that such processes can be modeled as transmission of information through a noisy channel that allows substitution, insertion and erasure errors.

The systematic study of fundamental properties, like unique decipherability, decipherability with finite delay and synchronizability of variable length codes in presence of errors was begun by the Coding Group at Park Mathematical Laboratories, [Hartnett (1974)]. Their most interesting result was the elaboration of constructive tests for the above properties when only substitutions errors are allowed. A more general approach has been recent proposed in Sato (1979). Sato introduced multivalued encodings, that is, variable length encoding systems in which not a single codeword but a set of codewords may correspond to each source symbol. It is clear, then, that if we consider each codeword belonging to the set associated to a source symbol as a possible noisy version of the original encoding of that symbol, multivalued encodings constitute a wellsuited approach to the problem of modeling the transmission over a channel that suffers of insertion, erasure and substitution errors. Unfortunately, since for multivalued encodings unique decipherability is not equivalent to unique decomposability (i.e., a code message might be parsed in two different ways both giving the same deciphering in terms of source symbols), the extention of fundamental properties for ordinary codes is not straightforward, neither it does appear possible to use existing methods to test whether a multivalued encoding posses such properties. Necessary and sufficient conditions for a multivalued encoding to be uniquely decipherable and decipherable with finite delay have been given, [Capocelli (1982), Capocelli et al. (1983) (1984)], and algorithms are known for the construction of decoders. In this paper we are concerned with the characterization of synchronizable multivalued encodings, that is encodings that allow the deciphering of any code message without knowing neither its beginning nor its end, exception made for a finite part of it. In particular we establish a necessary and sufficient condition that a multivalued encoding has to satisfy in order to have the property of synchronizability.

NOTATIONS AND DEFINITIONS

Let X be a finite nonempty set and let X^+ and X^* be the free semigroup and the free monoid generated by X, respectively. We recall that the free semigroup $X^+ = X^* - \lambda = \cup_{n=1}^{\infty} X^n$ denotes the set of all finite words over X, where λ and X^n are the empty word and the set of all strings of length n, respectively. Given $w \in X^+$ and $p, r, s \in X^*$, if $prs = w$ then p is a prefix of w, r is an infix of w and s is a suffix of w.

Given a finite set A of source symbols, a multivalued encoding is any mapping $F : A \rightarrow 2^{X^+}$ from the source alphabet A into the set of subsets (the power set) of X^+, denoted by 2^{X^+}. In order to define the encoding of strings of source symbols we expand the domain of F from A to A^* in the following way:

i) $F(\lambda) = \lambda$;

ii) for each $x \in A^*$ and for each $y \in A$ $F(xy) = F(x)F(y) = \{\alpha\beta | \alpha \in F(x)$ and $\beta \in F(y)\}$.

For each string of source symbols $x \in A^*$, $F(x)$ denotes the set of all possible encodings of the string x.

We give now the following definitions.

A multivalued encoding F is *uniquely decipherable* if and only if for any $x; y \in A^*$ $(x \neq y) \Rightarrow (F(x) \cap F(y) = \emptyset)$, i.e., there do not exist two sequences x and y having a common encoding.

Example 1. Given $A = \{0, 1\}$ and $X = \{a, b\}$, consider the multivalued encoding

$F(0) = \{ab, aba\}$, $F(1) = \{abb, bb\}$. It is uniquely decipherable, even though the code message $ababb$ can be parsed in two different ways, namely $ab|abb$ and $aba|bb$. In fact, since $ab, aba \in F(0)$ and $abb, bb \in F(1)$, both parsings give the same decipheration 01.

A uniquely decipherable multivalued encoding F is *decipherable with finite delay* P if and only if the non deterministic generalized sequential machine (gsm) M that implements F has an inverse machine M^{-1} such that the serial connection $M^{-1}M$ of M and M^{-1} in the initial states amounts to a delay machine with maximum delay P. A non deterministic gsm D is called a delay machine with maximum delay P if and only if for any arbitrary input $x \in A^+$, $l(x) > P$, any associated output y is a prefix of x, with $l(x) - l(y) \leq P$. In other words, F is decipherable with delay P if and only if the individuation in a message of $P + n$ initial consecutive codewords suffices to determine the first n symbols of the source sequence that generated the message.

Example 2. Consider the multivalued encoding of Example 1 and the infinite message $ab(bb)^n \ldots$. Since it results $ab(bb)^n \ldots = abb(bb)^{n-1}b \ldots$, $ab(bb)^n \in F(01^n)$ and $abb(bb)^{n-1} \in F(1^n)$, the prefix of code message $ab(bb)^n \ldots$ has two different decipherations. It follows that one cannot start deciphering the message without knowing its end, that is, the multivalued encoding is not decipherable with finite delay.

Finally we define a uniquely decipherable multivalued encoding F *synchronizable with delay* Q if and only if the nondeterministic gsm M that implements F has a self-synchronizing inverse machine M^{-1} such that the serial connection $M^{-1}M$ of M in its initial state and M^{-1} in any state is a self-synchronizing delay machine with maximum delay Q; moreover the serial connection $M^{-1}M$ of M in its initial state and M^{-1} in its initial state amounts to a delay machine with maximum delay Q. A non deterministic gsm D is called a self-synchronizing delay machine with maximum delay Q if and only if for any initial state, for any input $x \in A^+$, $l(x) > Q$, any associated output y has a suffix z such that $x = x_1 z x_2$ and $l(x_1) + l(x_2) \leq Q$. In other words, F is synchronizable with delay Q if and only if the individuation in a message of $Q + n$ consecutive codewords suffices to determine an infix of length greater or equal to n of the source sequence that generated it, in particular if the $Q + n$ codewords are the beginning of a message, one can find a prefix of length not less than n of the source sequence that generated it.

It follows from the definition that if a multivalued encoding $F : A \to 2^{X^+}$ is synchronizable then the encoding and its reversal, i.e., the encoding obtained by reversing in each set $F(a)$, $a \in A$, the order of letters in each codeword, are both decipherable with finite delay. This condition is not sufficient, in fact a multivalued encoding may not be synchronizable even though it and its reversal are both decipherable with finite delay.

Example 3. Given $A = \{0, 1\}$ and $X = \{a, b, c\}$, the multivalued encoding $F(0) = \{aa, abc\}$, $F(1) = \{aba, bba, bc\}$ is decipherable with finite delay $P = 0$, but it is not synchronizable. Indeed, it results $\ldots (aa)^n abc \ldots = \ldots a(aa)^n bc \ldots$, with $(aa)^n abc \in F(0^{n+1})$ and $a(aa)^n bc \in F(0^n 1)$. It follows that there exists an arbitrary long code message having $(aa)^n abc$ as infix, which has two decipherations differing on prefixes and suffixes of unbounded length. Therefore the multivalued encoding is not synchronizable.

A NECESSARY AND SUFFICIENT CONDITION FOR SYNCHRONIZABILITY

A multivalued encoding F can be represented [Sato (1979)] as a transduction by a finite state nondeterministic machine and formalized as a relation between input sequences and output sequences by using a parameter set P^+: $F = \{(\sigma(p), \tau(p)) | p \in P^+\}$ $\sigma(P) = A$, $\tau(P) = F(A) = \cup_{x \in A} F(x)$, $\sigma(P^+) = A^+$, $\tau(P^+) = F(A^+) = \cup_{x \in A^+} F(x)$, where, $\sigma : T^+ \to A^*$ and $\tau : T^+ \to X^+$ are mappings onto the input (a shortening homomorphism) and onto the output (a length-preserving homomorphism), respectively; T represents the set of the transitions in the state diagram of the finite state machine that realizes F; P represents the set of all elementary successfull transition sequences (i.e., of all transition sequences that produce a codeword), and P^+ is the set of all transition sequences.

Example 4. Let $A = \{0,1\}$ and $X = \{a,b,c\}$. Consider the multivalued encoding $F(0) = \{aa, bb\}$, $F(1) = \{abb, cab, bca\}$ and denote by T the set $T = \{t_1, t_2, t_3, t_4, t_5, t_6, t_7, t_8, t_9, t_{10}, t_{11}, t_{12}, t_{13}\}$. The two homomorphism σ and τ are defined by $\sigma(t_1) = \sigma(t_3) = 0$, $\sigma(t_5) = \sigma(t_8) = \sigma(t_{11}) = 1$, $\sigma(t_i) = \lambda$ $i \neq 1,3,5,8,11$ $\tau(t_1) = \tau(t_2) = \tau(t_5) = \tau(t_9) = \tau(t_{13}) = a$, $\tau(t_3) = \tau(t_4) = \tau(t_6) = \tau(t_7) = \tau(t_{10}) = \tau(t_{11}) = b$, $\tau(t_8) = \tau(t_{12}) = c$. Finally, the set P is given by $P = \{t_1t_2, t_3t_4, t_5t_6t_7, t_8t_9t_{10}, t_{11}t_{12}t_{13}\}$.

Let $\Gamma = T \times T$. Define the sets $R = \{x \in \Gamma^+ | \alpha_1(x) \in P^+ \wedge \alpha_2(x) \in P^+ \wedge \tau\alpha_1(x) = \tau\alpha_2(x)\}$ and $S = \{x \in \Gamma^+ | \sigma\alpha_1(x) = \sigma\alpha_2(x)\}$ where α_1 and α_2 are two (projection) homomorphism defined by $\alpha_1(t_i, t_j) = t_i$ and $\alpha_2(t_i, t_j) = t_j$, $(t_i, t_j) \in \Gamma$. Notice that F is uniquely decipherable if and only if $R \subset S$, Sato (1979). Consider now the minimal-state finite automaton $A_P = \{K_P \cup \{q_\emptyset\}, T, \delta_P, q_0, F_P\}$ that accepts the regular set P^+, where $q_\emptyset$ is a $\emptyset$ state, that is $\delta_P(q_\emptyset, t_i) = q_\emptyset$ for each $t_i \in T$; δ_P is the transition function, and the final state set $F_P = \{q_0\}$. Construct the incompletely defined automaton $A = \{K_P \times K_P, \Gamma, \delta, (q_0, q_0), F_P \times F_P\}$, where $\delta((q_i, q_j), (t_h, t_k)) = (\delta_P(q_i, t_h), \delta_P(q_j, t_k))$ if $\delta_P(q_i, t_h) \neq \emptyset$, $\delta_P(q_j, t_k)) \neq \emptyset$ and $\tau(t_h) = \tau(t_k)$, and is not defined otherwise. Here $(q_i, q_j) \in K_P \times K_P$, $(t_h, t_k) \in \Gamma$. It is easily seen that A accepts R. Note that A is not generally a minimal-state machine, and states can exist that either cannot be reached from the initial state or from which the final state cannot be reached. We shall give necessary and sufficient conditions for a multivalued encoding to be synchronizable. For each $x \in \Gamma^+$ define the remainder of x with respect to $(a,b) \in A^* \times A^*$ as follows

$$Rem(x;(a,b)) = \begin{cases} (y,\lambda) & \text{if } a^{-1}\sigma_1(x) = b^{-1}\sigma_2(x)y, \\ (\lambda, y) & \text{if } a^{-1}\sigma_1(x)y = b^{-1}\sigma_2(x), \\ (\lambda,\lambda) & \text{if } a^{-1}\sigma_1(x) = b^{-1}\sigma_2(x), \\ \emptyset & \text{otherwise} \end{cases}$$

where $\sigma_1 = \sigma\alpha_1$ and $\sigma_2 = \sigma\alpha_2$. Recall that if α, β are two strings on the same alphabet, $\alpha^{-1}\beta$ is equal to γ if $\beta = \alpha\gamma$, is not defined if α is not a prefix of β. Let $(a,b) \in A^* \times A^*$ fixed, it is easy to show the following properties of the function Rem

a) for each $x, y \in \Gamma^+$ $(Rem(x;(a,b)) = Rem(y;(a,b))) \Rightarrow$ for each $z \in \Gamma^+$ $(Rem(xz;(a,b)) = Rem(yz;(a,b)))$

b) for each $x \in \Gamma^+$ $(Rem(x;(a,b)) = \emptyset) \Rightarrow$ for each $z \in \Gamma^+$ $(Rem(xz;(a,b)) = \emptyset)$

c) for each $x,y,z \in \Gamma^+$ $(Rem(xz;(a,b)) = Rem(yz;(a,b)) \neq \emptyset) \Rightarrow (Rem(x;(a,b)) = Rem(y;(a,b)) \neq \emptyset)$

The reason for which we have introduced this rather messy sequence of notations and definitions is the following. To each input sequence x of the automaton A corresponds, by means of the homomorphisms τ, α_1 and α_2, an infix of code message given by $\tau\alpha_1(x) = \tau\alpha_2(x)$. In this way, studying the properties of the input sequences of A, we can study the properties of the decipherations of code messages, which is our primary concern. The tool to accomplish the study of these decipherations is the function Rem. In fact, if $Rem(x;(a,b)) = (\lambda, y)$, for instance, this means that the two decipherations of the infix of code message $\tau\alpha_1(x) = \tau\alpha_2(x)$ are one az and the other bzy for some $z \in A^*$.

Let now m be the number of internal states of A. Denote by $A(Q_i,Q_j)$ the set of all strings that bring A from the state Q_i to the state Q_j and by $A_k(Q_i,Q_j)$ the set of all strings in $A(Q_i,Q_j)$ of length not greater than k.

Definition 1. We say that $x, xy \in \Gamma^+$ $(xy \neq \lambda)$ represents (is) a loop of states in the state diagram of A if and only if $x, xy \in A(Q_i,Q_j)$, for some Q_i,Q_j states in A. We say that the loop x,xy has length n if $l(xy) = n$.

Definition 2. Let $x,xy \in A(Q_i,Q_j)$ and $\alpha,\alpha\beta \in A(Q_i,Q_r)$ be loops of states in A. We say that x,xy and $\alpha,\alpha\beta$ are linked if the following conditions hold:

i) $\alpha = x_1x_2$, $x = x_1x_3$, $l(x_1) \leq m$, $(x_1,x_2,x_3 \in \Gamma^*)$;

ii) $x_1 \in A_m(Q_i,Q_k) \Rightarrow A_m(Q_k,Q_r) \neq \emptyset$.

Definition 3. Let D be the set $D = \{(a,b) \in (A^* \times A^*) | \exists Q_i,Q_j$ states of A, $\exists x,xy \in A_m(Q_i,Q_j)$ such that $\sigma_1(xy) = az$ and $\sigma_2(xy) = bzc$ or $\sigma_1(xy) = azc$ and $\sigma_2(xy) = bz, z,c \in A^*\}$.

Definition 4. We say that $x,xy \in A(Q_i,Q_j)$ is a loop of states $n-$*favorable* if there exists $(a,b) \in D$ such that $Rem(x;(a,b)) = Rem(xy;(a,b)) \neq \emptyset$, and for each loop of states $\alpha,\alpha\beta$ $(l(\alpha\beta) \leq n)$ linked to the loop of states x,xy $Rem(\alpha;(a,b)) = Rem(\alpha\beta;(a,b)) \neq \emptyset$. We say that a loop of states $x,xy \in A(Q_i,Q_j)$ is favorable if it is n-favorable whatever n.

We need the following technical lemma.

Lemma 1. For each Q_i,Q_j, if each loop of states $x,xy \in A_m(Q_i,Q_j)$ is m-favorable then for each $\alpha \in A_m(Q_i,Q_j)$, $\alpha\beta \in A(Q_i,Q_j)$ the loop of states $\alpha,\alpha\beta$ is favorable.

Proof. The proof is by inductive argument. Let $k \geq m$. We will show that if for each Q_i,Q_j, for each $\alpha \in A_m(Q_i,Q_j)$ and for each $\alpha\beta \in A_k(Q_i,Q_j)$ the loop of states $\alpha,\alpha\beta$ is k-favorable, then for each Q_i,Q_j, for each $x \in A_{k+1}(Q_i,Q_j)$ and for each $xy \in A_{k+1}(Q_i,Q_j)$ the loop of states x,xy is (k+1)-favorable. Let $x \in A_m(Q_i,Q_j)$ and $xy \in A_{k+1}(Q_i,Q_j) - A_k(Q_i,Q_j)$. Since $l(xy) \geq m+1$, it is possible to find x_1,x_2,x_3, $l(x_1) \leq m-1$ and $x_2,x_3 \in \Gamma^+$, such that $xy = x_1x_2x_3$, $\delta(Q_i,x_1) = Q_l = \delta(Q_l,x_2)$ and $\delta(Q_i,x_1x_3) = Q_j = \delta(Q_i,x_1x_2x_3)$ where Q_l is not necessarily distinct from Q_j. In order to show that the loop of states x,xy is (k+1)-favorable, it is convenient to distinguish the following cases a) and b).

Case a). $l(x) \leq l(x_1)$.

We can write $x_1 = xa$, with $a \in \Gamma^*$. Therefore $x_1x_3 = xax_3$ and $x \in A_m(Q_i, Q_j)$, $xax_3 \in A_k(Q_i, Q_j)$, $x_1 = xa \in A_m(Q_i, Q_l)$, $xax_2 \in A_k(Q_i, Q_l)$. By inductive hypothesis the loop xa, xax_2 is k-favorable. Since the loop x, xax_3 is linked to the loop xa, xax_2 and $l(xax_3) \leq k$, it follows that there exists $(a, b) \in D$ such that

$$Rem(xa; (a,b)) = Rem(xax_2; (a,b)) \neq \emptyset, \tag{1}$$

$$Rem(x; (a,b)) = Rem(xax_3; (a,b)) \neq \emptyset \ . \tag{2}$$

From (1) and (2), using the properties of Rem, one gets

$$Rem(xax_3; (a,b)) = Rem(xax_2x_3; (a,b)) = Rem(x_1x_2x_3; (a,b)) \neq \emptyset.$$

$$Rem(x; (a,b)) = Rem(x_1x_2x_3; (a,b)) = Rem(xy; (a,b)) \neq \emptyset.$$

Case b). $l(x) > l(x_1)$.

We can write $x = x_1a$ with $a \in \Gamma^+$. It is convenient to distinguish the following two cases i) and ii).

i) $l(a) \leq l(x_2)$. We can then write $x_2 = ab$, $y = bx_3$ with $b \in \Gamma^*$ and $x_1 \in A_m(Q_i, Q_j)$, $x_1x_2, x_1x_3b \in A_k(Q_i, Q_l)$. By inductive hypothesis, the loop x_1, x_1x_2 is k-favorable. Since the loop x_1, x_1x_3b is linked to the loop x_1, x_1x_2 and $l(x_1x_3b) \leq k$ one has that there exists $(a, b) \in D$ such that

$$Rem(x_1; (a,b)) = Rem(x_1x_2; (a,b)) = Rem(x_1x_3b; (a,b)) \neq \emptyset. \tag{3}$$

From the right equality of (3) and the properties of Rem, one gets

$$\begin{aligned} Rem(x_1x_3b; (a,b)) = Rem(x_1x_2x_3b; (a,b)) &= Rem(x_1abx_3b; (a,b)) \\ &= Rem(xyb; (a,b)) \neq \emptyset. \end{aligned}$$

and then $Rem(xb; (a,b)) = Rem(xyb; (a,b)) \neq \emptyset$. From which $Rem(x; (a,b)) = Rem(xy; (a,b)) \neq \emptyset$.

ii) $l(a) > l(x_2)$. We can then write $a = x_2b$, $x_3 = by$ with $b \in \Gamma^*$, $x_1b \in A_m(Q_i, Q_j)$, $x_1x_2 \in A_k(Q_i, Q_l)$, $x_1by \in A_k(Q_i, Q_j)$. By inductive hypothesis, the loop x_1b, x_1by is k-favorable. Since the loop x_1, x_1x_2 is linked to the loop x_1b, x_1by and $l(x_1x_2) \leq k$, it follows that there exists $(a, b) \in D$ such that

$$Rem(x_1; (a,b)) = Rem(x_1x_2; (a,b)) \neq \emptyset$$

$$Rem(x_1b; (a,b)) = Rem(x_1by; (a,b)) = Rem(x_1x_3; (a,b)) \neq \emptyset.$$

From which, one has

$$Rem(x_1x_3; (a,b)) = Rem(x_1b; (a,b)) = Rem(x_1x_2b; (a,b)) = Rem(x; (a,b)) \neq \emptyset,$$

$$Rem(x; (a,b)) = Rem(x_1x_2x_3; (a,b)) = Rem(xy; (a,b)) \neq \emptyset.$$

We have proved that for each Q_i, Q_j, $x \in A_m(Q_i, Q_j)$, $xy \in A_{k+1}(Q_i, Q_j)$ there exists $(a, b) \in D$ such that $Rem(x; (a,b)) = Rem(xy; (a,b)) \neq \emptyset$. We shall show that for each Q_i, Q_j, $x \in A_m(Q_i, Q_j)$, $xy \in A_{k+1}(Q_i, Q_j)$ among all $(a, b) \in D$ such that $Rem(x; (a,b)) = Rem(xy; (a,b)) \neq \emptyset$ there exists $(a', b') \in D$ such that for each $\alpha \in A_m(Q_i, Q_p)$, $\alpha\beta \in A_{k+1}(Q_i, Q_p)$, where $\alpha, \alpha\beta$ is a loop linked to x,xy, it holds $Rem(\alpha; (a,b)) = Rem(\alpha\beta; (a,b)) \neq \emptyset$. Let $\alpha_1, \alpha_1\beta_1; \ldots; \alpha_r, \alpha_r\beta_r$ ($\alpha_i \in A_m(Q_i, Q_{p_i})$,

$\alpha_i\beta_i \in A_{k+1}(Q_i, Q_{p_i}))$, be all loops of states linked to x,xy. Suppose that $\alpha_i\beta_i \in A_k(Q_i, Q_{p_i})$, $1 \leq i < p \leq r$ and for xy the decomposition described in case a) holds, that is $xy = x_1x_2x_3$, $\delta(Q_i, x_1) = Q_l = \delta(Q_l, x_2)$, $\delta(Q_i, x_1x_3) = Q_j = \delta(Q_i, x_1x_2x_3)$ and $x_1 = xa$, $a \in \Gamma^*$. It is easily seen that each loop $\alpha_i, \alpha_i\beta_i$. $1 \leq i \leq p$ is linked to the loop xa, xax_2. Consider now a loop $\alpha_i, \alpha_i\beta_i$, $p \leq i \leq r$, with $\alpha_i \in A_m(Q_i, Q_{p_i})$ and $\alpha_i\beta_i \in A_{k+1}(Q_i, A_{p_i}) - A_k(Q_i, Q_{p_i})$. As above seen, it is possible to decompose each loop of states of length $k+1$ in loops of length not greater than k. Therefore

$$\alpha_i\beta_i = x_{1_i} x_{2_i} x_{3_i} \tag{4}$$

with $\delta(Q_i, x_{1_i}) = Q_{r_i} = \delta(Q_{r_i}, x_{2_i})$, $\delta(Q_i, x_{1_i}x_{3_i}) = Q_{p_i} = \delta(Q_i, x_{1_i}x_{2_i}x_{3_i})$ and Q_{r_i} is not necessarily distinct from Q_{p_i}. As for the loop x,xy, for the decomposition (4) one can distinguish the case a) and b). Suppose case a) true for the loop x,xy. It is easily seen that the loops of length not greater than k in which we decompose $\alpha_i, \alpha_i\beta_i$, are linked to xa, xax_2. It follows that there exists $(a,b) \in D$ such that $Rem(x;(a,b)) = Rem(xy;(a,b)) \neq \emptyset$ and for each $\alpha_i \in A_m(Q_i, Q_{p_i})$, $\alpha_i\beta_i \in A_{k+1}(Q_i, Q_{p_i})$ with $\alpha_i, \alpha_i\beta_i$ linked to x,xy it holds $Rem(\alpha_i;(a,b)) = Rem(\alpha_i\beta_i;(a,b)) \neq \emptyset$. Suppose case b) true for the loop x,xy. Again, one has that there exists a loop of states, say x',x'y', such that each loop of length greater than k linked to x,xy is also linked to it. Moreover,if $\alpha, \alpha\beta$ is a loop of length $k+1$ linked to x,xy, each loop of the decomposition of $\alpha, \alpha\beta$ is linked to x',x'y'. In each case, by using the inductive hypothesis, it is possible to show that the loop x,xy, $x \in A_m(Q_i, Q_j)$, $xy \in A_{k+1}(Q_i, Q_j)$, is $(k-1)$-favorable.

Q.E.D.

Let now $w \in \Gamma^*$, $\delta(Q, w) = Q'$ for some Q, Q' states of A. It can be shown that it is possible to write $w = x_0^{h_0} x_1 x_2^{h_1} x_3 \ldots x_{2n-1} x_{2n}^{h_n} x_{2n+1}$ $n \geq 0$, $h_j \geq 0$, $0 \leq j \leq n$. $x_0, x_{2n+1} \in \Gamma^*$, $x_i \in \Gamma^+$, $1 \leq i \leq n$. in such a way that $l(x_1x_3 \ldots x_{2n+1}) \leq m - 1$ and

$$\delta(Q, x_0) = Q,\ \delta(Q, x_1) = Q_1 = \delta(Q_1, x_2), \ldots,\ \delta(Q, x_1x_3 \ldots x_{2n-1}) = Q_n = \delta(Q_n, x_{2n})$$

with $Q_i \neq Q_j$ if $i \neq j$. Further, if for any Q_i, Q_j, for any $x, xy \in A_m(Q_i, Q_j)$ the loop of states $x.xy$ is favorable, from Lemma 1 one has that there exists $(a,b) \in D$ such that

$$\begin{aligned} &Rem(\lambda;(a,b)) = Rem(x_0;(a,b)) \neq \emptyset \quad (if\, x_0 \neq \lambda) \\ &Rem(x_1;(a,b)) = Rem(x_1x_2;(a,b)) \neq \emptyset \\ &\ldots \\ &Rem(x_1x_3 \ldots x_{2n-1};(a,b)) = Rem(x_1x_3 \ldots x_{2n-1}x_{2n};(a,b)) \neq \emptyset \end{aligned} \tag{5}$$

We need now a result from Capocelli et al. (1986).

Theorem 1. A multivalued encoding F is synchronizable if and only if, given any infix of code message β, all possible decipherations of β have a common infix and differ, at most, for prefixes and suffixes of total length bounded by a constant Q.

Since the number of code messages is infinite, the above theorem does not provide a finite procedure for testing whether a multivalued encoding is synchronizable. The following theorem, which is our main result, provides the desired test.

Theorem 2. A necessary and sufficient condition for a multivalued encoding F to be synchronizable is that any loop of states in the state diagram of A of length not greater than m is m-favorable, where m is the number of states of A.

Proof. Sufficiency. Let us consider $w = x_0^{h_0} x_1 x_2^{h_1} x_3 \ldots x_{2n-1} x_{2n}^{h_n} x_{2n+1} \in \Gamma^+$, with $\delta(Q,w) = Q'$, $Q, Q' \in K_P \times K_P$. One has that there exists $(a,b) \in D$ such that $Rem(\lambda;(a,b)) = Rem(x_0;(a,b)) \neq \emptyset$, and, from the properties of Rem, one has $Rem(\lambda;(a,b)) = Rem(x_0^{h_0};(a,b)) \neq \emptyset$. From (5) from the properties of Rem, one has then $Rem(x_1;(a,b)) = Rem(x_0^{h_0} x_1;(a,b)) \neq \emptyset$, $Rem(x_1;(a,b)) = Rem(x_0^{h_0} x_1 x_2^{h_1};(a,b)) \neq \emptyset$ and so on up to get $Rem(x_1 x_2 \ldots x_{2n-1};(a,b)) = Rem(x_0^{h_n} x_1 x_2^{h_1} x_3 \ldots x_{2n-1} x_{2n}^{h_n};(a,b)) \neq \emptyset$. By definition of Rem one has that $\sigma_1(w)$ and $\sigma_2(w)$ either coincide or have a common infix. The same holds for $\sigma_1(x_1 x_3 \ldots x_{2n-1})$ and $\sigma_2(x_1 x_3 \ldots x_{2n-1})$. In case $\sigma_1(w)$ and $\sigma_2(w)$ do not coincide, the prefixes and the suffixes for which they differ are the same for which $\sigma_1(x_1 x_3 \ldots x_{2n-1})$ and $\sigma_2(x_1 x_3 \ldots x_{2n-1})$ differ. Since $l(x_1 x_3 \ldots x_{2n-1}) \leq m-1$, the length of these prefixes and suffixes are themselves not greater than $m-1$. Then one has that for each $w \in \Gamma^+$ such that $\delta(Q,w)$ is defined, for some state Q of A, and then for each infix of code message $\tau\alpha_1(w) = \tau\alpha_2(w)$, it is possible to determine an infix of the source sequence that generated the message with resulting (undeciphered) prefix and suffix having bounded length.

Necessity. Let us suppose that there exists a loop of states $x, xy \in A_m(Q_i, Q_j)$ that is not m-favorable. It is possible to distinguish the following two cases.

Case a). For each (a,b) $Rem(x;(a,b)) = Rem(xy;(a,b)) = \emptyset$ or $Rem(x;(a,b)) = Rem(xy;(a,b))$. We will show that for each $R > 0$ there exists k such that $\sigma_1(xy^k)$ and $\sigma_2(xy^k)$ either do not have a common infix or differ for prefixes and suffixes of length greater than R. Indeed, if this is not true, one has $\sigma_1(y) = cd$ and $\sigma_2(y) = dc$ $c,d \in A^*$ from which it follows $Rem(x;(\sigma_1(x),\sigma_2(x)\beta)) = Rem(xy;(\sigma_1(x),\sigma_2(x)\beta)) = (\beta,\lambda) \neq \emptyset$, that contradicts the hypothesis, because $(\sigma_1(x),\sigma_2(x)\beta) \in D$. Then, there exists a code message such that either its decodings do not have a common infix or they differ for prefixes and suffixes of unbounded length, and the encoding is not synchronizable.

Case b). There exists $\alpha, \alpha\beta \in A_m(Q_i, Q_k)$ loop of states linked to x,xy such that there does not exist $(a,b) \in D$ for which $Rem(x;(a,b)) = Rem(xy;(a,b)) \neq \emptyset$ and $Rem(\alpha;(a,b)) = Rem(\alpha\beta;(a,b)) \neq \emptyset$. Since the loop of states $\alpha, \alpha\beta$ is linked to the loop of states x,xy, one has $x = x_1 x_2$ $\alpha = x_1 x_3$ and $x_1 \in A_m(Q_i, Q_r)$ $x_4 \in A_m(Q_k, Q_r)$. It is possible to show that for each $R > 0$ there exists h_1, h_2, h_3 such that $\sigma_1(x_1(x_3\beta^{h_1}x_4)^{h_2}x_2y^{h_3})$ and $\sigma_2(x_1(x_3\beta^{h_1}x_4)^{h_2}x_2y^{h_3})$ differ for prefixes and suffixes of length greater than R. Indeed, if this should not be true, one would have that for each $h \geq 0$ $\sigma_1(x_3\beta^h x_4) = a(h)b(h)$ and $\sigma_2(x_3\beta^h x_4) = b(h)a(h)$, $a(h), b(h) \in A^*$, i.e., the two decodings of the cycle $x_3\beta^h x_4$ must be one a cyclic permutation of the other. In addition $\sigma_1(\beta) = cd$ and $\sigma_2(\beta) = dc$ $c,d \in A^*$ Hence, it follows that $\sigma_1(x_3\beta^h x_4) = ab(h)$ and $\sigma_2(x_3\beta^h x_4) = b(h)a$ [Resp. $\sigma_1(x_3\beta^h x_4) = a(h)b$ and $\sigma_2(x_3\beta^h x_4) = a(h)b$]. This implies for some $z \in A^+$ either $c^{-1}\sigma_1(x_3)z = \sigma_2(x_3)$ and $c^{-1}\sigma_1(x_3\beta)z = \sigma_2(x_3\beta)$, or $c^{-1}\sigma_1(x_3) = \sigma_2(x_3)z$ and $c^{-1}\sigma_1(x_3\beta) = \sigma_2(x_3\beta)z$. Moreover, if $\sigma_1(x_1(x_3\beta^{h_1}x_4)^{h_2}x_2y^{h_3})$ and $\sigma_2(x_1(x_3\beta^{h_1}x_4)^{h_2}x_2y^{h_3})$ differ for prefixes and suffixes of bounded length, it must hold that for each h_1, h_2, h_3 either $c^{-1}\sigma_1(x_2y)s = \sigma_2(x_2y)$ and $c^{-1}\sigma_1(x_2y^2)s = \sigma_2(x_2y^2)$, or $c^{-1}\sigma_1(x_2y) = \sigma_2(x_2y)s$ and $c^{-1}\sigma_1(x_2y^2) = \sigma_2(x_2y^2)s$.

It follows that $Rem(x_1x_3;(\sigma_1(x_1)c,\sigma_2(x_1)) = Rem(x_1x_3\beta;(\sigma_1(x_1)c,\sigma_2(x_1)) \neq \emptyset$ and $Rem(x_1x_2;(\sigma_1(x_1)c,\sigma_2(x_1)) = Rem(x_1x_2y\beta;(\sigma_1(x_1)c,\sigma_2(x_1)) \neq \emptyset$ that contradicts the hypothesis. Then one gets that there exists an arbitrary long infix of a sequence of codewords $\tau_1(x_1(x_3\beta^{h_1}x_4)^{h_2}x_2y^{h_3})$ such that its decodings either do not have a common infix or differ for prefixes and suffixes of unbounded length. Q.E.D.

REFERENCES

Bahl L.R. and Jelinek F. (1975): Decoding for Channels with Insertions, Deletions and Substitutions with Applications to Speech Recognition. In: IEEE Trans. Inform. Theory, IT-21, 404-411.

Capocelli R.M. (1982): A Decision Procedure for Finite Decipherability and Synchronizability of Multivalued Encodings. In: IEEE Trans. Inform. Theory, IT-28, 307-318.

Capocelli R.M., Gargano L. and Vaccaro U. (1986): A Model for Communication over Noisy Channels. In: Physics of Cognitive Processes, E.R. Caianiello, Editor, World Publishing, Singapore.

Capocelli R.M. and Vaccaro U. (1983): Finite Decipherability of Multivalued Encodings. In: Proc. of Twenty-first Annual Allerton Conf. on Communication, Control and Computing, 528-536.

Capocelli R.M. and Vaccaro U. (1984): Structure of Decoders for Multivalued Encodings. Submitted.

Hartnett W.E., Editor (1974): Foundations of Coding Theory. Boston MA: Reidel.

Jelinek F., Bahl L.R. and Mercer R.L. (1975): Design of a Linguistic Statistical Decoder for the Recognition of Continuous Speech. In: IEEE Trans. Inform. Theory, IT-21, 250-256.

Jelinek F., Bahl L.R. and Mercer R.L. (1983): A Maximum Likelihood Approach to Continuous Speech Recognition. In: IEEE Trans. Pattern Analysis and Machine Intelligence, PAMI-5, 179-190.

Sato K. (1979): A Decision Procedure for the Unique Decipherability of Multivalued Encodings. In: IEEE Trans. Inform. Theory, IT-25, 356-360.

Dipartimento di Informatica ed Applicazioni
Università di Salerno
84100 Salerno, Italy

AUTOMATIC CONTROL OF GAS TRANSPORT LINES RESISTING TO RANDOM NOISE

Jiří Čermák, Miloslav Driml

Prague

Key words: gas network control conception, automatic control algorithms

ABSTRACT

To the gas transport problems to be solved in the conception of the very fast acting network lines in nuclear energy producing units, chemical plants, gas works, there belongs the fully automatic real-time computer control, fulfilling the tasks in this area, without participation of the human factor in the closed control loops. In the contribution, there is a short information about some disposition of a new simple but effective type of control algorithms which fulfil many claims in the automatic transport problems and also resist well to the disturbing parasitic random noise on the signals delivered to the controlling centre. Also some results obtained on the Hewlett-Packard computer are described in a very concise form in order to show that the conception of our solution is promising.

1. In the area of chemical plants, gas networks and nuclear energy producing units there will be requested in the near future a real time computer-aided and fully automatic control of the fast acting network lines. The mechanism of the control of these objects will be performed without any participation of a human factor in the closed control loops, because in assumed high rates control actions a human being could be neither physically nor

mentally able to operate as a member of these control loops.

In our contribution, we bring a very short information on the qualities and characteristic dispositions of a simple and effective type of algorithms fulfilling many of the claims in the modern automatic control of gas transport. An important property of these algorithms is their well-resistance to parasitic random noise on the signals delivered from the measured points of the plant to the controlling centres.

The function of algorithms is based before all on the knowledge of repeatedly measured border values of pressures at the ends of each gas conduit section. It means that a homogeneous basic set of measured "intensive" physical state parameters is applied. These values enter the mathematical description of the control algorithms as boundary conditions of the solved problem.

The control algorithms use both the special type of feed-forward (with the state parameters reconstruction) and the feed-back principle. The values of the "feed-forward" line is determined from the knowledge of some basic structure parts: simple static and dynamic model of the controlled system, autostability and special predictive ability of the control system. The measured values of the pressures at both ends of every conduit section are mutually independent and form in certain sense a minimum possible set of state values which must be used for the control.

Two simple examples obtained by the mathematical simulation of control processes on Hewlett-Packard computer are shown in this paper. The investigations and the results affirm that the conception of such an analysis and solution of the given problematics is promising.

2. The automatically controlled gas transport system could be hypothetically realized by the gas pressure measuring system together with a control computer complex. It can be demonstrated schematically as shown in Fig. 1. On this figure there are shown: a compressor station, a set of two gas conduit elements - section (A), (B) as a transport system with three nodes (1,2,3) (where the gas pressures (Y_1, Y_2, Y_3) are repeatedly measured) - and a computer complex. The values of measured pressures are transmitted (by

a long distance transmission) into the computer centre, where the signals can be effectively exploited for the computer-aided control actions.

The recommended value $^{*}Y_1$ for the compressor station (with the momently existing pressure Y_1 at the output) is repeatedly evaluated by the computer complex according to the software algorithms in every sampling moment.

$\downarrow X_2$ is the mass-flow value of gas flowing at the end of the first section - conduit (A) - in the direction to the node 2.

$\uparrow X_2$ is the mass-flow value of gas going out of the node 2 at the beginning of the section (B).

$\downarrow X_3$ is the mass-flow value of gas flowing at the end of the second section conduit (B) in the direction to the node 3.
In our example the required (reference) value of the end pressure $Y_E \equiv Y_3$ of the whole system is Y_R.

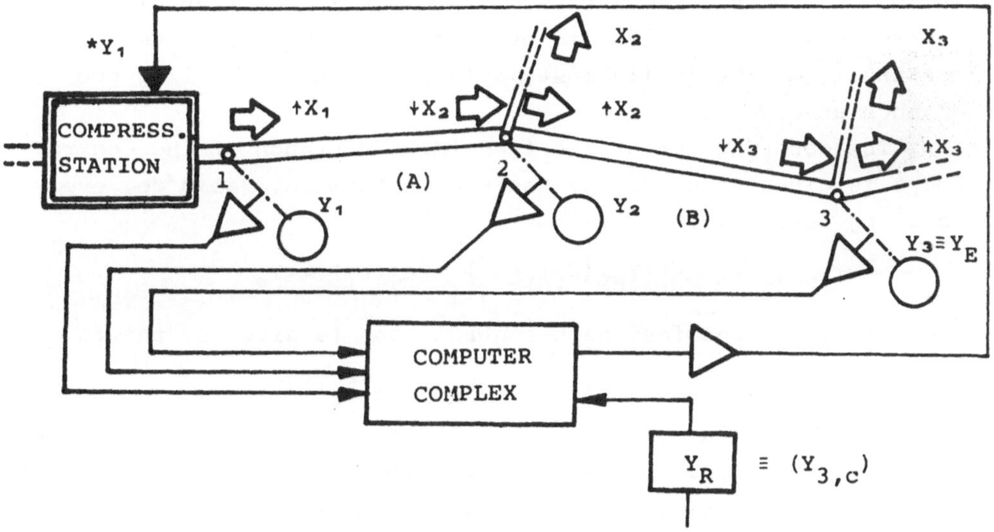

Fig. 1

3. In Fig. 2, there is given a short look into the computer software block modelling the control conception.
a) In this schema we see the blocks (A), (B), which are simple mathematical models for evolving and repeated computing of the values of the aforementioned quantities of $\downarrow X_2$, $\downarrow X_3$.
b) Afterwards, in the following prediction blocks PRED(A) and PRED(B), there are generated the fictive in advance expected values of pressure differences $*\Delta_A$, $*\Delta_B$, having the character of "feed-forward" - (better "state reconstruction") quantities, created in every sampling moment by the help of pseudo-static relations.
c) In the summing block SUM, there are both the "feed-forward" values $*\Delta_A$, $*\Delta_B$ of pressure differencies summed together simultaneously with the simple feed-back difference value $*\Delta_{FB}$. Thus, in every sampling moment we get repeatedly a new resulting value of the quantity $*Y_1$ "recommended" for the gas pressure Y_1 at the output of the compressor station.(In our analysis we suppose that the compressor station operates ideally, i. e. it is able to set the prescribed value $*Y_1 \equiv Y_1$ immediately, without any delay.)

The mentioned set of operations creates a sequence of recommended "gas pressure state reconstructions", which enables to the control computer complex (in every sampling moment) a very robust, continual and physically most rapid possible controlling process of such an object.

The construction of the feed-forward branch of the control loops is given generally by the simple formula, valid for every sampling interval:

$$*Y_{FF}(t_i) = \sum_{j=1}^{N} *\Delta_j(t_i) + Y_R.$$

The branch for the feed-back control loop is given by the formula

$$*\Delta_{FB}(t_i) = (Y_R - Y_E(t_i)),$$

where $Y_E(t_i)$ is the end pressure of all the conduit system at the sampling moment t_i. The global recommended value of the gas pressure at the beginning of the conduit $*Y_1$ or at the beginning of the whole set of several conduits in series - is formed by the sum

$$*Y_1(t_i) = *Y_{FF}(t_i) + *\Delta_{FB}(t_i).$$

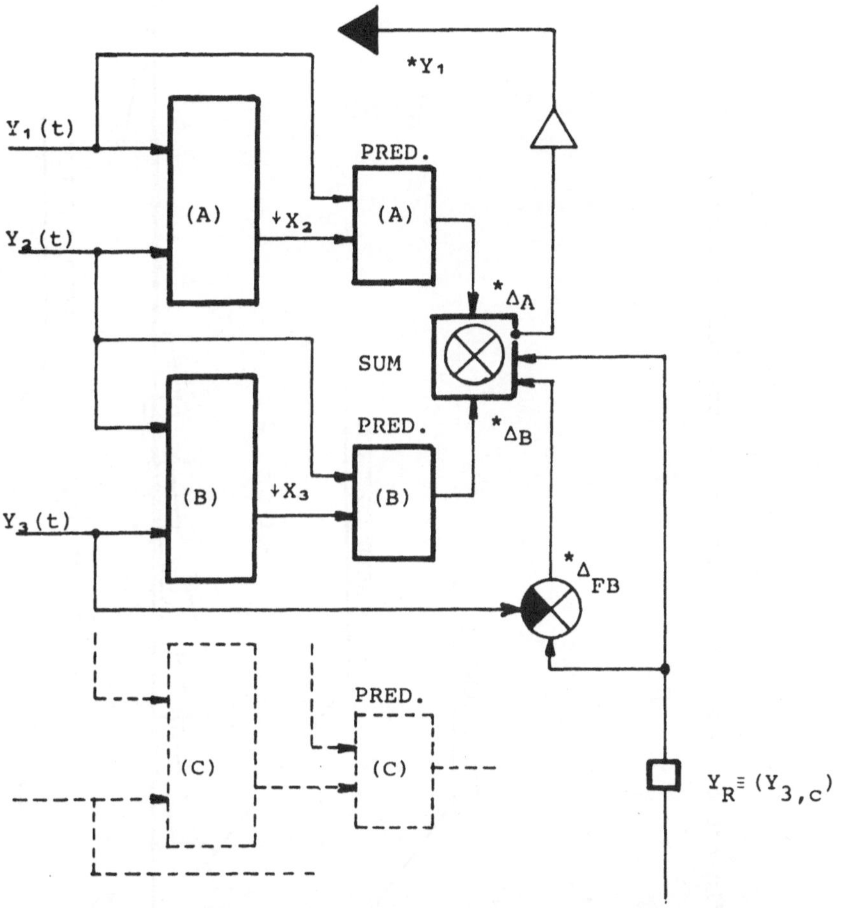

$Y_3(t) \equiv Y_E(t)$ (in the case of two conduit elements)

$*Y_{FF} = *\Delta_A + *\Delta_B + Y_R$

$*Y_1 = *Y_{FF} + *\Delta_{FB}$

Y_R = desired (reference) value of controlled pressure value $Y_{3,c}$ (at the end of the second (last) conduit element).

Fig. 2

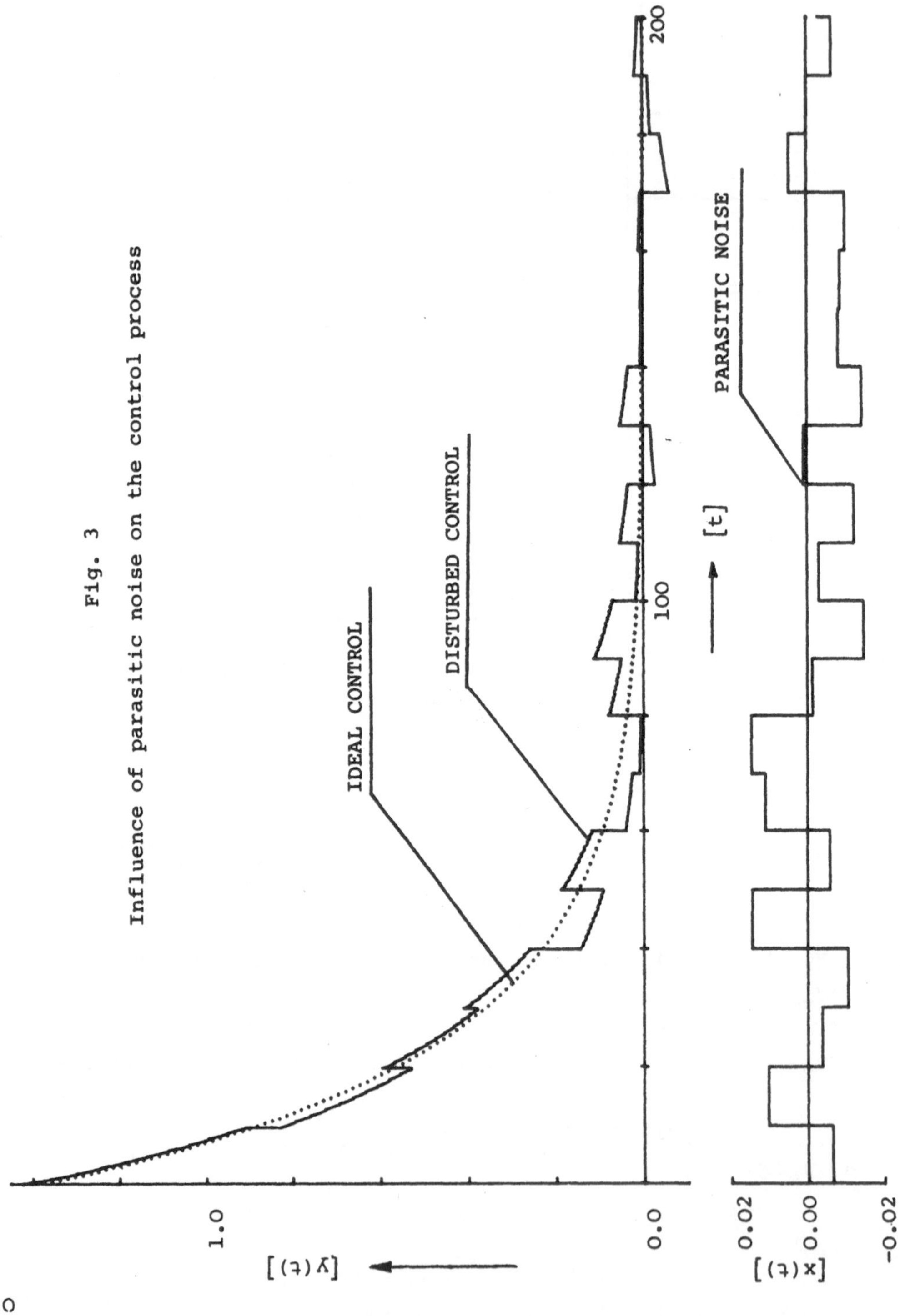

Fig. 3

Influence of parasitic noise on the control process

4. The main principle of the conception of solution of our problem is shown in Fig. 1 and especially on the block scheme in Fig. 2. The corresponding algorithms were investigated with different mathematical models of gas transport lines and all the obtained results proved a good applicability.

Now, we study the influence of a parasitic random noise applied to the measured quantity in the course of the control process. In regard to the complexity of the general problem, we have chosen the most simple dynamic model of the gas transport conduit, i. e. object with lumped parameters. This leads to a linear differential equation of first order. At the same time, we have used the relative deviation form of considered quantities, i. e. we have only considered relative (dimensionless) deviations near to some stable regime. In this formulation, all the considered variables ($\frac{\Delta X(t)}{X_0} = x(t)$, $\frac{\Delta Y(t)}{Y_0} = y(t)$, etc.) are equal to zero at the stable regime. Figs. 3 and 4 are examples of the control process under additive random noise at the gas flow output. The course of the control process in both figures is characterized by these parameters:

i) the basic time constant of the controlled object $T_0 = 100$ time units;

ii) the relation of the end pressure to the starting pressure at the gas transport element $\varepsilon = 0.3$;

iii) the relative change of the gas outflow (consumption) $x_2(t) = -0.3$.

The additive disturbing noise applied to the gas outflow $x_2(t)$ has two forms. In Fig. 3 it is a step function changing its values in intervals of 10 time units according to a sequence of independent random variables uniformly distributed over the interval [-0.015,0.015]. It means that the maximum amplitude of the noise represents 10 % of the basic relative change of the gas outflow.

The noise in Fig. 4 is characterized by a sequence of dependent random variables with the change interval 1 time unit. The random variables are moving averages of 10 subsequent independent variables uniformly distributed over the interval [-0.1,0.1].

In both figures, the dotted line represents an exponential course of the ideal non-disturbed control process. The full line

Fig. 4

Influence of parasitic noise on the control process

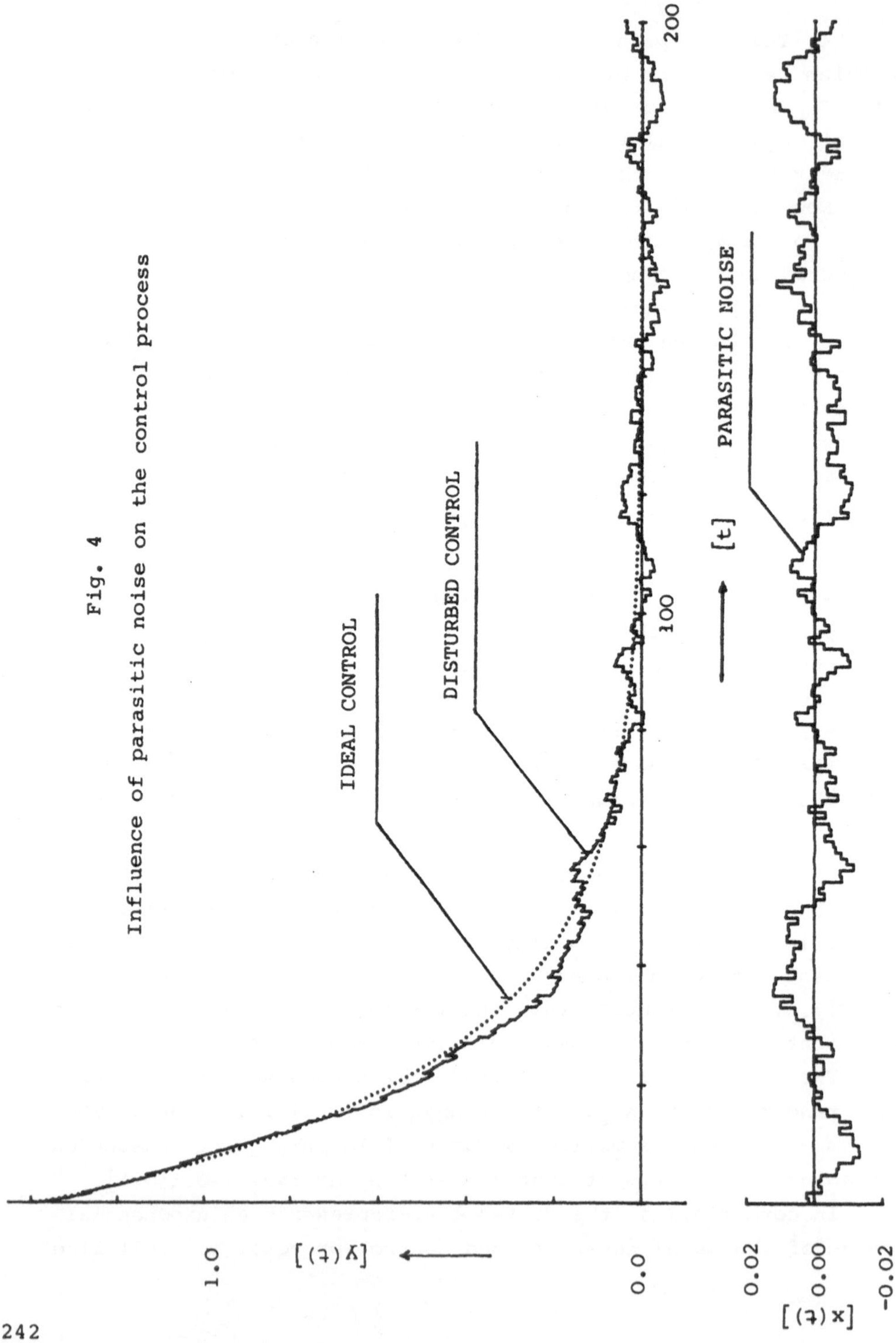

represents the control process under additive parasitic noise applied to the relative change of the gas outflow. In both the cases, the course of the noise is shown in the lower part of the figures.

The investigation of the disturbed control process has been satisfactory up to now and we suppose to continue studying the more complicated cases.

REFERENCES

Čermák J. (1983): Automatic supervising and control of large scale gas transit systems. In: Fourth Formator Symposium on Mathematical Methods for the Analysis of Large-Scale Systems, Academia, Prague 1983, 129-140.

Čermák J. (1985): Automatic control of gas transport (in Czech). Czechoslovak State Patent Office, Prague 1985. Patent No. 224787.

Czechoslovak Academy of Sciences
Institute of Information
Theory and Automation
Pod vodárenskou věží 4
182 08 Prague 8
Czechoslovakia

JOINT ROBUST ESTIMATES
OF LOCATION AND SCALE PARAMETERS

N.I.Chernov, G.A.Ososkov

Dubna

Key words: robust estimation, regression models, Monte Carlo study

ABSTRACT

Robust estimates of regression parameters are studied in linear models for heavy contaminated distribution of errors with uniformly distributed noise. Maximum likelihood approach to joint estimating location and scale parameters leads to an algorithm for computation of regression parameters. This algorithm was tested by Monte Carlo method in experimental data models of a particle track detector.

INTRODUCTION

We consider the robust estimation of a location parameter or regression coefficients in some models arising in the particle track recognition problems of high energy physics. We use the gross-error model of the contaminated distribution of errors

(1) $$f(x)=(1-\varepsilon)\varphi(x)+\varepsilon h(x)$$

with $\varphi(x)=(2\pi\sigma^2)^{-1/2}\exp(-x^2/2\sigma^2)$ and some long-tailed noise distribution h(x) specified below.

In the case of automatic scanning the experimental data obtained from particle track detectors consist of a useful ("good") part related to the track to be found, as well as of signals of background tracks, fiducials and other noise points. The noise points are usually uniformly distributed. This is the reason why we suppose h(x) in (1) to be uniform: $h(x)=h_o$ in a sufficiently large interval I_h of the length $1/h_o \gg \sigma$ and $\varepsilon > 1/2$ (even close to 1).

These models are usually explored by the pattern recognition or clustering methods. The robust estimates are also applicable in these cases, but with certain modifications or auxiliary means. We propose one of these modifications and show its high efficiency in the regression model by Monte Carlo method.

THE CHOICE OF THE WEIGHT FUNCTION FOR M-ESTIMATION

It is convenient to begin with a one-parameter model of estimating the location parameter a=Ex from a sample $x_1, x_2, \dots, x_n$, where $x \sim a+(1-\varepsilon)\varphi(x)+\varepsilon h(x)$, the functions φ,h are described above. We use Huber's M-estimates

(2) $$L(a,\sigma) = \sum_i \rho\left(\frac{x_i - a}{\sigma}\right) \to \inf_a$$

or

(3) $$a = \frac{\sum w_i x_i}{\sum w_i}$$

with the weights $w_i = w((x_i - a)/\sigma)$, where $w(t) = \rho'(t)/t$ is the weight function of the estimator. The usual requirements on (2) are: the function $\rho(t)$ must be even, C^2-smooth, not decreasing for $t > 0$, $\rho(0)=0$, $\rho(t) \sim t^2/2$ as $t \to 0$ (i.e. w(0)=1), the function $w(t) \geqslant 0$ and does not increase for $t > 0$, and the estimate (2) must be shift- and scale-invariant, as well as the estimate of σ if its value is unknown.

The problem of the choice of the function $\rho(t)$ (or w(t)) is widely discussed in the literature on robust statistics. Unbounded convex functions $\rho(t)$ provide the uniqueness of the estimate (2), its consistency, asymptotic normality in some models and a certain minimax efficiency - see Huber (1981), Yohai and Maronna (1979). But these estimates are practically unsuitable for heavy contaminated data models with $\varepsilon > 1/2$ and asymmetric, not unimodal function h(t).

The M-estimators with bounded function $\rho(t)$ are very robust in these cases, but there are many difficulties in their use. The first one is that there is almost no theoretical foundation for the use of such functions. In particular, they all are obtained by their authors heuristically. In any case there are certain objections against their application - see Huber (1981).

We shall demonstrate that the maximum likelihood estimation in the framework of our model straightforwardly leads to a bounded function $\rho(t)$ in (2). Evaluating the corresponding likelihood equation

$$\frac{\partial}{\partial a} \sum \ln\left(\frac{1-\varepsilon}{\sqrt{2\pi\sigma^2}} \exp(-(x_i - a)^2/2\sigma^2) + \varepsilon h_o\right) = 0$$

we obtain $a=\sum w_i x_i / \sum w_i$, where $w_i=w((x_i-a)/\sigma)$ with the weight function

(4) $$w(t)=w_U(t)=\frac{1+c}{1+ce^{t^2/2}}$$

with $c=\sqrt{2\pi}\,\sigma h_o \varepsilon/(1-\varepsilon)$ (the factor $1+c$ is introduced in (4) to fulfil $w(0)=1$). The weight function (4) corresponds to the bounded function

(5) $$\rho(t) = (1+c)\ \ln \frac{c+1}{c+e^{-t^2/2}} \ .$$

The function (4) has no scale parameter ($w(t)\neq w_o(t/c)$). The only parameter c is the ratio of the mean number of noise observations within an interval of the length $\sqrt{2\pi}\,\sigma$ to the mean number of useful observations in the sample. It is determined by the contamination of data not in the whole range of the sample but within its essential part where all useful observations are practically concentrated (for instance, in the interval $(a-3\sigma, a+3\sigma)$). The value of c is often approximately known in experimental models.

The upper bound of (5) $(1+c)\ln(1+1/c)$ increases without limit as $c\to 0$ (with the noise diminishing). Hence the boundedness of this function is significant only for $c>0.1$ which corresponds to heavy contamination. Fig.1 shows the function (4) with $c=0.2$ compared to Tukey's bi-square weight (see Tukey 1974)

(6) $$w(t)=w_T(t)=\left\{(1-(t/c_T)^2)^2 \text{ for } |t|<c_T \text{ and } 0 \text{ for } |t|\geq c_T\right\}$$

with $c_T=4$. These functions are close to each other, but (6) is more preferable due to faster computations. We shall suppose that $w(t)=w_T(t)$ in our further considerations.

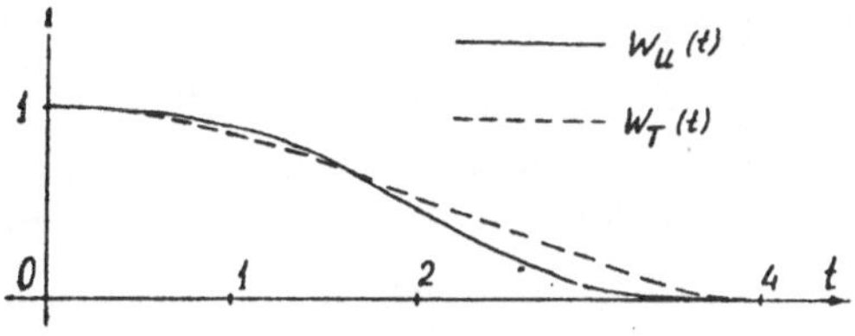

Fig. 1

ESTIMATION OF σ

Another problem caused by the use of bounded functions $\rho(t)$ is connected with the non-uniqueness of the estimate (2). The function $L(\underline{a},\sigma)$ often has several minima. It is difficult to find them all and to choose one of them for estimating a. The number of minima and their location depends on the value of σ, i.e., the problem of M-estimating a and σ are closely related. J.O.Ramsay (1977) also notes that separate procedures for estimating a and σ are unwise. Our approach is based upon the joint M-estimates of a and σ.

It is difficult problem to estimate the parameter σ in our model. The common robust estimate $\hat{\sigma} = \text{const}\cdot\text{med}\{|x_i - a|\}$ is unavailable for $\varepsilon > 1/2$ as well as other estimates based on the order statistics. From the likelihood equation for σ

$$\frac{\partial}{\partial\sigma}\sum \ln\left(\frac{1-\varepsilon}{\sqrt{2\pi\sigma^2}}\exp\left(-\frac{(x_i-a)^2}{2\sigma^2}\right)+\varepsilon h_o\right) = 0$$

we obtain

(7)
$$\sigma^2 = \frac{\sum w_i (x_i-a)^2}{\sum w_i}$$

with w_i defined in (4). This estimate is applicable for $\varepsilon > 1/2$, too. It was proposed by J.O.Ramsay (1977), S.A.Aivazyan et al. (1985) with different functions w(t). It also satisfies Huber's definition of M-estimate of σ through the solution of the following equation (see Huber 1981)

(8)
$$\sum_i \chi\left(\frac{x_i-a}{\sigma}\right) = 0$$

with an even function $\chi(t)$. The estimate (7) corresponds to (8) when $\chi(t) = t^2 w(t) - w(t)$. Therefore it is shift- and scale-invariant. Using the same weight function w(t) in (3) and (7) we can consider a- and σ-estimating as a single problem.

SOME SPECIAL GEOMETRIC PROPERTIES OF M-ESTIMATES (3),(7)

Let us consider the function $L(a,\sigma)$ in (2) as a two-parameter function. The set of local conditional minima of $L(\underline{a},\sigma)$ for all fixed $\sigma > 0$ form a finite collection of smooth curves in the semi-plane $\{(a,\sigma), \sigma > 0\}$. Denote them $\gamma_1, \gamma_2, \ldots, \gamma_m$. There is $\sigma_1 > 0$ such that the semi-plane $\{(a,\sigma), \sigma > \sigma_1\}$ contains only one of these curves which is infinite and has the asymptote $a = (x_1 + x_2 + \ldots + x_n)/n$ as $\sigma \to \infty$ (we denote this curve by γ_1).

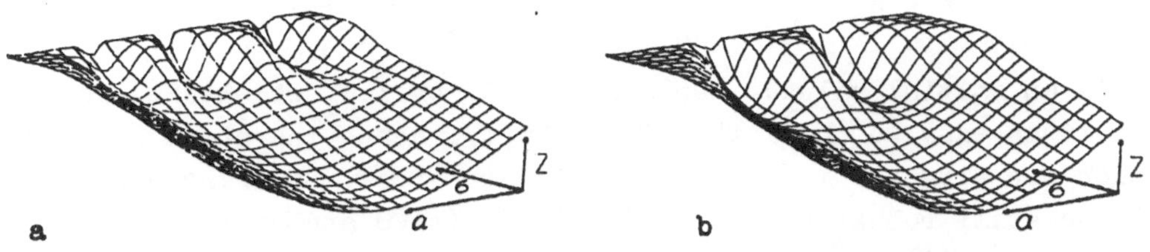

Fig.2

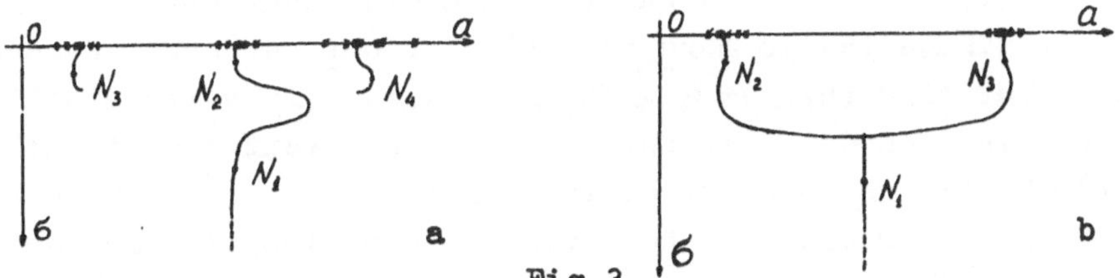

Fig.3

In fig.2 we give two examples of the surface $z=L(a,\sigma)$ for the samples each containing 20 points grouped into three and two clusters, correspondingly. On these surfaces one can see "ravines" (sometimes quite curved), the number of which increases as $\sigma \to 0$.

Fig.3 represents the curves $\gamma_1, \gamma_2, \dots, \gamma_m$ which are the "bottoms" of these ravines (the corresponding samples are marked with asterisks). These curves are always disconnected except the special case of symmetric samples like 3b, which lead to branching γ_1 into two different curves. The upper ends of the curves $\gamma_1, \gamma_2, \dots, \gamma_m$ usually lay on the axis $\sigma=0$, but there are some exceptions. At the lower ends of these curves $\partial^2 L/\partial a^2=0$, i.e., the tangents at these points are parallel to the axis $\sigma=0$.

In order to find the estimate (7) on the curves $\gamma_1, \gamma_2, \dots, \gamma_m$ consider the function $M(a,\sigma)=\sigma^2\sum w_i - \sum w_i(x_i-a)^2$. It is easy to show that $M(a,\sigma)>0$ for the curve γ_1 for large σ ($\sigma >$ const) and $M(a,\sigma)<0$ at the endpoints of these curves on the axis $\sigma=0$. However it does not mean the existence of a solution $M(a,\sigma)=0$ in any of the curves $\gamma_1, \dots, \gamma_m$ and, morover, there are even examples of absence of such solutions on these curves.

This annoying situation can be overcome if we replace (7) by

$$\sigma^2 = \frac{\sum w_i (x_i-a)^2}{\sum \varkappa_i} , \qquad (9)$$

where $\varkappa_i = \varkappa((x_i-a)/\sigma)$, $\varkappa(t)=\rho''(t)$. The idea of substituting $\sum \varkappa_i$ for $\sum w_i$ has come up in calculating the local $L(\underline{a},\sigma)$ minimum by Newton's iteration method

$$a = a_o - \frac{\partial L/\partial a}{\partial^2 L/\partial a^2} = a_o + \frac{\sum w_i(x_i - a_o)}{\sum \varkappa_i}$$

(a_o is an initial value). Replacing $\sum \varkappa_i$ by $\sum w_i$ we obtain exactly (3). It is easy to check that $\varkappa(t) \leq w(t)$, hence the above substitution can slightly increase the estimate of σ . Nevertheless, it will be still robust because $\varkappa(t) \equiv 0$ if $w(t) \equiv 0$ and $\varkappa(t) \to 1$ as $t \to 0$. The estimate (9) also satisfies the definition (8), where $\chi(t) = t^2 w(t) - \varkappa(t)$, hence it is shift- and scale-invariant.

The estimate (9) is more suitable for our purposes, because the behaviour of $M_1(a,\sigma) = \sigma^2 \sum \varkappa_i - \sum w_i(x_i - a)^2$ is quite similiar to $M(a,\sigma)$ mentioned above, but with the important exception: $M_1(a,\sigma) < 0$ at each endpoint of the curves $\gamma_1,\dots,\gamma_m$ (both for $\sigma = 0$ and $\sigma > 0$). Therefore the curve γ_1 contains at least one solution $M_1(a,\sigma) = 0$. Moreover, there must be the solution with $\partial M_1/\partial \sigma > 0$ as the necessary condition of the σ-estimate (otherwise the part of the sample in the interval $(a - c_T\sigma, a + c_T\sigma)$ is concentrated at the ends of the considered interval, where $\varkappa((x-a)/\sigma) < 0$ which is in contradiction with the normal distribution of useful observations).

The obtained joint estimate (2),(9) can be defined by the set of equations containing the function $L(a,\sigma)$ only:

(10)
$$\begin{cases} \dfrac{\partial L}{\partial a} = 0 \\[2ex] \dfrac{\partial^2 L}{\partial a^2} + \dfrac{1}{\sigma}\dfrac{\partial L}{\partial \sigma} = 0 \end{cases}$$

with the following conditions: $\partial^2 L/\partial a^2 > 0$, $\partial/\partial\sigma\,(\partial^2 L/\partial a^2 + 1/\sigma\ \partial L/\partial \sigma) > 0$. Note that (10) defines just the maximum likelihood estimate in the classical case of $\varepsilon = 0$ and $\rho(t) = t^2/2$.

The estimates obtained from (10) are denoted by $N_1, N_2, \dots$ in fig.3. Studying many different samples simulated by the computer we noticed the cases of just one estimate (10) in the whole area $\sigma > 0$ as well as the cases with many simultaneous estimates (9) on several curves $\gamma_1, \gamma_2, \dots$. The multiplicity of solutions (10) corresponds to the existence of clusters in the sample $x_1,\dots,x_n$.

ALGORITHM FOR COMPUTATION OF M-ESTIMATE

The existence of at least one estimate (10) on the curve γ_1 suggests the following algorithm. Starting from some point (a_o,σ_o) on the curve γ_1 with sufficiently large σ_o and $a_o \approx (x_1+\ldots+x_n)/n$ we move along the curve γ_1 decreasing σ and looking through all solutions of (10). The one closest to the axis $\sigma=0$ must be chosen as the M-estimate.

In our model $\sigma \ll 1/h_o$, i.e. σ is much less than the range of the sample. Therefore we can simplify our algorithm by moving along γ_1 without looking for solutions of (10) but stopping when we reach a small threshold $\sigma = \sigma_{min}$.

MONTE CARLO RESULTS

We studied a linear regression model y=ax+b with a uniform contamination arising in special emulsion data processing in high energy physics (see Bencze and Soroko 1985). The sample consisted of N_o "good" points $y_i=ax_i+b+\varepsilon_i$ and N_1 noise points uniformly distributed in the square $(0,1)\times(0,1)$. The factors x_i were uniformly distributed in the intervals $((i-1)/N_o, i/N_o)$, $i=1,2,\ldots,N_o$ and $\varepsilon_i \sim N(0,10^{-6})$. The parameters a,b were uniformly distributed in the domain $\{(a,b):\ 0<b<1\ ,\ 0<a+b<1\ ,\ |a|<\operatorname{tg} 30^\circ\}$.

The parameters a,b were estimated by the iterational reweighted least-square procedure

$$\begin{cases} a^{(k)}\sum w_i x_i^2 + b^{(k)}\sum w_i x_i = \sum w_i x_i y_i \\ a^{(k)}\sum w_i x_i + b^{(k)}\sum w_i = \sum w_i y_i \end{cases}$$

with $w_i = w_T((y_i-a^{(k-1)}x_i-b^{(k-1)})/\sigma^{(k-1)})$ by c=3.5. Starting with $\sigma^{(o)}=1$ we diminished σ very slowly from iteration to iteration: $\sigma^{(k)}=(1-\delta)\sigma^{(k-1)}$ ($\delta=0.05$), in order to retain the point $(a^{(k)}, b^{(k)}, \sigma^{(k)})$ in the neighbourhood of γ_1. The procedure stopped at $\sigma^{(k)}<\sigma_{min}=0.001$. The obtained estimates $\hat{a},\hat{b}$ were considered as correct if they differed from the "true" values a,b by less than 0.001 .

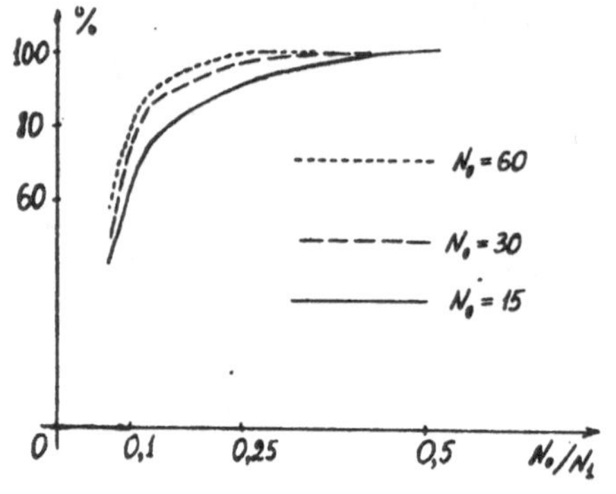

Fig.4 shows the percentage of the cases of correct esimating to the total number of simulated samples. One can note rather high efficiency of the algorithm even for $N_1=10N_0$ (i.e. $\varepsilon=0.9$ in the model (1)). The efficiency increases when the sample size augments and N_1/N_0 remains fixed, that indicates the possible consistency of the obtained estimate for any $\varepsilon < 0.9$.

The authors are greatful to V.K.Khoromskaya and S.V.Kunyaev for valuable help in preparing the manuscript.

REFERENCES

Aivazyan S.A., Yenyukov I.S., Meshalkin L.D. (1985): Applied statistics, Fin. i Statist., Moscow (in Russian).

Huber P.J. (1977): Robust methods of estimation of regression coefficients. Mathem. Operationsforschung und Statist., Ser. Statist. 8, No. 1, 41-53.

(1981): Robust statistics, J.Willey and Sons, New York.

Ramsay J.O. (1977): A comparative study of several robust estimates of location, slope, intercept and scale in linear regression, J.A.S.A. 72, No 359, 608-615.

Bencze Gy.L., Soroko L.M. (1985): Event searching algorithms in mesooptical Fourier transform microscope for nuclear research emulsion, JINR communication P13-85-137, Dubna (in Russian).

Tukey J.W. (1974): Introduction to today's data analysis. In: Critical evaluation of chemical and physical structural information, Nat. Acad. Science, Wash., 3-14.

Yohai V.J., Maronna R.A. (1979): Asymptotic behaviour of M-estimators for the linear model, Annals of Statist. 7, 258-268.

Joint Institute for Nuclear Research
Head Post Office, P.O. Box 79, Moscow

THE MD-METHOD, AN INTERACTIVE MULTI-CRITERIA GROUP DECISION-MAKING PROCEDURE

László Cserny dr.
Budapest

Key words: multi-criteria decision-making, group decision-making, interactive decision-making, rank odering

ABSTRACT

The MD-method is a two-phases interactive multi-criteria decision-making procedure, which is able to order variants in accordance with the group interests. In the first phase of the procedure the decision-makers determine the weights of the criteria /decision variables/, then in the second phase the ranks of the variants are evaluated. In both cases the participants have the possibility to study the structure of the decision-makers' group.

1. THE MAIN PRINCIPLES

The MD-method is usable in the cases of

- rank ordering of variants on the base of one or more criteria by one or more experts;
- multi-criteria rank ordering of the variants, taking the various importance of the criteria and that of the decision-makers too into consideration;
- bringing to light the internal structure of the group of the decision-makers.

2. THE DETAILED INTERPRETATION OF THE METHOD

2.1. Rank odering of the criteria

For further dicuss let the set of the criteria be denoted with $N = \{1,2,\ldots,n\}$, the set of the experts with $M = \{1,2,\ldots\ldots,m\}$ and the set of the variants with $L = \{1,2,\ldots\ldots,\ell\}$.

a./With the aid of the method of pair comparisons let us determine the preference matrix

$$(1) \qquad P = (p_{ijk}) \quad \text{where} \quad p_{ijk} = \begin{cases} 1 & \text{if the k-th expert prefers the i-th criterion to the j-th} \\ 0 & \text{in other cases} \end{cases}$$

b./For every expert let us compute the coefficient of consistency [5]

$$(2) \qquad q_k = 1 - \frac{z_k}{z_M} \qquad \forall k \in L$$

where

$$z_M = \begin{cases} n(n^2-1)/24 & \text{if } n \text{ is odd} \\ n(n^2-4)/24 & \text{if } n \text{ is even} \end{cases}$$

and

$$z_k = \frac{n(n-1)(2n-1)}{12} - \frac{1}{2}\sum_{i=1}^{n}\left(\sum_{j=1}^{n} p_{ijk}\right)^2 \qquad \forall k \in L$$

If $\exists q_k$, $k \in L$ so, that $q_k < q_o$ given, then the k-th expert must repeat the procedure from a./

c./Now let us determine the rank number matrix A, using the Borda-Kendall/or other suitable/ rank ordering function [3,4,6] β, in the following way:

$$(3) \qquad \beta : \sum_{j=1}^{n} p_{ijk} \longrightarrow a_{ik} \qquad \forall i \in N \text{ and } \forall k \in L$$

where $a_{ik} \in \{1,2,\ldots,n\}$ is the rank number of the i-th criterion according to the k-th expert and

$$a_{ik} < a_{pk} \iff \sum_{j=1}^{n} p_{ijk} > \sum_{j=1}^{n} p_{pjk}$$

d./In the knowledge of the rank number matrix A, let us compute the rank correlation matrix R, where

$$(4) \qquad r_{ij} = 1 - \frac{6\sum_{k=1}^{n}(a_{ki} - a_{kj})^2}{n(n^2 - 1)} \qquad \forall\, i,j \in L$$

is the Spearman-kind correlation coefficient between the orders of rank of the i-th and j-th experts.

e./For weighting the experts, let us cluster them hierarchically [2] on the base of the decreasing value of the Kendall-kind coefficient of concordance [5] w.

We can determine the value of w as the sum of the elements of the rank number matrix or as that of the rank correlation matrix:

$$(5) \qquad w = \frac{12\sum_{i=1}^{n}\left(\sum_{k=1}^{m} a_{ik} - \frac{m(n+1)}{2}\right)^2}{m^2 n(n^2-1)}$$

or

$$(6) \qquad w = \frac{\sum_{i=1}^{m}\sum_{j=1}^{m} r_{ij}}{m^2}$$

At the union of two clusters, the new value of the concordance is determined as it is following:

$$(7) \qquad w = \frac{m_1^2 w_1 + 2m_1 m_2 w_{12} + m_2^2 w_2}{m^2}$$

where w_1, w_2 are the coefficients of concordance of the experts to be clustered;

w_{12} is the coefficient of mutual concordance between the two clusters. /It is the average rank correlation between the members of the one and the other cluster./;

m_1, m_2 are the numbers of experts in the two clusters;

and $m = m_1 + m_2$

If the concordance of the whole group of experts is less than a given value w_o, then the group have to repeat the whole procedure from the begenning.

f./Using the formula /7/ we can determine the weights of the clusters/finally those of the experts/ in the following way:

$$(8) \qquad b_1 = \frac{m_1^2 w_1 + m_1 m_2 w_{12}}{m^2 w} b \qquad b_2 = \frac{m_2^2 w_2 + m_1 m_2 w_{12}}{m^2 w} b$$

where b is the weight of the cluster with the value w of concordance and

$b=1$ is the initial value of the cluster weights.

g./Knowing the weights $b_1, b_2, \ldots, b_m$ of the experts, we can compute the weights of the criteria too. Let us denote the preference frequency of a criteria by h_i, $i \in N$, then

$$(9) \qquad h_i = \sum_{j=1}^{n} \sum_{k=1}^{m} b_k p_{ijk} \qquad \forall i \in N$$

and finally thus we have

$$(10) \qquad c_i = \frac{h_i}{\sum_{j=1}^{n} h_j} \qquad \forall i \in N$$

as the weights of the criteria.

2.2. Rank ordering of the variants

The rank ordering of the variants is similar to that of the criteria.

a./Let us score the variants criterion by criterion on the scale $[0,10]$ so, that

(11) $T = (t_{ijk})$ where $t_{ijk} \in [0,10]$ is the score of the i-th variant according to the j-th criterion, given by the k-th expert.

b./The weighted scores give the co-ordinates of a vector $\underline{d}_{ik}$ in the n-dimensional decision space:

(12) $\underline{d}_{ik} = (c_1 t_{i1k}, c_2 t_{i2k}, \ldots\ldots, c_n t_{ink})$ $\forall i \in L$ and $\forall k \in M$

Comparing the variants pair by pair, we can get the following vector differences:

(13) $\Delta \underline{d}^k_{ij} = \underline{d}^k_i - \underline{d}^k_j$ $\forall i,j \in L$ and $\forall k \in M$

We have got a maximum value of the vector difference $\Delta \underline{d}$, if

(14) $\Delta \underline{d}_M = (c_1 t_{max}, c_2 t_{max}, \ldots\ldots, c_n t_{max})$

and here $t_{max} = 10$.

c./On the base of a preference matrix P/see below/, we can again cluster the experts in the same way, as it is mentioned previously/see paragraph a.-f. in 2.1./

(15) $P = (p_{ijk})$ where $p_{ijk} = \begin{cases} 1 & \text{if the scalar product } (\Delta \underline{d}^k_{ij} \cdot \Delta \underline{d}_M) > 0 \\ 0 & \text{in other cases} \end{cases}$

After proceeding the steps mentioned in the paragraph a.-f., we have get the weights b_i, $i \in M$ of experts.

d./Considering the weights b_k, $k\in M$ and the weights c_i, $i\in N$, let us determine the weighted score matrix G of the whole group of the experts in the following way:

$$(16)\qquad g_{ij}=\sum_{k=1}^{m}\frac{b_j+c_k}{2}\,t_{ijk}\qquad \forall i\in L \quad\text{and}\quad \forall j\in N$$

Then similarly to the way mentioned in the previous paragraph b., we determine the scalar product matrix $U=(u_{ij})$, as if it is following

$$(17)\qquad u_{ij}=(\Delta \underline{d}_{ij}\cdot\Delta\underline{d}_M)\qquad \forall i,j\in L$$

where $\Delta\underline{d}_{ij}=\underline{d}_i-\underline{d}_j$ and $\underline{d}_i=(g_{i1},g_{i2},\dots,g_{in})$ $\forall i,j\in N$

and

$$\Delta\underline{d}_M=(g_{M1},g_{M2},\dots,g_{Mn})$$

$$g_{Mj}=\sum_{k=1}^{m}\frac{b_j+c_k}{2}\,t_{max}\qquad \forall j\in N$$

e./Using the values of the precedence matrix $U=(u_{ij})$, we can produce the preference matrix P, in the same way as it is given in the formula /15/. Then applying the Borda-Kendall/or other suitable/ rank ordering function, we determine the final order of ranks of the variants.

REFERENCES

[1] Cserny,L./1982/: Többváltozós csoportdöntések problémái, alkalmazásai/The problems and applications of multi-criteria group decision-making/,Építésügyi Szemle,XXV.,No.11.,345-352

[2] Cserny,L./1983/:Rangsoroptimálás a csoportérdekek figyelembevételével/Rank ordering in accordance with group interests/,XIII.Magyar Operációkutatási Konferencia,Balatonfüred,29-30

[3] Fishburn,P.C./1973/:The Theory of Social Choice,Princeton Press

[4] Inoue,K.-Tanino,T.-Nakayama,H.-Sawaragi,Y./1981/: A Trial Towards Group Decisions in Structuring Environmental Science; in Morse,J.N.ed.: Organizations: Multiple Agents with Multiple Criteria, Springer Verlag,Berlin

[5] Kendall,M.G./1970/: Rank Correlation Methods,Griffin,London

[6] Sen,A.K./1970/: Collective Choice and Social Welfare,Holden-Day, San Francisco

"Ybl Miklós" College for Building Industry

Budapest,70. P.O.B.117.
H-1442, Hungary

EXTENDING FOSTER'S ERGODICITY CRITERIA TO CONTROLLED MARKOV CHAINS AND ANALYZING INTEGRATED SERVICE LOCAL AREA NETWORKS

Sándor Csibi

Budapest

Key words: Controlled Markov chains, ergodicity, Foster's criteria, cooperative queueing, integrated services, local area networks, piggy backing data onto speech transmission

ABSTRACT

A further extention of Foster's ergodicity criteria to controlled Markov chains is presented by developing the approach introduced by the same author (1985 a) further. An insight is obtained in this way into the possibilities of utilizing also the partial activity of busy talkers for data transmission; an essential agenda provided the maximum speech throughput is close to unity.. Stationary and Markovian on-off sequences are assumed for each talker, under a very mild decay of the correlation function. (Markovity appears in the present context as a fairly realistic assumption, because of the length of the speech packets adopted usually.)

PREREQUISITES

Let $\varrho=(\varrho_k;\ k=0,1,\ldots)$ be a countable homogeneous irreducable and aperiodic Markov chain. (Denote the states of ϱ by $i=1,2,\ldots$).

Theorem 1 (Foster/1953/) If ϱ is such that for some positive integer n_0 and some positive real d

$$E(\varrho_1-\varrho_0|\varrho_0=i)\leq -d \tag{1}$$

for $i > n_0$, and

$$E(\varrho_1 - \varrho_0 | \varrho_0 = i) < \infty \quad (2)$$

for $i \leq 0$, then ϱ is ergodic, i.e., recurrent positive. (E stands for expectation.)

Proof See Foster (1953) and Cohen (1982).

Application Tsybakov and Mikhalov (1978) realized first the significance of Theorem 1 for the study of stable block wise random access data communication: viz., in the course of which newly arriving data packets are accessed to some common medium (e.g., either to some bus, some surface radio or geostationary satellite medium) as soon as previously collided packets have been transmitted. It is well known that the throughput as well as the asymptotic average delay (for stable performance) can be appropriately estimated in this way.

Blockwise techniques offer definite advantages for integrated local area services, embedding a data packet flow randomly into the pauses left free by a scheduled flow of active speech packets. (Such sort of embeddings are usually called piggy backings, and are of interest also in other kinds of data services.) However, for studying the performance feature of such sort of services an extention of Foster's criteria to controlled Markov chains appears as obviously necessary, simple because of the very nature of the two flows handled simultaneously.

Theorem 2 (Csibi 1985) Let $\varrho = (\varrho_k ; k=0,1\ldots)$ be a controlled Markov chain defined by

$$\varrho_k = \sigma_k \gamma + \psi_k$$

(γ is standing for some positive integer, fixed at the outset. $\sigma = (\sigma_k ; k=0,1\ldots)$ is a sequence of countable non-negative valued and $\psi = (\psi_k ; k=0,1\ldots)$ another sequence of finite non-negative valued random variables. Let σ take the values i=1,2,...and ψ_k $\gamma = 1,\ldots\gamma$) Call ψ the control sequence and σ the controlled chain. Assume ϱ to be a homogeneous, irreducible aperiodic Markov chain with the following properties: There exists some positive integer n_0 and some positive real d such that

$$E(\sigma_1 - \sigma_0 | \varrho_0 = i) \leq -d \quad (4)$$

for $i > n_0$ and

$$E(\sigma_1 - \sigma_0 | \varrho_0 = i) < \infty \quad (5)$$

for $i > n_0$.

Then ς is ergodic.

Proof Annex 4 in Csibi (1985a)

Application By Theorem 2 the supremum of the stable data throughput for some maximum speech throughput can be underestimated, meaningful design relations being obtained in this way (Csibi/1985a/). Notice however that the study in Csibi (1985) has been confined for the sake of simplicity, by a crude exaggeration, taking all simultaneously busy talkers also simultaneously active. One excludes in this way the study of the additional data transmission possibilites offered by the actual partial (and of course also fluctuating) activity of the busy talkers.

These additional possibilities may be very essential provided the maximum speech throughput, corresponding to the simultaneous activity of all busy talkers, is close to unity.

Another negative consequence of the aforementioned appealing simplification is that one has got, in this case, only a rough underestimation for the asymptotic average length $Z_1 = \lim_{k \to \infty} E\Theta_k$ of the collision resolution intervals Θ_k . One obtains in this way, only a very loose upper estimation for the asymptotic average delay as for the latter an underestimation of Z_1 is also needed.

As a matter of fact it is of fair interests to develop some more tight estimation for the average length of the speech packets skipped by the superimposed data flow.

Getting rid of these difficulties is the main purpose of this paper.

A FURTHER EXTENTION FOR STUDYING PARTIAL ACTIVITY

For getting some useful insight also into the advantages of utilizing the partial activity of busy talkers, let us broaden our interest in seeking Foster type criteria not just in terms of the single step behavior of the considered controlled chain, but admitting also some reference to long run behavior. One can get rid, in this way, of the complications due to the fluctuating partial activity of the talkers.

Theorem 3 Assume a nonnegative valued, countable homogeneous, irreducible and aperiodic Markov chain (having states according to (2)) with the following properties: There exists some positive integer

n_0 , some positive real d and a pair of real valued sequences $\varepsilon = (\varepsilon_k ; k = 0,1 \ldots)$ and $\tilde{\varepsilon} = (\tilde{\varepsilon}_r ; r = 0,1,\ldots)$ of random variables such that

$$E(\sigma_{k+1} - \sigma_k \mid \rho_k = i) \leq -d + \varepsilon_k$$

for any $i > n_0$ and all $k = 0,1,\ldots$,

$$E(\sigma_1 - \sigma_0 \mid \rho_0 = i) < \infty$$

for any $i \leq 0$,

$$\frac{1}{r} \sum_{k=1}^{r} \varepsilon_k \leq \tilde{\varepsilon}_r$$

for any $r = 1,2,\ldots$, and $\tilde{\varepsilon}_r \to 0$.

Then the Markov chain ρ is ergodic.

APPLICATION

Let us adopt Theorem 3 to the data flow embedding model, introduced in Csibi (1985a) with the following single revision: Assume the binary on-off sequences $\eta^{(l)} = (\eta_j^{(l)} , j = 0,1,\ldots)$, describing by the state one the activity of the speech packet due to the l th actually busy talker, to be a stationary Markov chain with $E\eta_j^{(l)} = E\eta_j^{(0)}$ for any $l = 0,1,\ldots,N-1$. ($E\eta_j^{(0)}$ stands for the average activity of the talkers.) Considering the long run behavior of any $\eta^{(l)}$, what one really needs is the law of large numbers. We assume this to hold; which is the case, for any sequence $\eta = (\eta_n , n = 0,1,\ldots)$ the correlation $\rho_n = E\eta_n\eta_0$ of which is decaying as $n \to \infty$ so mildly as

$$\rho(n) = O(1/[(\log n)(\log_2 n)^{1+\delta}])$$

for $\delta > 0$. (See Masry and Győrfi /1986/).

Recollect that only blockwise collision resolution is considered for the data service, for which short packets of unit length are assumed. For conveying the messages of data users and also for the active speech segments of the actually busy talkers long packets of K length are taken into account.

It is to be noticed, this time again, that the rules for accessing were set up in Csibi(1985a) very carefully; to include, as far as possible, the practically really relevant features of the considered services. Constraints, introduced for the sake of simple mathematical analysis, were introduced intentionally in a way not to restrict really the scope of real life applications.

Admit the start of a short as well as a long packet only at nodes. Long packets of at most N simultaneously busy talkers are included after another, in frames of length M (where $N < M$). However the really relevant constraint sounds as follows: the start of a collision resolution period is admitted only at the outset of a frame. (Fig.1.)

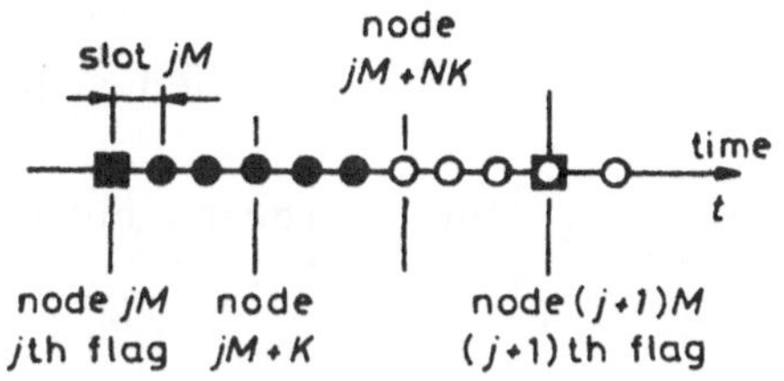

Fig.1. Slotting and flagging (for K=3, M=9 and N=2)

New demands for data transmission are assumed to arrive according to a homogeneous Poisson process with a data call rate $\lambda_D = S_D / K$. (S_D stands for the data throughput.) However, as the study is confined only to blockwise collision resolution, newcomers have to wait until previously collided data transmission demands have been actually served. Denote by ξ_k number of data transmission demands waiting at the outset of the kth collison resolution interval (i.e. at node kM).

Denote by $S_S = NK/M$ the maximum and by $\bar{S}_S = NKE\beta_0^{(0)}/M$ the average speech throughput. (S_S is describing the extremely unlikely instants when all busy talkers appear to be simultaneously active.) Call $\bar{S} = \bar{S}_S + S_D$ the average overall throughput.

Assume, when studying the performance, all active speech packets, occuring either within the Lth frame or next to this slot, to be lost. However, as the final estimates do not depend on L, one can finally consider $L \to \infty$. (I.e., the intermediate truncation to at most L speech packets per frame does not finally matter.)

Define the sequence of random variables

$\zeta = (\zeta_k \,,\ k = 0,1,\dots)$ by $\zeta_k = \xi_k \gamma + \tilde{\alpha}_k$

Here

$$\alpha_k = (\eta^{(\ell)}_{(\nu_k/M)+j}\,;\ 0 \le j \le \min(\nu_k, L-1),\ 0 \le \ell \le N-1)$$

$\tilde{\alpha}_k$ denotes the binary number, the jth bit of which is defined by $\eta^{(\ell)}_{(\nu_k/M)+j}$. $\gamma = \sum_{i=1}^{L} 2^{iN}$.

Under the assumptions and constraints considered, ζ is a countable, nonnegative valued, homogeneous, irreducable aperiodic Markov chain controlled by the finite valued sequence α of random variables.

<u>Theorem 4</u> For ζ the following constraint holds:

$$E\,(\zeta_{k+1} \mid \xi_k) < \xi_k + \varepsilon_k + B$$

Here

$$A = \frac{(\overline{TS} - \bar{S}_s)(1 + C_0/K)}{1 - \bar{S}_s} \ge 0$$

$$B = \frac{(\overline{TS} - \overline{\overline{TS}})(M/K)}{1 - \bar{S}_s} \ge 0$$

$$\varepsilon_k = E\,\Big(\sum_{j=0}^{\nu_k} \sum_{l=0}^{N-1} (\eta^{(\ell)}_{(\nu_k/M)+j} - E\,\eta_0 \mid \xi_k\Big)$$

$$\frac{1}{r} \sum_{k=1}^{r} \varepsilon_r \le \tilde{\varepsilon}_r$$

$$\tilde{\mathcal{E}}_r = KL \sum_{j=0}^{L-1} \sum_{l=0}^{N-1} \frac{1}{r} \sum_{k=1}^{r} \left[E(\eta^{(l)}_{(\nu_k/M)+j} \mid \xi_k) - E\eta_0 \right] \tag{6}$$

where $\tilde{\mathcal{E}}_r \to 0$ as $r \to \infty$ (C_0 = 8/3.) Then ϱ is ergodic.

Hint to the proof See Annex. (For the details see Csibi (1986).)

CONCLUSIONS

Readers familiar with Csibi (1985a) will immediately realize that from Theorem 4 relations in Secs. 6 and 7 in Csibi (1985a) follow for (i) stable performance (for $A < 1$)(ii) for the lower bound S_∞ of the supremum of the stable average overall throughput (for any L , i.e. for $L \to \infty$) and (iii) for the asymptotic average delay, provided S_s is replaced by $\overline{S_s}$. An insight is obtained in this way for utilizing the partial activity of the simultaneously busy talkers. In addition the asymptotic delay can be, overestimated in this case, also for the considered joint data and speech services, with the very same tightness as in Tsybakov and Mikhalov (1978).

While important questions, left open still in Csibi (1985a), could be answered by using Theorems 3 and 4, the interesting question of how to introduce reasonable contraints for controled Markov chains with (not finite but) countable control, and find meanigful Foster-type ergodicity criteria, is still left open (Csibi /1985b/).

ANNEX

The idea of proving Theorem 3 is to replace in (A.4.9), Annex 4, Csibi (1985a) d by $d-(r^{-2})^{-1}\sum_{k=2}^{r} \varepsilon_k$ and to overbound $r^{-1}\sum_{k=1}^{r} \varepsilon_k$ by $\tilde{\varepsilon}_r$. One arrives in this way, as $k \to \infty$

$$\pi_{n_0} > 0$$

for the limiting probability π_{n_0} of state n_0 . By this $\pi_i > 0$, for all i . Thus ξ is proved to be recurrent positive, i.e. ergodic.

The essential point of proving Theorem 4 is to replace (A.3.3.) in Csibi (1985a) by the following overbound:

$$E\,(\omega_k \mid \xi_k) \le K E\Big(\sum_{j=0}^{\nu} \sum_{l=0}^{N-1} E\eta_0^{(\bullet)} \mid \zeta_k\Big) + \varepsilon_k$$

Recollect from Theorem 4 that

$$\varepsilon_k = K E\Big[\sum_{j=0}^{\nu_k} \sum_{l=0}^{N-1} \Big(\eta^{(l)}_{(\nu_k/M)+j} - E\eta_0\Big) \mid \zeta_k\Big]$$

When one overestimates in (6) ν_k by $L-1$, certain members of the on-off sequences $\eta^{(l)}$ $(l=0,1,\ldots N-1)$ of the busy talkers are taken several times into account. However, this multiplicity will never exceed L. Thus

$$\frac{1}{r}\sum_{k=1}^{r} \varepsilon_k \le \tilde{\varepsilon}_r$$

for an $\tilde{\varepsilon}_k$ defined by (6).

By Masry and Győrfi, L. (1986): $\tilde{\varepsilon}_r \to 0$ as $r \to \infty$. Thus Theorem 3 holds.

REFERENCES

Foster (1953): Stochastic matrices associated with certain queueing problems. Ann. Math. Statist. Vol. 24, 355-360, Thm.2.

Cohen (1982): The single server queue (Rev'd, 2nd edn.) North Holland Publ. Comp.

Tsybakov-Mikhalov (1978): Synchronous free access of packets to a broadcast channel (in Russian) Problemy Peredachy Informatsii, Vol. 14, 32-59.

Csibi (1985a): On the stability of random access communication during the time left by speech packets. Pr. of Contr. and Inf. Theory, Vol. 14, 231-246.

(1985b): On the furthering of extended Foster type theorems to more general classes of controlled Markov chains. Open questions. Workshop on Multi-User Information Theory. U. Lund - U. Linköping. - IPPI. Supplement to appear

(1986): On the Stability of a data packet flow embedded randomly in a partially active, scheduled speech packet flow. Preprint. HEI, BME

ENTROPY AND BOUNDARY FOR RANDOM WALKS

ON LOCALLY COMPACT GROUPS

- THE EXAMPLE OF THE AFFINE GROUP -

Yves DERRIENNIC

BREST

Key words : Entropy, information, random walks, boundary, harmonic function.

ABSTRACT

Given a probability measure μ on a locally compact, separable group G, we consider the random walk $S_n = X_1 \ldots X_n$ where (X_n) is an i.i.d. sequence with common distribution μ. The mutual information $I(S_1, S_n)$ of the variables S_1 and S_n has a limit which is the "asymptotic entropy" h_μ of the random walk. This number is linked to the exit boundary of the random walk, that is the set of bounded μ-harmonic functions. The case where G is either the discrete dyadic affine group or the connected real affine group is particularly considered.

UNIVERSITE DE BRETAGNE OCCIDENTALE

Département de Mathématiques et Informatique

6 Avenue V. Le Gorgeu

29287 BREST CEDEX - FRANCE -

In the study of the asymptotic behaviour of a random walk on a locally compact group, the description of the *exit boundary* should precede the derivation of any limit theorem. The exit boundary corresponds to the space of the bounded harmonic functions : two different bounded harmonic functions yield two different final behaviours of the walk. The nature of the exit boundary is linked to the structure of the group. Let us recall that the exit boundary is trivial if the group is abelian. To study discrete non abelian groups, Avez introduced the notion of entropy of a random walk ([1], [2]). This direction of research was further developped by several authors. A rich account of this circle of ideas was given by Kaimanovich and Vershik ([6]).

In this paper we first explain how the *asymptotic entropy* of a random walk on a non discrete locally compact group can be defined and used to study the bounded harmonic functions. The general notion of mutual information plays in this definition a crucial rôle. The detailed proofs are not given here : they can be found in [4]. In a second part we study the random walks defined on the "affine group ax+b" which displays many interesting features.

I - THE GENERAL NOTION OF ASYMPTOTIC ENTROPY

We use freely the definitions and properties of the notions of entropy (denoted $H(\mu)$ or $H(X)$), relative entropy ($H(Q ; P)$), mutual information ($I(X,Y)$), conditionnal information ($EI(X,Y/Z)$), for general random variables as they are given in Pinsker's book ([7] ; a résumé is given in [4]).

Given three random variables Y_1, Y_2 and Y_3 they form a Markov chain if and only if $EI(Y_1, Y_3|Y_2) = 0$. Therefore, given a Markov chain $(Y_n)_{n\geqslant 0}$, we have

$$I(Y_1, Y_n) = I(Y_1, (Y_n, Y_{n+1}, \ldots)).$$

Obviously this sequence is decreasing. If it is finite we get

$$\lim_n I(Y_1, Y_n) = I(Y_1, \mathcal{A})$$

where $\mathcal{A}$ denote the σ-algebra of the asymptotic events of the chain (Y_n), i.e. the tail σ-algebra.

Definition : *The asymptotic entropy of a Markov-chain* $(Y_n)_{n\geq 0}$ *, starting at the point* $Y_0 = y$ *(of some measurable state space* F, $\mathcal{F}$ *) is the number* $h(y) = I_y(Y_1, \mathcal{A})$. *That is the mutual information of* Y_1 *, the position of the chain at time 1, and the tail* σ*-algebra* $\mathcal{A}$*, with respect to* P_y *the law of the chain starting at* $Y_0 = y$ *a s*

If the tail-σ-algebra $\mathcal{A}$ is trivial mod P_y, then $h(y) = 0$. In general, $h(y)$ depends on y. But there is an important case where $h(y)$ is constant : the case where the Markov chain is *spatially homogeneous* A Markov chain is spatially homogeneous if the state space $(F, \mathcal{F})$ is endowed with a family $\mathcal{T}$ of bijective bimeasurable transformations T which commute with the transition probabilities of the chain, that is

$$P(Ty,TB) = P(y,B)$$

i.e.

$$E(Y_{n+1} \in TB | Y_n = Ty) = E(Y_{n+1} \in B \,|\, Y_n = y)$$

$$= E(Y_1 \in B \,|\, Y_0 = y)$$

for any $y \in F$ and $B \in \mathcal{F}$, and act transitively on F.

Theorem 1 : *For a spatially homogeneous Markov chain* $(Y_n)_{n\geq 0}$ *, the asymptotic entropy is the number*

$$h = I(Y_1, \mathcal{A})$$

which is independent of the starting point (or the initial distribution) of the chain. The tail σ*-algebra* $\mathcal{A}$ *is trivial if and only if* $h = 0$.

The proof given in [**4**] for random walks extends without change to the more general situation considered here.

A random walk on a locally compact separable group G is defined as the chain $S_n = X_1 \dots X_n$, where $(X_n)_{n\geq 1}$ is an i.i.d. sequence of random variables taking values in G. The group law is denoted by multiplication because the group is not necessarily abelian. The common distribution of the X_n is a probability measure μ on G. For any $g \in G$, $g S_n$ is a Markov chain, whose starting point is g and transition probability is $P(y, B) = \mu(y^{-1}B)$. Obviously this Markov chain is spatially homogeneous. (If M is a homogeneous space under the action of G :

$$M \times G \longrightarrow M$$
$$(m,g) \longmapsto mg$$

then mS_n is also a spatially homogeneous Markov chain on M).

A real bounded measurable function f, defined on G, is called μ - *harmonic* if

$$f(x) = \int_G f(xy)\, d\mu(y)$$

for every $x \in G$.

Theorem 2 *Let the probability measure* μ *be "adapted" on* G *(that is : G is the smallest closed subgroup with* μ*-measure equal to* 1 *) and "spread out" (that is "étalée" : there exists* n *such that the convolution power* μ^n *is non singular) The bounded continuous* μ *-harmonic functions are constant if and only if* $h_\mu = 0$, *where* h_μ *is the asymptotic entropy of the random walk induced by* μ

For a random walk, the asymptotic entropy can be computed as a limit, knowing the "entropies" of the convolution powers μ^n of μ .

Theorem 3 *If* G *is a discrete, denumerable group, if the entropy*

$$H(\mu) = - \sum_{x\in G} \mu(x) \log \mu(x)$$

is finite, then the asymptotic entropy of the random walk induced by μ *is :*

$$h_\mu = \lim_n \left[H(\mu^n) - H(\mu^{n-1})\right] = \lim_n \frac{1}{n} H(\mu^n)$$

If G *is a separable, locally compact group with left Haar measure* m *, if* μ *has a density* $\varphi = \frac{d\mu}{dm}$ *, if the "differential entropies"*

$$\tilde{H}(\mu^n) = \tilde{H}(\varphi^{*n}) = - \int_G \varphi^{*n}(x) \log \varphi^{*n}(x)\, dm(x)$$

are finite, then the asymptotic entropy is :

$$h_\mu = \lim_n \left[\tilde{H}(\varphi^{*n}) - \tilde{H}(\varphi^{*(n-1)})\right] = \lim_n \frac{1}{n} \tilde{H}(\varphi^{*n}).$$

II - THE EXAMPLE OF THE RANDOM WALKS ON THE AFFINE GROUP

A. We first consider the affine group of the dyadic-rational line. This is the group $G = Aff(\mathbb{Z}[1/2])$ of matrices $\begin{pmatrix} a & b \\ 0 & 1 \end{pmatrix}$ where $a = 2^n$, $b = \frac{k}{2^\ell}$ with $k, \ell, n \in \mathbb{Z}$. It is determined by the generators $\alpha = \begin{pmatrix} 2 & 0 \\ 0 & 1 \end{pmatrix}$ and $\beta = \begin{pmatrix} 1 & 1 \\ 0 & 1 \end{pmatrix}$ and the relation $\beta^2\alpha = \alpha\beta$. It is solvable of length 2 and has exponential growth. In [6], p. 484, Kaimanovich and Vershik asserts, without details, that the boundary of this group G is trivial for every finitary symmetric measure μ on G. We first prove this fact with the help of an elementary argument hinted by an unpublished manuscript of H. Carnal.

Proposition 1 *: Let* μ *be the following measure on* G *:* $\mu(\alpha) = \mu(\alpha^{-1}) = \frac{1}{2}p$, $\mu(\beta) = \mu(\beta^{-1}) = \frac{1}{2}q$ *with* $p > 0$, $q > 0$, $p+q = 1$. *Every bounded* μ*-harmonic function on* G *is constant. (i.e. the exit boundary is trivial ; i.e. the asymptotic entropy* h_μ *is zero).*

Proof : Let $a(S_n) = a(X_1)\dots a(X_n)$ be the first component of S_n. We have $a(S_n) = 2^{v_1+\dots v_n}$ where $v_1+\dots+v_n$ is a symmetric random walk on $\mathbb{Z}$. This random walk is recurrent and the time T of first return at 0 is a.s. finite. If f is a bounded μ-harmonic function we get $f\begin{pmatrix} a & b \\ 0 & 1 \end{pmatrix} = E\left[f\begin{pmatrix} a & b \\ 0 & 1 \end{pmatrix} S_T\right]$. Therefore, for every a, we have

$$f\begin{pmatrix} a & b \\ 0 & 1 \end{pmatrix} = \sum_{j\in D} f\begin{pmatrix} a & b+j \\ 0 & 1 \end{pmatrix} P\left[ab(S_T) = j\right]$$

where D is the abelian group of the dyadic rational numbers. Since the distribution of $ab(S_T)$ is adapted on D the Choquet-Deny theorem implies that f is constant with respect to b. But, with respect to a, we have

$$f(a) = \frac{1}{2}\, p\, (f(2a) + f(\tfrac{a}{2})) + qf(a)$$

and again Choquet-Deny theorem yields that f is constant.

It is clear that this argument would give the same result for a measure μ on G such that the first component $a(S_n)$ be recurrent and the second $b(S_n)$ "adapted enough" on D. In [1] Avez asked the following question : does there exist a nondegenerate finitary measure with zero asymptotic entropy on a discrete group with exponential growth ? A positive answer was given in [6]. Obviously the preceding proposition answers also positively this problem because of theorem 2.

Proposition 2 *: Let* μ *be the following probability measure on* G *:* $\mu(\alpha) = p$ *and* $\mu(\beta) = q$ *with* $p > 0, q > 0$, $p+q = 1$. *The asymptotic entropy* h_μ *is strictly positive (i.e. there exists non constant bounded* μ*-harmonic functions ; i.e. the exit boundary is non trivial).*

Proof : Let T_k be the time of the kth occurence of α in the sequence (X_n). The 2nd component of S_{T_k} is $b(S_{T_k}) = r_0+2r_1+\dots+2^{k-1}r_{k-1}$ where $(r_i)_{i\geq 0}$ are i.i.d.r.v. with the geometric law of parameter p (i.e. $P(r_i=n)=q^n p$, $n\geq 0$). From the Markov property $I(S_1, S_{T_k}) = I(S_1, (S_{T_k}, S_{T_{k+1}}, \dots))$. Thus this sequence decreases to h_μ with k. On the other hand $I(S_1, S_{T_k}) \geq I(S_1, b(S_{T_k}) \bmod 2)$, and $b(S_{T_k}) = r_0 \bmod 2$. As $I(S_1, r_0 \bmod 2) = H(X_1) - P(r_0 = 0 \bmod 2)\, H(X_1 | r_0 = 0 \bmod 2)$ an easy computation yields $h_\mu \geq p \log(2-p) + \left(\frac{q-1}{q+1}\right) q \log q > 0$. We used the obvious fact that, knowing $r_0 = 1 \bmod 2$, necessarily we have $X_1 = \beta$.

The computation of the exact value of h_μ seems intricate. We shall come back to this question at the end.

B. We consider now the connected real affine group. This is the group G of matrices $\begin{pmatrix} a & b \\ 0 & 1 \end{pmatrix}$ with $a > 0$, $b \in \mathbb{R}$. It is solvable with exponential growth, non unimodular : a left Haar measure is $dm(a,b) = \frac{1}{a^2}\, da\, db$ and the module of the group is $\Delta(g) = 1/a(g)$.

Let μ be a probability measure on G with bounded density and compact support. We recall that every bounded continuous μ-harmonic function is constant if $\int_G \log a(g)\, d\ (g) \geqslant 0$. If $\check{\mu}$ denotes the image of μ by $g \to g^{-1}$, there is the following general relation between h_μ and $h_{\check{\mu}}$:

$$h_{\check{\mu}} = h_\mu - \int_G \log \Delta(g)\, d\mu(g).$$

(for the easy proof see [4], part VI). These facts together yield :

Proposition 3 *: Under the preceding assumptions*

$$h_\mu = \max\left(0, \ - \int \log a(g)\, d\mu(g)\right)$$

This implies the existence of bounded continuous nonconstant μ-harmonic functions when $\int \log a(g)\, d\mu(g) < 0$. This fact was proved in [3] by a direct method which we sketch now. Under the assumption $\int \log a(g)\, d\mu(g) < 0$, there exists on $\mathbb{R}$ a unique probability measure ν which is μ-invariant, that is : $\iint_{G\times\mathbb{R}} \Psi(g.x)\, d\mu(g)\, d\nu(x) = \int_{\mathbb{R}} \Psi(x)\, d\nu(x)$ for any function Ψ , where $g.x = a(g)x + b(g)$. Then, for any bounded continuous function Ψ on $\mathbb{R}$, the function $f(g) = \int_{\mathbb{R}} \Psi(g.x)\, d\nu(x)$ is a bounded continuous μ-harmonic function on G. From the exact value of h_μ given by proposition 3, it is interesting to deduce the following result, already proved by a different argument in [5].

Propositon 4 *Under the preceding assumptions, if* $\int \log a(g)\, d\mu(g) < 0$, *all bounded measurable* μ*-harmonic functions on* G *are of the type*

$$f(g) = \int_{\mathbb{R}} \psi(g.x)\, d\nu(x)$$

where ψ *is any bounded measurable function and* ν *the unique* μ*-invariant probability measure on* $\mathbb{R}$ *(i e.* $(\mathbb{R}, \nu)$ *is the exit boundary or the Poisson boundary of* (G,μ)*).*

Proof From the entropy criterion given in [4], part V, it suffices to show :

$$h_\mu = - \int_G d\mu(g) \int_{\mathbb{R}} (\log \frac{dg^{-1} * \nu}{d\nu}(x))\, d\nu(x),$$

where we know that the integral is well defined. Since μ has a density, ν has a density φ too ; $\frac{d\, g * \nu}{d\nu}(x) = \frac{1}{a(g)} \varphi\left(\frac{x-b(g)}{a(g)}\right) / \varphi(x)$ and a direct computation, using the μ-invariance of ν, shows that the integral equals $-\int \log a(g)\, d\mu(g)$.

Let us come back to the discrete affine dyadic group. It is unimodular and $h_\mu = h_{\check{\mu}}$, at least if μ has finite support. Thus the situation is rather different from what we observed on the connected real affine group Yet there is a μ-invariant probability measure ν on $\mathbb{R}$ as soon as $\Sigma\, a(g)\, \mu(g) < 0$, and the following inequality holds :

$$h(\mu) \geqslant - \Sigma \mu(g) \int_{\mathbb{R}} (\log \frac{dg^{-1} * \nu}{d\nu}(x))\, d\nu(x).$$

A natural question is : when is it an equality ?

BIBLIOGRAPHY

[1] (1972).: **AVEZ A.** - Entropie des groupes de type fini. C.R. Acad. Sc. Paris 275 A 1363-1366

[2] (1976) : **AVEZ A.** - Croissance des groupes de type fini et fonctions harmoniques. L.N. in Math. Springer n° 532, 35-49.

[3] (1970] : **AZENCOTT R.** - Espaces de Poisson des groupes localement compacts. L.N. in Math. Springer n° 148

[4] (1986) : **DERRIENNIC Y.** - Entropie, théorèmes limite et marche aléatoire. L.N. in Math. Springer (to appear). Probability measures on groups VIII

[5] (1978) : **ÉLIE L.** - Fonctions harmoniques positives sur le groupe affine. L.N. in Math. Springer n° 706, 96-110.

[6] (1983) : **KAIMANOVICH V.A. et VERSHIK A.M.** - Random walks on discrete groups : boundary and entropy. The annals of Proba. Vol. 11, n° 3, 457-490.

[7] (1964) : PINSKER M.S. - Information and information stability of random variables and processes. HOLDEN-DAY.

-ooooooooooooooo-

ESTIMATION OF THE SIGNAL'S APPEARING MOMENT

Rositsa Dodunekova

Sofia

Key words: signal, nose, minimax estimator

ABSTRACT

In a white Gaussian noise a signal appears at a random moment with an unknown distribution function. The linear minimax estimator of the moment of signal's appearing and the linear minimax risk have been found and investigated.

1. Let $\eta(t)$, $0 \leq t \leq T$, be a random process defined by

$$(1) \qquad \begin{aligned} d\eta(t) &= \Theta(t,\omega)dt + \varepsilon dW(t), \\ \eta(0) &\equiv 0. \end{aligned}$$

Here $\Theta(t,\omega) = I_{\{\Theta(\omega) \geqslant t\}}(\omega) = \begin{cases} 1, & \text{if } t \geq \Theta(\omega) \\ 0, & \text{if } t < \Theta(\omega), \end{cases}$

$\Theta(\omega)$ is a random variable with values in $[0,T]$ and such that it's distribution function is unknown, $W(t)$ is a standard Wiener process, independent on Θ, and ε is a given positive constant.

The following problem is considered in this paper: To estimate (in some reasonable sense) the value of the random variable Θ, using the observation $\eta(t)$, $0 \leq t \leq T$.

Let us note, that the model (1) describes the following real situation. In a white Gaussian noise with an intensity ε a unit signal appears at a random moment. The additive sum, noise + signal, is integrated by an integrator and the result goes to the output, which is observable. Having the observation (a continuous trajectory) we have to estimate the moment of the appearing of the signal.

There is a lot of papers, dealing with problems close to the one formulated here (see, e.g. references) and they all differ in the a priori information about Θ.

As in our case the distribution of Θ is unknown, it seems quite natural to apply the minimax approach.

We are interested in the class M of all linear estimators of the form

(2) $$\hat{\Theta} = \int_0^T l(t)d\eta(t) - \alpha,$$

where the weight function $l(t)$ belongs to the space $L_2[0,T]$ and α is a real constant.

D e f i n i t i o n 1. The linear minimax (quadratic) risk Δ^* is defined by

(3) $$\Delta^* = \inf_{\hat{\Theta} \in M} \sup_{\Theta} E\{|\hat{\Theta} - \Theta|^2\}$$

Here sup is taken over all the random variables Θ with values in $[0,T]$.

D e f i n i t i o n 2. The linear estimator Θ^* is called minimax in M, if for every estimator $\hat{\Theta} \in M$ it holds

$$\sup_{\Theta} E\{|\hat{\Theta}-\Theta|^2\} \geq \sup_{\Theta} E\{|\Theta^*-\Theta|^2\},$$

i.e., Θ^* establishes inf in (3).

Our purpose now is to find Θ^* and Δ^* and to investigate their properties.

2. It can be shown, that if $\hat{\Theta}$ has the form (2), then

$$E\{|\hat{\Theta}-\Theta|^2\} = \int_{0^-}^{T} (\int_{t}^{T} l(s)ds-t-\alpha)^2 dF_{\Theta}(t)+\varepsilon^2||l||^2,$$

where $F_{\Theta}(t)$ denotes the distribution function of Θ.
From this and (3) it follows, that

$$(4) \qquad \Delta^* = \inf_{l,\alpha} \sup_{F}\{\int_{0^-}^{T} (\int_{t}^{T} l(s)ds-t-\alpha)^2 dF(t) + \varepsilon^2||l||^2\},$$

where sup is taken over all the distribution functions with a support in $[0,T]$ and inf is taken over $l\ L_2[0,T], \alpha\in R$.

Let $t(l,\alpha)$ be a point of the absolute maximum in $[0,T]$ of the function

$$(\int_{t}^{T} l(s)ds-t-\alpha)^2.$$

It is obvious that sup in the right hand side of (4) is given by the function

$$F_{l,\alpha}(t) = \begin{cases} 0, & \text{if } t < t(l,\alpha) \\ 1, & \text{if } \geq t(l,\alpha) \end{cases}$$

and then the finding of Θ^* and Δ^* has reduced to the following variation problem:

$$(5) \quad \varepsilon^2\|l\|^2+(\int_{t(l,\alpha)}^{T} l(s)ds - t(l,\alpha) - \alpha)^2 \to \inf_{l\in L_2[0,T],\alpha\in R}.$$

Here we furmulate the main results of this paper.

T h e o r e m 1. The solution of (5) is given by

$$l^*(t) \equiv -\frac{T}{T+4\varepsilon^2}, \quad \alpha^* = -\frac{T^2+2T\varepsilon^2}{T+4\varepsilon^2}.$$

T h e o r e m 2. The linear minimax estimator and the linear minimax risk in problem (1) are respectively

(6) $$\Theta^* = -\frac{T}{T+4\varepsilon^2}\eta(T) + \frac{T^2+4T\varepsilon^2}{T+4\varepsilon^2},$$

$$\Delta^* = \frac{\varepsilon^2 T^2}{T+4\varepsilon^2}.$$

3. a) From (6) it is seen, that the linear minimax estimator depends only on the terminal value $\eta(T)$. Actually, from (1) it is clear, that the "weights" on the trajectory $\eta(t)$ given by the weight function $l(t)$ ought to be one and the same for every moment $t < \Theta$. The same is valid for $t \geq \Theta$. And as for Θ we in fact have no a priori information, this is reached with $l(t) = \text{const}$.

b) It can be checked, that if the intensity $\varepsilon \to 0$, then $\Theta^* \to \Theta$ a.s. and in a square mean. When $\varepsilon = 0$ (no noise) then $\Theta^* = \Theta$.

c) Let $T \to \infty$ and $\varepsilon = T^{-\alpha}$, $\alpha > 0$. In this case

$$\Delta^* \underset{T\to\infty}{\sim} T^{2(\frac{1}{2}-\alpha)}$$

Hence if $\alpha > \frac{1}{2}$ (i.e., $\varepsilon \leq \frac{1}{\sqrt{T}}$), then $\Delta^* \underset{T\to\infty}{\to} 0$

and $\Theta^* \to \Theta$ a.s. and in a square mean.

REFERENCES

Shiryaev A.N. (1963): On optimum methods in quickest detection problems. In: Theory Prob. Appl., 8, 1, 22-46.

Ибрагимов И.А.
Хасьминский Р.З. (1979): Асимптотическая теория оценивания. Наука, Москва.

Вострикова Л.Ю. (1981): Обнаружение "разладки" винеровского процесса. Теория вероятн. и ее примен., 26, 2, 362-368.

Institute of Mathematics
Bulgarian Academy of Sciences
P.O.B. 373
Sofia

SUFFICIENT OPTIMALITY CONDITIONS FOR SEMI-MARKOV DECISION PROCESSES WITH INCOMPLETE STATE-INFORMATION: UNDISCOUNTED CASE

Maria Drăguţ
Bucharest

Key words: sufficient statistics, finitely transient policies, piecewise monotonic transformations

ABSTRACT

This paper presents a particular case of a Bayesian dynamic controlled model, namely an infinite-horizon partially observed one, where the core process is a controlled finite state space, discrete-time, semi-Markov process.

We assume that the times of the control reset and the noise corrupted observations of the core process occur at times of core process transitions. The control employed each time of the core process transitions is allowed to be functionally dependent on the sample path of the core process only through the history of the corrupted observations.

Based on the construction of a sufficient statistic, the reduction of the considered controlled models with incomplete-information to controlled models with completely state-information was established and conditions for the average cost optimality criterion were stated in Drăguţ (1983).

In order to outline algorithms to calculate the optimal control policy and optimal payoff functions, new conditions must be imposed on the state structure of the partially observed semi-Markov process.

DEFINITION AND NOTATION

In this section we develop the definitions and notation for a class of semi-Markov decision processes with incomplete state information SMDPi in a way similar to Ross (1970), Drăguț (1981).

A Borel set X is a Borel subset of a complete separable metric space. We denote by P(X) the set of all probability measures on X and by F(X) the set of all real valued, bounded Borel measurable function on X.

If X and Y are non-empty Borel sets, the set of all conditional probabilities on Y given X is denoted by Q(Y/X) and the Cartesian product of X and Y is denoted by XY.

A SMDPi can be written in the following form: $(M, S, N, A, q, r, q_0, \bar{\tau})$ where :

1) M, S, A are finite sets: $N=\{0,1,\dots\}$; $S=\{1,\dots,N_S\}$; MN is the observed state space, S is the concealed state space, A is the action space.

2) $q \in Q(MSN/MSNA)$; $q(m_n, s_n, \tau'_n, a_n; m_{n+1}, s_{n+1}, \tau'_{n+1}) = k(s_n, a_n; s_{n+1}) \cdot \tau(s_n, a_n, s_{n+1}; \tau'_{n+1}) \cdot l(s_{n+1}; m_{n+1})$

where $k \in Q(S/SA)$, $\tau \in Q(N/SAS)$ and $l \in Q(M/S)$

3) $q_0 \in P(MSN)$

4) $r \in F(SA)$; $r(s_n, a_n) = \sum_{m_n \in M} \sum_{s_{n+1} \in S} l(s_n; m_n) \cdot k(s_n, a_n; s_{n+1}) \cdot$

$\left[c(s_n, m_n, a_n, s_{n+1}) + d(s_n, a_n, m_n) \cdot \sum_{\tau'_{n+1} \in N} \tau'_{n+1} \tau(s_n, a_n, s_{n+1}; \tau'_{n+1})\right]$

where $c \in F(SMAS)$; $d \in F(SAM)$

5) $\bar{\tau} \in F(SA)$

$\bar{\tau}(s_n, a_n) = \sum_{s_{n+1} \in S} k(s_n, a_n; s_{n+1}) \sum_{\tau'_{n+1} \in N} \tau'_{n+1} \cdot \tau(s_n, a_n, s_{n+1}; \tau'_{n+1})$

CONDITION 1. $\tau(s, a, s'; 0) = 0$ for all $(s, a, s') \in SAS$ and $q_0(m_0, s_0, \tau'_0) = 0$ for all $(m_0, s_0, \tau'_0) \in MSN$ with $\tau'_0 \neq 0$.

A semi-Markov decision process with complete state information (SMDPc) can be written in the form $((YN, \mathcal{Y}), (A, \mathcal{A}), p'_0, p', \tau, s)$ (Ross (1970)) where:

1) Y, A are Borel sets and $\mathcal{Y}, \mathcal{A}$ are the corresponding σ-fields of Borel subsets : $N=\{0,1,\dots\}$

2) $p'_0 \in P(Y)$, $p' \in Q(Y/YA)$, $\tau \in Q(N/YAY)$ and $s \in F(NYA)$

Let's consider a SMDPc defined by $((YN, \mathcal{Y}), (A, \mathcal{A}), p'_0, p'_0, \tau, s)$.

DEFINITIONS 1.A policy π' for a SMDPc is a sequence of transitions probabilities $\pi'_n \in Q\,(A/(YAN)^{n-1}YN)$ with the property $\pi'_n(y_0,t_0,a_0,\ldots,y_n,t_n;A)=1$.

Let's denote by Δ' the set of all policies for a SMDPc. $\pi'\in\Delta'$ is a Markov policy if $\pi'_n(y_0,t_0,a_0,\ldots,y_n,t_n;\cdot)=\pi'_n(y_n;\cdot)$. $\pi'\in\Delta'$ is a stationary policy if it is a Markov policy and $\pi'_n=\pi'$ for all $n\in N$ and there exists a measurable map $f:Y\to A$ with the property $\pi'(y;\{f(y)\})=1$.

Let's consider a SMDPi defined by (M,S,N,A,q_0,q,r,τ).

DEFINITION 2.A policy π' for a SMDPi is a sequence of transitions probabilities $\pi_n\in Q\,(A/(MAN)^{n-1}MN)$ with the property $\pi_n(m_0,t_0,a_0,\ldots,m_n,t_n;A)=1$.

Let's denote by Δ the set of all policies for a SMDPi. $\pi\in\Delta$ is an admissible policy if there exists a sequence of functions $[\,y(t);t=0,1,\ldots]$ $y(t)=y(t_n)$, $t_n\le t<t_{n+1}$, $y(t_n):H_n\to A,[y(t);t=0,1,\ldots]$ is a sufficient statistic for SMDPi, where $H_n=\{h_n=[p_0,a_0,t_1,m_1,\ldots,a_{n-1},t_{n-1},a_n]$, $p_0\in P(S),a_i\in A,t_i\in N,m_i\in M\}$.

$\pi=f\in\Delta$ is a stationary policy if there exists a measurable function $f:H_n\to A$ with the property $\pi_n(h_n;\{f(h_n)\})=1$, $(\forall)\,h_n\in H_n$.

According to the theorem of Kolmogorov, the probability measure $P_\pi=q_0\otimes\bigotimes_{n\in N}(\pi_n\otimes q)$ can be defined on the space $(MNSA)^N$ for all $\pi\in\Delta$.

We consider the average expected cost: $G_\pi=\lim_{n\to\infty}\dfrac{\sum_{i=0}^{n-1}E_{P_\pi}(r_i)}{\sum_{i=0}^{n-1}E_{P_\pi}(\tau_i)}$

and look for plans which minimize G_π.

For any $\pi'\in\Delta'$ we can also define $Q_\pi=p'_0\otimes\bigotimes_{n\in N}(\pi'_n\otimes p')$ which is a probability measure on $(YNA)^N$.

CONDITIONS FOR OPTIMALITY

Let's consider the process $[y(t);t=0,1,\ldots]$,$y(t)=y(t_n)$ for $t_n\le t<t_{n+1}$ where $y(t_n)=\tilde{y}(n,h_n)=[\tilde{y}_1(n,h_n),\ldots,\tilde{y}_{N_S}(n,h_n)]$, $\tilde{y}_i(n,h_n)=P[s(t)=i\,/\,h_n,t=t_n]$, $i\in S=\{1,2,\ldots,N_S\}$ and where by $[\,s(t);t=0,1,\ldots]$ is denoted the core process.

THEOREM 1.(Drăguţ (1983)).A SMDPi defined by (M,S,N,A,q_0,q,r,τ) can be transformed into an equivalent in the sense of Rhenius (Rhenius (1974)) SMDPc defined by $((YN,\mathcal{Y}),(A,\mathcal{A}),\ p_0,\ p,\ \overline{\tau},\ s)$ with respect to average cost criterion ($Y=(y_i,i\in S,\ 0\le y_i\le 1,\ \sum_{i\in S} y_i=1)$.

REMARK. From Th. 1 we have that

$$\tilde{y}_j(n+1,h_{n+1})=P[s(t_{n+1})=j/z(t_{n+1})=\theta,\ t_{n+1}-t_n=m,\ h_n,\ a(t_n)=a]=$$
$$=T_j(\theta,m,\tilde{y}(n,h_n),a)=\Big[\sum_{s\in S} k(s,a;j)\cdot\tau(s,a,j;m)\cdot \ell(j;\theta)y_s\Big]/V(\theta,m,y,a);$$
$$V(\theta,m,y,a)=\sum_{i\in S}\sum_{j\in S} k(s,a;j)\,\tau(s,a,j;m)\,\ell(j;\theta)y_s$$

Let's denote by $T(\theta,m,y,a)=[T_1(\theta,m,y,a),\dots,T_{N_S}(\theta,m,y,a)]$

THEOREM 2. (Drăguț (1983), White (1976)). The process $[y(t);t=0,1,\dots]$ is sufficient statistic in the sense of Hinderer (Hinderer (1970)) for the SMDPi defined by (M,S,N,A,q_0,q,r,τ).

THEOREM 3. (Drăguț (1983)). Let $\pi'\in\Delta'$ be a stationary policy for the equivalent SMDPc of Th. 1. Then $[y(t);t=0,1,\dots]$ is a semi-Markov process and: $P[y_n\in W,\ t_{n+1}-t_n=m/\mathcal{B}_n]=$
$=P[y_n\in W,\ t_{n+1}-t_n=m/y_{n-1}=y]=\sum_{\theta\in L} V(\theta,m,y,\alpha)$
where $L=\{l\in M\ /\ T(l,m,y,a)\in W\}$, $\mathcal{B}_n$ is the σ-fields generated by $\{(y_0,t_0=0),(y_1,t_1-t_0),\dots,(y_n,t_n-t_{n-1})\}$ and W is a Borel subset of Y.

CONDITION 2. $E_{Q_\pi}[T/\ y_0=y]<\infty$ for all $\pi'\in\Delta'$.

CONDITION 3. There exists a real number M_0 such that $r(s,a)\le$ $\le M_0\cdot\overline{\tau}(s,a)$.

CONDITION 4. There exists $M_1\in N^*$ such that $\tau_{ij}(m,a)=0$ for all $m\ge M_1$.

The controller is assumed to know times of the core process transitions t_n, and also the realizations of the random variable $z(t_n)$ (the process $[z(t);t=0,1,\dots]$; $z(t)=z(t_n)$ for all $t_n\le t<t_{n+1}$ is called the observable process. Let's denote by $u(t)$ the control vector for the interval $(t_n,t_{n+1}]$. Associated with the core process transition times and the observation process, there are known time independent conditional probabilities

$$c_{ij}(m,a)=P[t_{n+1}-t_n=m,\ s(t_{n+1})=j/s(t)=i,\ u(t_n)=a]$$

and $q_{jk}(a)=P[z(t_{n+1})=k\ /s(t_{n+1})=j;\ u(t_n)=a]$

for all i,j,k and a. It is convenient to define the matrices

$$C(m,a)=\{c_{ij}(m,a)\},\quad R_k(a)=diag\{q_{jk}(a)\}$$

i.e. the jth main diagonal term of $R_K(a)$ is $q_{jK}(a)$ and all off-diagonal terms are 0 and $D_K(m,a)=C(m,a)R_K(a)$.

Let's denote by $A=\max_{i,j,m,a} c_{ij}(m,a)$; $B=\min_{i,j,m,a} c_{ij}(m,a)$; $C=\max_{j,K} q_{jK}$; $D=\min_{j,K} q_{jK}$.

CONDITION 5. For $\lambda = \frac{BD}{AC}$ we have :

$$\sum_{j\in S}\sum_{m\in N}\max_{(a,a',i,k)} | c_{ij}(m,a)-\lambda c_{kj}(m,a') | = \alpha_0 < \lambda$$

THEOREM 4. (see Drăguţ (1981)). Suppose Conditions 1→5 are verified. Then there exists a bounded continuous function f(y) and a constant g satisfying:

$$(1)\quad f(y)=\min_{a\in A}\{r(y,a)+\sum_{m=1}^{M_1}\sum_{k\in M} V(k,m,y,a) f(T(k,m,y,a)) - g\cdot\sigma(y,a)\}$$

and for any stationary policy π^* such that where in state y selects an action minimizing the right hand side of (1) we have $G_{\pi^*}(y)=g=\min_{\pi\in\Delta} G_\pi(y)$ for all $y\in Y$.

FINITELY TRANSIENT POLICIES AND OPTIMALITY CONDITIONS

In order to use Howard-like policy iteration algorithm developed by Sondik (1978) and extended in Drăguţ (1983) to SMDPi, there exists two difficulties. One of them, the calculation of the total expected cost for a given policy is overcome by using finitely transient policies. Sufficient conditions for the existence of the gain for every finitely transient policy and also for the existence of an ε -optimal gain in the class of finitely transient stationary policies are obtained.

DEFINITION 3. A partition $\{Y_i\}_{i=1}^m$ of Y is called simple if each Y_i is a convex polyedral set, where a convex polyedral set is the solution set of a finite system of linear inequalities:

$$Y_i=\{y\in Y \mid \alpha_{ij}\cdot y<0;\ i=1,\dots,n_i\},\quad \alpha_{ij}\in R^{N_S\cdot M_1}$$

and $\alpha_{ij}\cdot y$ is the inner product of α_{ij} and y.

DEFINITION 4. A stationary policy $\pi\in\Delta$ is called simple with respect to a simple partition $\{Y_i\}_{i=1}^m$ if $\pi(y)=a_i$ on Y_i , i=1,2,..,m.

DEFINITION 5. A real valued function C on Y is called piecewise linear if $C(y)=\alpha_i y$ on Y_i, i=1,..,m and $\alpha_i\in R^{N_S\cdot M_1}$.

For a simple policy define $D^K=\bigcup_{\theta,m}\{y/T(\theta,m,y,\pi(y))\in D^{K-1}\}$ where $D_0=\bigcup_i \Delta_i^a=\bigcup_{i,j}\{y\in Y/\alpha_{ij}y=0\}$

which forms the boundary set of the partition $\{Y_i\}_{i=1}^{m}$ corresponding to the simple policy Π.

DEFINITION 6. A simple policy is called finitely transient if there exists an integer $K<\infty$ such that $T(V_j^K/\theta,m,\pi) \subset V_{\nu(j,\theta,m)}^K$ for all $(\theta,m)\in M\times\{0,\dots,M_1\}$ where $T(V/\theta,m,\pi)=\{T(\theta,m,y,\pi(y)),\ y\in V\}$, $\nu(j,\theta,m)$ is the index of the set containing $T(\theta,m,y,\pi(y))$ for $y\in V_{jK}^K$ and $V^K=\{V_j^K\}_{j=1}^{m'}$ is the collections of sets whose boundaries are $\bigcup_{\ell=0} D^\ell$.

THEOREM 5. If Π is a finitely transient policy of index K for the SMDPc defined by $((YN), A, p_0', p', \tau, s)$ and $Y^K=\{Y_i^K\}_{i=1}^{L}$ is its associated partition, then:

a) there exists a map $\nu:\{1,2,\dots,L\}\times\{1,2,\dots,M_1\}\times M\rightarrow\{1,2,\dots,L\}$ with the property that $T(\theta,m,y,\pi(y))\in Y_{\nu(j,\theta,m)}^K$ for all $y\in Y_j^K$;

b) for all $y\in Y_j^K$ the expected reward for n-units of time associated with policy Π, denoted by $C_{n\Pi}(y)$ equals $y\alpha_j^n$, where α_j^n is the unique bounded solution to the vector equation:

$$\alpha_j^n = g(a_j)+\sum_{m=1}^{M_1}\sum_{k\in M} D(k,m,a_j)\,\alpha_{\nu}^{n}(j,k,m) \tag{2}$$

where a_j and α_j^n are the action and respectively the vector associated with Y_j^K;

c) for every transient policy there exists an unique finite state space equivalent semi-Markov process;

d) the gain g_Π associated with a finitely transient policy, if it exists, satisfies the following vectorial equation:

$$g_\Pi\bar{\tau}+\bar{\alpha}=\bar{r}+\sum_{m=1}^{M_1}\bar{D}(m)\,\bar{\alpha} \tag{3}$$

where $\bar{\alpha}$, $\bar{r}$, $\bar{D}(m)$ are quantities associated with the semi-Markov process of c).

PROOF. For details see Drăguţ (1983). a) and b) easily follow from the definition and the properties of a finitely transient policy. In order to demonstrate c) we must introduce the $LN_{S'}$-dimensional vectors $\bar{\alpha}$, $\bar{\tau}$ and $\bar{g}$, where $\bar{\alpha}=[\alpha_1,\dots,\alpha_L]^T$; $\bar{g}=[g(\pi(1)),\dots,g(\pi(L))]^T$, $\bar{\tau}=[\bar{\tau}(\pi(1)),\dots,\bar{\tau}(\pi(L))]$ where $\pi(j)$ is the action for every $y\in Y_j^K\in\{Y_i^K\}_{i=1}^{L}$

Let's define $\bar{D}(m)=\sum_{k\in M} D(k,m,\pi(j))$ as follows:

$$\overline{D}_{ij}(m) = \begin{cases} \sum_{k\in M} D(k,m,\Pi(j)) & \text{for all k with the property:} \\ 0 & \text{otherwise} \end{cases}$$

where $i,j \in \{1,\dots,L\}$. Taking limit as $n \to \infty$ in 2) and supposing that g_Π exists, equation 3) is easily obtained. q.e.d.

For the convenience of the reader, in the following, we shall treat only the case $N_{S'}=2$. Similar results can be obtained for the case $N_S \geq 2$.

Let's denote by $E = M\cdot\{0,1,\dots,M_1\}\cdot A$. For each $e \in E$, let's consider the map $T_e(\cdot) = T(\theta,m,\cdot,a)$ as in Th.1 which maps $I=[0,1]$ onto the closed interval $I_e \subset I$. For any $e^{(n)}=(e_1,\dots,e_n) \in E^n$, $n \geq 1$, let's put $T_{e^{(n)}} = T_{e^1} \circ \dots \circ T_{e^n}$ (denote composition of functions). Clearly $T_{e^{(n)}}$ maps I onto the closed interval $I_{e^{(n)}}$ with endpoints $T_{e^{(n)}}(0)$ and $T_{e^{(n)}}(1)$.

Let's consider the usual Markov process X_t associated with the semi-Markov process Y_t defined in Th.3. For any stationary policy Π let $q^n_\Pi(W,y)$ with $y \in Y$ and $W \in \mathcal{Y}$ be the n-step transition probability for the Markov chain X_t.

CONDITION 6. $\sup_{x\in I} |T'_{e^{(n)}}(x)| / \inf_{x\in I} |T'_{e^{(n)}}(x)| \leq C$

for any $e^{(n)} \in E^n$, $n \geq 1$, where C is a constant ≥ 1.

THEOREM 6. Conditions 1→6 are sufficient for the existence of a unique bounded solution to the functional equation (1) in Th.4 and also for each finitely transient policy Π there exists a constant g_Π as a gain.

PROOF. The first assertion is from Th.4. For the second one, we observe that T_e has an inverse. Let's denote it by U_e. It maps I_e onto I. We can construct a continuous map $U: G \to I$ where G is an open set with $\lambda(G)=1$ (λ is the Lebesgue measure. We also observe that $\bigcup_{e\in E} I_e \supset G$ and for any $e \in E$ the set $I_e \cap (I-G)$ consists exactly of the endpoints of I_e and that the restriction of U to $I_e \cap G$ is strictly monotonic and extends to a $C^1(I)$ function on I ($C^1(I)$ means the set of all real valued functions defined on I which are continuous and have continuous derivatives). So a $C^1(I)$ monotonic piecewise transformation can be constructed. Under Condition 6 (see Iosifescu), there exists a unique absolutely continuous (with respect to λ) U-invariant probability measure ρ. The probability density $r = d\rho/d\lambda$ of ρ satisfies almost everywhere on I the inequalities $1/c \leqq r \leqq c$ and r and U are invariants amounts to the equation Pr=r almost everywhere in I when P is the Frobenius Perron operator associated with h defined as: $Ph(x) = \sum_{e \in E} |T_e(x)| \, h(T_e(x))$, $x \in I$

Transforming the SMDPC into an equivalent Markov decision process (Ross(1970)), for the equivalent Markov decision process the Conditions $A_1 \to A_5$ of Th.2 in Tijms are satisfied.
The considered Markov chain associated with a finitely transient policy π satisfies the so-called Doeblin condition which assures the existence of a constant g_π as a gain. q.e.d.

REFERENCES

Drăguţ M. (1981): Non-discounted semi-Markov decision processes with incomplete state-information. In: Proc. of the Sixth Brasov Conference, 1979, Publishing House of the Academy of RSR, 403-411.

(1983): On semi-Markov decision processes with incomplete state-information. Doctoral Thesis, Centre of Mathematical Statistics, Bucharest, Romania.

Hinderer K. (1970): Foundations of non-stationary dynamic programming with discrete time parameter. In: Lecture Notes in Oper. Research and Mathem. Systems, vol. 33, Springer-Verlag.

Iosifescu M.: On invariant densities for piecewise monotonic transformation. To be published.

Kurano M. (1976): On the existence of an optimal stationary I-policy in non-discounted Markovian decision process with incomplete state-information. Bull. Math. Statist. 17, No. 3-4, 75-81.

Rhenius D. (1974): Incomplete information in Markov decision models. Ann. Statist. 2, No. 2, 1327-1334.

Ross S.M. (1970): Average cost semi-Markov decision processes. J. Appl. Probab. 7, No. 3, 649-656.

Sondik E. (1978): The optimal control of partially observable Markov processes over the infinite horizon: discounted costs. Oper. Res. 26, No. 2, 282-304.

Tijms H.C.: On dynamic programming with arbitrary state space, compact action space and the average return as criterion. To be published.

White C.C. (1976): Procedures for the solution of a finite horizon partially observed semi-Markov optimization problems. Oper. Res. 24, No. 2, 348-358.

Central Mathematical Institute
Centre of Mathematical Statistics
174 Stirbei Voda Street
77 104 Bucharest, 6
Romania

ON OPTIMIZATION WITH RANDOM SEEKING*

Ewa Dudek-Dyduch, Tadeusz Dyduch

Kraków

Key words: iterative algorithm, non-deterministic procedure, two-level optimization

ABSTRACT

Two level iterative optimization algorithms are considered. The random seeking in such algorithms is discussed. Finally an example is given which shows in detail the way of improvement of the algorithm effectiveness by modifying the probability distributions utilized in seeking.

INTRODUCTION

Random elements arise in the most of optimization algorithms. While the effectiveness of algorithms is compared then the random set of instances is defined. Random graphs described in Karp (1976) and utilized to compare the abilities of graph algorithms for finding Hamilton circuts or maximum independent set of nodes are good example of it.

Moreover, when an algorithm gives approximately optimal solution, it is often started many times for the same data from different starting points chosen in random. It is done in order to avoid an influence of starting point on a solution.

At last there are many algorithms based on probabilistic rules as for example ones based on Monte Carlo methods.

* The paper has been written as a part of subject "Resortowy Program Badań Podstawowych R.P.I.02."

In this paper we consider some iterative optimization algorithms for both continuous and discrete problems, named two level algorithms with random seeking.

TWO LEVEL ITERATIVE OPTIMIZATION ALGORITHM

An iterative optimization procedure is necessary when the calculus problems are very complicated or no solving method is known except trial-and-error method. From the other hand, it is applicable in a case of mixed, continuous-discrete optimization problem, when that problem, to say in general

$$\text{minimize} \quad f(x) : X \to R$$

is regular, that is in X set a metric $\varrho(x_i,x_j)$ is defined and following relation occurs

$$\forall \varepsilon > 0 \quad \exists d(\varepsilon) > 0 \quad \exists x_i, x_j \in X : \varrho(x_i,x_j) < \varepsilon \Rightarrow |f(x_i) - f(x_j)| \leq d(\varepsilon)$$

The two level iterative algorithm may be formally described as follows. The aim

$$\text{find } \hat{x} \text{ such that } f(\hat{x}) = \min_{x \in X} f(x)$$

is achieved by means of sequence of iterations. Let n denotes the number of iteration and $m(n)$, $w(n)$ - values computed in n-th iteration.

$$f(\hat{x}) \cong \lim_{n \to \infty} f(m(n), w(n)) = \lim_{n \to \infty} \left\{ \min_{m \in M} \left[\min_{w \in W_m} f(m(n-1), w) \right] \right\}$$

where $(m,w) \in M \times W = X$, $W_m \subset W$ is a subset determined by fixed $m(n-1)$.

The more detail description of above problem has been given in Dudek-Dyduch (1983).

EVOLUTIONARY ALGORITHM

In Wala (1979) the idea of evolutionary algorithm is given. The algorithm belongs to the class characterized above. Each iteration consists of three steps:

1) a step from the last achieved base point in direction chosen at random. (So called upper level of the algorithm).
2) finding a local optimum inside the defined neighbourhood of the randomly chosen point. (So called lower level of the algorithm).

3) moving the base point to the point found in 2) step, when a value of performance index is better than previous one.

The used name - evolutionary algorithm - seems to be very right, in spite of its similarity to equations of evolution, which may mislead mathematicians. Following Darvin's meaning of evolution of species, it is characterized by:

- stochastic disturbances in genetic code (genotype) given descendants from parents,
- competition of descendants in their ability to meet the terms of environment,
- greater amount of descendants from better adapted parents.

Similarity to the evolutionary algorithm is easy to see.

Nevertheless, it should be explained why random seeking is utilized in the upper level of the evolutionary algorithm. Deterministic procedures using, for example, the choice according to lexicographical order give random effects for different real problems. However in worse cases these procedures have a tendency to fall into loops (do not terminate). This disadvantage does not occur in algorithms with a random choice.

We should stress probabilistic assumptions under which some deterministic optimization algorithms work. Let us take the gradient method for continuous functions. A step made over a surface along a gradient computed in the starting point gives the best result with probability 0 . (If it is not a trivial case of plane) Nevertheless, if there is no additional data and a function changes not too fast, a gradient points the most perspective, in probabilistic meaning, direction of searching.

EXAMPLE OF EVOLUTIONARY ALGORITHM

The authors have worked out several evolutionary algorithms, utilizing following assumptions:

- optimization problem is regular,
- an evolutionary method is applicable, it is to say that assuming uniform distribution of random choices on the upper level the algorithm should reach the solution with probability 1 .

Let us consider the possibility of improvement of algorithms efficiency. It may be achieved by means of utilizing more efficient optimization procedures of the lower level, or by the modification of probability distributions of the upper level procedure, which should improve a frequency of successful steps. In order to modify the probability distribution the additional analysis of the lower level results must be done. It can be seen more detaily in the example of the algorithm. It has been developed by author Dudek-Dyduch (1984), partially based on the algorithm given in Kornai, Liptak (1965). It is an application of evolutionary algorithm in some case of discrete programming problem.

The problem is as follows:
find K set and $\bar{x}$, $\bar{y}$, $\bar{z}$ vectors which minimize

$$f = \bar{c}_2 \circ \bar{x} - \bar{c}_1 \circ \bar{y} + \bar{c}_3 \circ \bar{z}$$

under constrains

$$A\,\bar{z} + \bar{x} - \bar{y} = \bar{b}$$

$$z_i \leq h_i\, v_i \quad , \quad i \in J$$

$$\bar{q}(K) \circ \bar{v} \leq Q$$

$$\bar{x}\ ,\ \bar{y}\ ,\ \bar{z} \geq \bar{0}$$

where J - set of indexes of $\bar{z}$ vector elements,

$K \subset J$ - subset of J ,

$\bar{q}(K)$ - given vector function of K subsets,

$v_i = \begin{cases} 0 & \text{for } i \in J \setminus K \\ 1 & \text{for } i \in K \end{cases}$ - Boolean variables

Let us notice that if the subset K is fixed then the limitations for the linear programming task are fixed. These limitations should be determine in such a way to solution of the linear programming task be minimal.

The evolutionary algorithm involves random choice of subsets K(n) (by means of modifying previous ones) on the upper level and utilizes linear programming procedure with decomposition as it is described in Kornai, Liptak (1965) on the lower level. The modification of K sets is performed due to dual solutions obtained on the lower level.

Let us assume for a while that vectors $\bar{h}$ and $\bar{q}$ can change continuously. Thus you can estimate the gradient as follows:

$$\forall\, i \in J\ ; \quad \frac{\partial f}{\partial q_i} \simeq \frac{\partial f}{\partial h_i}\frac{dh_i}{dq_i} = \lambda_i \frac{dh_i}{dq_i} \simeq \lambda_i \frac{h_i}{q_{mi}} \stackrel{df}{=} e_i$$

where λ_i - a value of dual variable for i-th inequality,

h_i - a value of i-th constraint,

q_{mi} - a mean of $q_i(K)$ function.

Searching of a new subset $K(n+1)$ consists in replacing some part of $K(n)$ elements by other elements from $J \setminus K(n)$ set. In order to do it the vector $\bar{s}$ (defined below) is computed on a baiss of estimated gradient $\bar{e}$.

$$s_i(K(1)) = e_i(K(1))$$

$$s_i(K(n)) = (1 - r)\, s_i(K(n - 1)) + r\, e_i(K(n))$$

where $r \in (0,1)$ - given parameter,

$i \in J$

The new set $K(n+1)$ is generated as follows:

1. Probabilistic choice of numbers that are to be removed from $K(n)$ set. The probability that i-th element of $K(n)$ will stay in $K(n+1)$ is proportional to $s_i(K(n))$.
2. Probabilistic choice of new numbers which will belong to $K(n+1)$ The probability that i-th element of $J \setminus K(n)$ will belong to $K(n+1)$ is also directly proportional to $s_i(K(n))$.
3. Solving of linear programming task. Computing value of function f denoted as $f(K(n+1))$.
4. If $f(K(n+1))$ is better than previous ones then set $K(n+1)$ replaces $K(n)$ and algorithm goes to 1. If performance index did not improve, the sampling is repeated.

The searching is stopped after a fixed number of iterations, when the best value of performance index does not change.

r parameter is introduced to improve the stability of the algorithm.

CONCLUSION

An elasticity is a great advantage of the discussed algorithm. It gives a chance to fulfill different, even non-analitical constrains of the solved problem. On the other hand, it enables to take advantage of different ways of improving the performance of the algorithm. These ways may be suggested by experts or be any heuristics.

Unfortunately, heuristics introduce probabilistic dependencies that are extremaly difficult to analyze.

REFERENCES

Dudek-Dyduch E. (1983): On evolutionary algorithm. In: Proc. 1-st IASTED Symp. on Applied Informatics. Lille, 1983, vol.III, 37.

(1984): The evolutionary method for planning preparatory works in mines. Postępy Cybernetyki vol.4, 1984, 131-139.

Karp R. M. (1976): The probabilistic analysis of some combinatorial search algorithms. In: Algorithms and complexity. Directions and recent results. Ed. by Traub J.F. 1976, Academic Press,

Kornai J.,Liptak T. (1965): Two-level planning. Econometrica 1965, 141-163.

Wala K. (1979): Symulacyjne metody optymalizacji dyskretnych procesów produkcyjnych. (Simulation methods for optimization of discrete production processes). Zeszyty Nauk. AGH, Kraków 1979, automatyka z.21.

Institute of Automatic Control
University of Mining and Metallurgy
al.Mickiewicza 30
30-059 Kraków, Poland.

A NOTE ON STOCHASTIC APPROXIMATION USING ISOTONIC REGRESSION

Václav Dupač, Ulrich Herkenrath

Prague, Bonn

Key words: stochastic approximation, isotonic regression, delayed observation

ABSTRACT

Mukerjee's stochastic approximation by observations on a lattice using isotonic regression is shown to retain its convergence property even when observations are realized only after a time-delay.

INTRODUCTION

Mukerjee (1981) proposed a new stochastic approximation procedure for estimating the zero point (zero interval) of a regression function. His procedure requires fitting the sample isotonic regression at each stage and is thus, computationally, more time-consuming than standard stochastic approximation. On the other hand, it is as good as equivalent to standard procedures as to the convergence properties and, from the practical viewpoint, it is expected to behave more smoothly than the standard procedures do; especially, it is certainly less vulnerable to outlying observations.

In this note, we establish another positive feature of Mukerjee's method, namely, that it retains its convergence properties (under slightly modified assumptions) even when observations are realized only after a time-delay.

PROCEDURE

Assume that function $m:\mathbb{R}\to\mathbb{R}$ and reals $\gamma\le\delta$ satisfy

$$\sup_{x\le t} m(x) < 0 \quad \forall t<\gamma, \qquad \inf_{x\ge t} m(x) > 0 \quad \forall t>\delta;$$

$$m(\gamma)\le 0, \quad m(\delta)\ge 0, \quad m(x) = 0 \quad \forall \gamma<x<\delta \;;$$

γ and δ may be equal.

Let $L=(d_i=d_0+ih: i\in\mathbb{Z})$, for some $d_0\in\mathbb{R}$, $h>0$, be a discrete lattice.

We want to estimate $m^{-1}(0)$ by a point, using stochastic approximation on L. The estimate at time n will be denoted by θ_n.

One observation is taken at each time-instant $-K\le j<+\infty$. The initial observations are taken at fixed (nonrandom) points x_j, $-K\le j\le 0$; further observation points x_j, $1\le j<+\infty$, are determined sequentially by a rule described later.

The results of observations are $y_j=m(x_j)+z_j(x_j)$, where $(z_j(d),\ d\in L)_{-K\le j<+\infty}$ is a sequence of independent identically distributed random functions on L, the $z_j(d)$'s being centred and uniformly stochastically dominated by an integrable random variable; i.e.,

$$Ez_j(d) = 0, \qquad P(|z_j(d)|\ge t)\le P(|z|\ge t),$$

for all $-K\le j<+\infty$, $d\in L$, $t\ge 0$ and for some z with $E|z|<+\infty$. ($z_j(d)$ is the potential error of an observation of m, when taken at point d at time j; the distribution of $z_j(d)$ may depend on d, not on j.)

Observations are realized only after a time-delay. The delay t_j of an observation taken at time j equals 0, if the result of the observation becomes known before time j+1; it equals 1, if the result becomes known before time j+2 but not before j+1, etc.

The t_j's are assumed to be (integer valued) random variables, not necessarily independent; however, the independence between $(t_j)_{-K\le j<+\infty}$ and $(z_j(d), d\in L)_{-K\le j<+\infty}$ will be assumed.

Denoting by I(.) the indicator of the event in parentheses, we introduce, for $r,s\in L$, $r\le s$, and for $n\in\mathbb{N}$,

$$n(r,s) = \sum_{j=-K}^{n} I(r\le x_j\le s)\, I(j+t_j\le n),$$

the number of observations taken in the closed interval $[r,s]$, whose outcomes have been realized before time n+1. Denote by

$A_n(r,s)$ the average of these observations, i.e. put

$$A_n(r,s) = \begin{cases} \sum_{j=-K}^{n} y_j \, I(r \le x_j \le s) \, I(j+t_j \le n) \, / \, n(r,s), & \text{if } n(r,s)>0, \\ 0, & \text{if } n(r,s) = 0. \end{cases}$$

Denote by X_n the set of all points at which observations have been taken up to time n (irrespective of the time of their realization). For $n \in \mathbb{N}_0$ and $x \in X_n$ define

$$\text{(1)} \qquad m_n(x) = \max_{r \le x} \min_{s \ge x} A_n(r,s) .$$

If we restricted the domain of $m_n(x)$ to those observation points only, where observations have been realized before time n+1, then $m_n(x)$ would be the sample isotonic regression of the realized observations y on the corresponding points x. By (1), however, $m_n(x)$ is defined for the remaining $x \in X_n$ as well, its isotony property being automatically preserved.

Extend $m_n(x)$ to a continuous, piecewise linear function on $\mathbb{R}$, constant on the left of x_{nm} and on the right of x_{nM}, the minimal and maximal points in X_n. Further put

$$\theta_n = \begin{cases} x_{nm}-h, & \text{if } m_n(x) > 0 \text{ for all } x \in \mathbb{R}, \\ x_{nM}+h, & \text{if } m_n(x) < 0 \text{ for all } x \in \mathbb{R}, \\ \text{the middlepoint of the (possibly degenerate)} & \\ \text{interval } m_n^{-1}(0) \cap [x_{nm}, x_{nM}], & \text{otherwise.} \end{cases}$$

If $\theta_n \in L$, take it for x_{n+1}; if not, randomize for x_{n+1} between the two nearest lattice points, $\max(d \in L: d < \theta_n)$ and $\min(d \in L: d > \theta_n)$, with equal probability.

ASYMPTOTIC RESULT

Denote $d_m = \max(d \in L: m(d) < 0)$, $d_M = \min(d \in L: m(d) > 0)$.

THEOREM. We have

$P(d_m \le x_n \le d_M \text{ eventually}) = 1$ and $P(d_m \le \theta_n \le d_M \text{ eventually}) = 1$ for the described procedure.

Proof. Reorder the observation points x_j, observations y_j and error functions $z_j = (z_j(d), d \in L)$ increasingly according to time-instants $j+t_j$ of their realization; or, in case of equality, according to the time-instants j, when they were taken. Let the reordered points, observations and error functions be

$x_{j_k}, y_{j_k}, z_{j_k}, 1 \leq k < +\infty$; i.e., y_{j_1} is the observation that has realized as the first one, x_{j_1} the point at which it has been taken, etc. More formally, $(j_1, j_2, \ldots)$ is a random permutation of the sequence $(j)_{-K}^{\infty}$, defined by $j_1 = \min \operatorname{argmin}_j (j+t_j)$, $j_{k+1} = \min \operatorname{argmin}_{j \neq j_1, \ldots, j_k} (j+t_j)$, $k \geq 1$.

It is easy to show, that the sequence of random functions $(z_{j_k})_{1 \leq k < +\infty}$ possesses all the properties that have been assumed about the sequence $(z_j)_{-K \leq j < +\infty}$. Let us verify only the i.i.d. property: For Borel cylinders $A_1, \ldots, A_r$ in $\mathbb{R}^L$, we have

$$P(z_{j_1} \in A_1, \ldots, z_{j_r} \in A_r) =$$
$$= \sum_{a_1=-K}^{\infty} \cdots \sum_{a_r=-K}^{\infty} P(z_{a_1} \in A_1, \ldots, z_{a_r} \in A_r) P(j_1 = a_1, \ldots, j_r = a_r) = Q(A_1) \ldots Q(A_r),$$

where Q is the equidistribution (on $\mathcal{B}^L$) of the z_j's. Similarly,

$$P(z_{j_k} \in A_k) = \sum_a P(z_a \in A_k) P(j_k = a) = Q(A_k);$$

hence the i.i.d. property follows. We made use of the independence between $(j_k)_1^{\infty}$ and $(z_j)_{-K}^{\infty}$, which in turn follows from the independence between $(t_j)_{-K}^{\infty}$ and $(z_j)_{-K}^{\infty}$.

The observational point x_{j_1} belongs to the set $(x_{-K}, \ldots, x_0)$, as follows from the algorithm. Consider the observational point x_{j_k}, for some $k > 1$. It has been constructed by the algorithm by means of all the x_j, $j < j_k$, by observations that have already realized, i.e. by y_j, $j+t_j < j_k$, or, equivalently, by their errors $z_j(x_j)$, $j+t_j < j_k$, and finally, by the randomizing device, say R. The dependence on the x_j's reduces, however, to the dependence on the nonrandom initial values $x_{-K}, \ldots, x_0$, as all the others are determined by the initial ones, by the $z_j(x_j)$, $j+t_j < j_k$, and by R: an x_j, which follows after a time-instant at which no new realization occurred, remains equal to the preceding value x_{j-1}, or to its neighbour in L, according to R. The $z_j(x_j)$'s, $j+t_j < j_k$, are the $z_{j_1}(x_{j_1}), \ldots, z_{j_a}(x_{j_a})$, for some $a \leq k-1$. As the possibly remaining $z_{j_{a+1}}(x_{j_{a+1}}), \ldots, z_{j_{k-1}}(x_{j_{k-1}})$ do not influence the construction of x_{j_k}, we get finally, that x_{j_k} is measurable with respect to the σ-field $\mathcal{F}_{k-1} = \sigma(z_{j_1}(x_{j_1}), \ldots, z_{j_{k-1}}(x_{j_{k-1}}); R)$.

Taking into account the properties of random functions z_{j_k}, we can easily see, that the sequence $(z_{j_k}(x_{j_k}), \mathcal{F}_k)_{k=1}^{\infty}$ is a martingale difference sequence, satisfying conditions under which the strong law of large numbers applies (Loève (1955), Theor. 29.1E).

Let B be a fixed subset of lattice L. Consider a subsequence of $(z_{j_k}(x_{j_k}))_1^{\infty}$ consisting of those members for which $x_{j_k} \in B$ and completed by infinitely many zeros, if there is only a finite number of such x_{j_k}'s. According to the optional skipping theorem, see Doob (1953), p. 310, this is a martingale difference sequence as well. The strong law of large numbers applies again, which can be rewritten as

$$\sum_{j=-K}^{n} z_j(x_j) I(x_j \in B) I(j+t_j \leq n)/n_B \to 0 \text{ a.s. on } (n_B \to +\infty),$$

where $n_B = \sum_{j=-K}^{n} I(x_j \in B) I(j+t_j \leq n)$.

Starting from this point, the proof is (despite the slightly different meaning of symbols) formally identical with that given in Mukerjee (1981), the variant mentioned at the end of his Remark B. Hence, it will be omitted here.

ACKNOWLEDGEMENT

The authors acknowledge the support of their research by the Deutsche Forschungsgemeinschaft, SFB 72.

REFERENCES

DOOB, J.L. (1953): Stochastic Processes. Wiley, New York.

LOÈVE, M. (1955): Probability Theory. Van Nostrand, New York.

MUKERJEE, H.G. (1981): A stochastic approximation by observations on a discrete lattice using isotonic regression. Ann. Statist. 9, 1021-1025.

Dept. of Prob. & Math. Statist.
Charles University
Sokolovská 83, 18600 Prague 8
Czechoslovakia

Institut f. Angewandte Mathematik
der Universität Bonn
Wegelerstr. 6, 5300 Bonn 1
Federal Republic of Germany

POINT ESTIMATION IN CASE OF SMALL DATA SETS

Zdeněk Fabián

Praha

Key words: Point estimation, robust methods

ABSTRACT

Two new, promising methods of point estimation are briefly discussed and a relation between them is established.

INTRODUCTION

Current users of statistical methods are conscious nowadays of existence of some robust methods for the estimation of statistical characteristics of observed data. There are two frequent questions arising in connection with using robust estimators:

In which case to use them ?

What estimator is the most appropriate one ?

The answer to the first question seems to be quite easy: In any case.

The answer to the second question is unfortunately not easy at all. A number of point estimators have been developed, which are robust in the sense of insensitivity to outlier values and even to incorrectly assumed distribution function.

Robust estimation in "current use" (c.f. Andrews (1972), Huber (1981)), weighting outlier observations by means of, to a certain extent arbitrary weight functions, does not take into consideration any possible apriory information or assumption about the character of data. Due to the lack of appropriate criteria a comparison of quality" of different estimators is a difficult task. The problem of proper estimation is crucial particularly in case of small data sets, in which a construction of empirical distribution function losses a sense.

Recently, two new, promissing approaches to the estimation problem appeared. Both of them use inherent weights, which depends only on accepted assumptions about the data model (and, of course, on the data itself). In the first part of the paper both the methods are briefly discussed. In the second part, a relation between both the classes of estimators is established and some recomendations for the practical use are given.

2. DISTORTED MAXIMUM LIKELIHOOD ESTIMATORS

A broad class of D-estimators was developed by Vajda (1984a, b, c). The class includes various minimum distance estimators as well as some of well-known robust estimators in "current use".

Let $\mathcal{P}$ denotes the family of all probability distributions on a Borel space $(X,\mathcal{B})$, dominated by a measure λ. Vajda defined f-divergence of $P, Q \in \mathcal{P}$ with Randon-Nikodym densities p, q by

$$D_f(P,Q) = E_\lambda q f(p,q) \quad ,$$

where f is a real valued function continuous and convex on $(0,\infty)$, strictly convex at 1, $f(1)=0$. The special choise of the functions $f(u)=\alpha^{-1}(1-u^\alpha)$, $\alpha \in (0,1)$ yields D^α -divergences

$$D^\alpha(P,Q) = \alpha^{-1}(1-E_\lambda\, p^\alpha q^{1-\alpha}). \tag{2.1}$$

From the concept of f-divergence derived Vajda some classes of D-estimators: weak, normal, directed D-estimators and distorted maximum likelihood estimators (DMLE). We shall briefly discuse the last ones (Vajda (1986)).

Let Θ be an arbitrary parameter space, $\mathcal{P}_\theta = \{P_\theta : \theta \in \Theta\}$ parametrized subfamily of $\mathcal{P}$ dominated by a measure λ. Let $\mathrm{P} \in \mathcal{P}$ denotes subclass of all empirical distributions. By Θ^* we denote one-point compactification of Θ. Let $p_\theta = dP_\theta/d\lambda$ be positive, continuous and bounded on $\Theta \times X$ with $\lim p_\theta(x) = 0$ for $\theta \to \theta^*$ for every $x \in X$. Let $\dot{p}_\theta = (d/d\theta)\, p_\theta(x)$, $\ddot{p}_\theta = (d/d\theta)^T \dot{p}_\theta$ be continuous, $p_\theta^\alpha \cdot \ddot{p}_\theta / p_\theta$ bounded on int $\Theta \times X$ for $\alpha \in (0,1\rangle$.

If there exist ~~unique~~ solution $\theta^\alpha = \theta^\alpha(P)$ of the equation

$$E_p\,(f_\alpha\,(p_\theta)) = 0 \qquad (2.2)$$

on int Θ, where f_α is f-divergence then $T^\alpha(P) = \theta^\alpha(P)$ is DMLE.

It has been proved in Vajda (1986), that $T^\alpha(P)$ as well as corresponding influence curves and asymptotic variances are continuous at $\alpha = 0$. DMLE therefore posseses all the nice asymptotic properties of maximum likelihood estimates (strong consistency, asymptotic normality), except for the efficiency. From the other hand, DMLE has been shown to be robust in the sense of gross-error sensitivity and of limited sensitivity of asymptotic estimates to a violated assumption on generating probabilities P_θ.

By the use of D^α-divergence (2.1) equation (2.2) has the form

$$E_p\,\left(\frac{p_\theta'}{p_\theta}\,p_\theta^\alpha\right) = 0 \quad . \qquad (2.3)$$

Parameter α increasing from $\alpha = 0$ (maximum likelihood estimate) to $\alpha = 1$ (mean likelihood estimate) controls decreasing efficiency and increasing robustness of the estimate.

Given sample vector $\tilde{X} = (x_1,\dots,x_n) \in X^n$, the practical estimation consists in three steps:

1. Assumption on possible type of sample distribution P_θ.
2. Choice of appropriate probe value of α (using numerical experiments of Andrews et.al. (1972) Vajda recommended interval $\alpha \in \langle 0.1,\ 0.3\rangle$).
3. Solution of the system of equations

$$\sum_{i=1}^{n} \frac{1}{p_\theta(x_i)}\,\frac{\partial p_\theta(x_i)}{\partial \theta_j}\,p_\theta^\alpha(x_i) = 0,\ j = 1,\dots,k, \qquad (2.4)$$

where $\partial p_\theta(x_i)/\partial\theta_j = \partial p_\theta(x)/\partial\theta_j|_{x=x_i}$,

which gives the estimate $T^\alpha(P_\theta) = (\hat\theta^\alpha{}_1,\dots,\hat\theta^\alpha{}_k)$.

3. GNOSTICAL ESTIMATORS

Completely independently off concepts and mathematical tools of statistics developed Kovanic (1984a, b, c) his gnostical theory. He even refused the concept of random value dealing with "uncertainties" in data items. His theory is based on Riemannian geometrie approach and physical analogies. We shall follow some important steps of his reasoning, which lead to the estimator of the "parameter of location", which means some "center of mass" or "ideal value", denoted by z_o.

Starting from the observed finite set of positive-valued data $Z = (z_1,\ldots,z_n)$, Kovanic introduces model of one point of data in the form

$$z_j = z_o \exp(s\,\Omega_j) \qquad z_j, z_o, s \in R_+,\ \Omega_j \in R_1, \tag{3.1}$$

where s is a scale factor and $\Omega_j, (j=1,\ldots,n)$ describes influence of individual uncertainties. From the model he derived certain groups of quantifying and estimating transformations. By the use of the concept of "dissimilarity of events" he developed the measure of this dissimilarity in the form of the tensor

$$K_j^2 = \begin{bmatrix} f_j & h_j \\ h_j & f_j \end{bmatrix} ,$$

where

$$f_j = \frac{2}{\left(\frac{z_j}{z_o}\right)^{2/s} + \left(\frac{z_j}{z_o}\right)^{-2/s}} \tag{3.2}$$

$$h_j^2 = 1 - f_j^2 .$$

Values f_j ("fidelity") and h_j ("estimating irrelevance") playes the central role in Kovanic´s theory, having the meaning of weight and error in the data item. Data composition law, introduced as an axiom of the theory, claims, that K_c^2 of a "composite event" is a normalized result from the addition of individual K_j^2. Hovewer, as we noticed, for the estimation of z_o, Kovanic does not actually use of data composition law. In fact, he maximizes sums of some types of individual characteristies of data items. Of eight estimators proposed by Kovanic in (1984b) only last two are relevant. They have the form

$$\frac{d}{dz_o} \sum_{j=1}^{n} f_j^m = \frac{d}{dz_o} \sum_{j=1}^{n} f^m (z_j, z_o, \hat{s}) = 0 \tag{3.3}$$

where m = 1,2, $\hat{s}$ is an estimate of the scale parameter (or probe value) and f_j is given by (3.2).

Despite the actual simplicity of this part of the gnostic

theory leading to the estimation procedure (not apparent from the original papers), practical results of estimation of "ideal value" z_0 are comparable with results of estimation of the location parameter by means of robust statistical estimators in "current use", being sometimes (in case of gross outliers) better in the sense of smaller variance of the gnostic estimates (Kovanic, Novovičová (1986)).

4. RELATION BETWEEN GNOSTIC ESTIMATOR AND DMLE

Let (R_1, x, P_θ) be probability space, $\theta \varepsilon (R_1 \times (0, \infty))$. Let $p_\theta(x) = \frac{dP_\theta}{d\lambda}$ be Randon-Nikodym density in the form

$$p_\theta(x) = p_{(\mu,\sigma)}(x) = \frac{1}{\sqrt{2\pi}\,\sigma} \cosh^{-2}\left(\frac{x-\mu}{\sigma\sqrt{\pi/2}}\right) \qquad (4.1)$$

or, setting $s = \sqrt{2\pi}\,\sigma$

$$p_{(\mu,s)}(x) = \frac{1}{s} \cosh^{-2}\left(\frac{2(x-\mu)}{s}\right)$$

After substitution $z = \exp(x)$, $z_0 = \exp(\mu)$,

$$\frac{2(x-\mu)}{s} = \ln\left(\frac{z}{z_0}\right)^{2/s}$$

and

$$p_{(z_0,s)}(z) = \frac{1}{sz} \cosh^{-2}\left(\ln\left(\frac{z}{z_0}\right)^{2/s}\right) = \qquad (4.2)$$

$$= \frac{1}{sz}\left(\frac{2}{\left(\frac{z}{z_0}\right)^{2/s} + \left(\frac{z}{z_0}\right)^{-2/s}}\right)^2 ,$$

or, according to formula (3.2),

$$p_{(z_0,s)}(z) = \frac{1}{sz} f^2(z, z_0, z) .$$

Thus, the fidelity of one data item can be considered as likelihood function of the data item in the probability model with density in the form (4.2) (or, following Kovanic's procedure of exponencialising the data, by (4.1)).

Then equations for gnostic estimators (3.5) can be writen in the form

$$\sum_{i=1}^{n} \frac{d}{dz_0} p^{m/2}_{(z_0,s)} (z_i) = 0 ,$$

or

$$\sum_{i=1}^{n} \frac{p'_{(z_0,s)}(z_i)}{p_{(z_0,s)}(z_i)} p^{m/2}(z_i) = 0 ,$$

which is identical with formula (2.4) for DMLE for k=1, $\alpha=m/2$ = = {0.5, 1} and p_θ given by (4.2) (or, for "additive-error" data, by (4.1)).

We arrived at an interesting and, moreover, stimulating fact, that two quite different theories, apparently without any common feature, give the same formulae for point estimation, assigning the same, entirely differently motivated inherent weights to individual items of the data set (the former as the latest statistical result in the point estimation, the letter as a first attempt).

Remark: There is another density function, namely

$$\tilde{p}_{(z_0,s)} = f^2(z.z_0,s) ,$$

which is apparently used by Kovanic for practical purposes. Transformation in the "additive-error domain" leads to the density

$$\tilde{p}_{(\mu,\sigma)}(x) = \frac{1}{q(\mu,\sigma)} \cdot \frac{\exp(x)}{\cosh^2\left(\frac{x-\mu}{\sigma\sqrt{\pi/2}}\right)} ,$$

which has no "location-scale" structure and is highly nonsymetric in favour of positive valued errors. It may compensate in some cases influence the too high robustness of gnostical estimators.

5. CONCLUSION

From the foregoing discussion we conclude that the gnostic estimator (G-estimator) may be interpreted as an DMLE estimator of the parameter of location and that it possesess all the nice asymptotical properties as it has been proved by Vajda (1986) for common DMLE (p_θ given by (4.1) satisfies all requirements). On the other hand, there are justifiable doubts on acceptability of the use G-estimator for arbitrary observed data, even if we did not discuss a somewhat different type of G-estimator introduced in a further paper of Kovanic (1986).

DMLE opens a fairly new look in the point estimation problem. Instead of using different types of robust estimators, we can now choose different types of assumed densities and construct trajectories of the estimates for a continuous scale of reasonable robustness. We expect that G-estimator based on density of the form (4.1) (or (4.2) for "multiplicative-error" data) can be extremly useful. Using the comparison of normal density curve and "additive-error gnostical" density (4.1) (both for $\mu = 0, \sigma = 1$) in Fig 1., the latter seems to be more appropriate for the practical purposes because of its (may be more realistic) heavier tails.

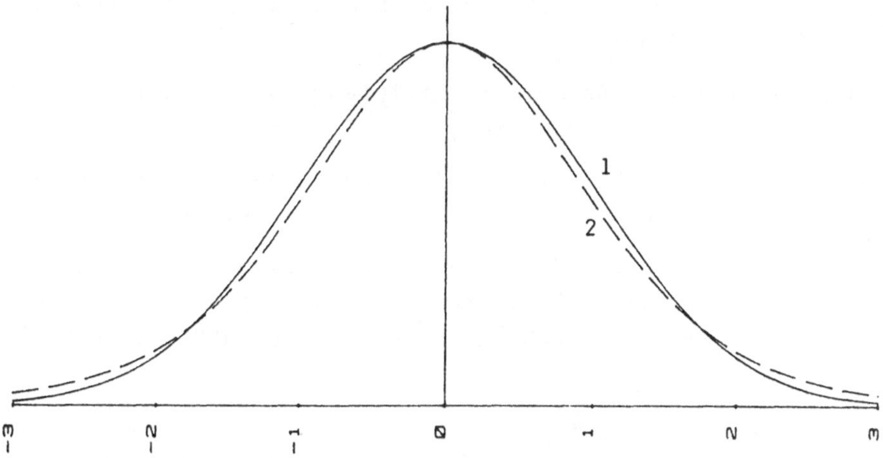

Figure 1. Comparizon of normal density curve (1) and density given by (4.1) for $\mu = 0, \sigma = 1$.

REFERENCES

Andrews D.F., Bickel P.J., Hampel F.R., Rogers W.H., Tukey J.W. (1972): Robust Estimates of Location. Survey and Advances. Princeton University Press.

Huber P.J. (1981): Robust Statistics. J.Wiley and Sons, New York.

Kovanic P. (1984a): Gnostical Theory of Individual Data. Problems of Control and Information Theory (PCIT) 13(4), 259-274.

Kovanic P. (1984b): Gnostical Theory of Small Smaples of Real Data. PCIT 13(5), 303-319.

Kovanic P. (1984c): On Relations between Information and Physics. PCIT 13(6), 383-399.

Michálek J., Vajda I., Víšek J.A. (1984): New Topics in Robust Statistics with Applications. Proceedings in computational statistics, COMPSTAT 1984, 73-83:

Stiegler S.M. (1977): Do Robust Estimators work with Real Data? Ann.Statist. 6, 1055-1098.

Vajda I. (1984a): Motivation,Existence and Equivariance of D-estimators. Kybernetika, 20, 189-208.

Vajda I. (1984b): Consistency of D-estimators. Kybernetika, 20, 283-303.

Vajda I. (1984c): Asymptotic Efficiency and Robustness of D-estimators. Kybernetika, 20, 358-375.

Vajda I. (1986): Efficiency and Robustness Control via Distorted Maximum Likelihood Estimation. Kybernetika, 22, 47-67.

Kovanic P., Novovičová J. (1986): Comparizon of Statistical and Gnostical Estimates of Parameter of Location on Real Data. Proceedings of conference ROBUST, JČSMF (in Czech).

General Computing Centre
Czechoslovak Academy of Sciences
Pod vodárenskou věží 2
182 07 Praha 8

OPTIMAL DESIGNS FOR SPATIALLY-AVERAGED OBSERVATIONS

Valeri Fedorov

International Institute for Applied Systems Analysis
Laxenburg, Austria

ABSTRACT

The methods of optimal design of experiments are considered for the regression problem when the observations are some averages of a response function over an interval whose length can be controlled. It is shown that the problem is closely related to the classical problem of Markov's moments.

INTRODUCTION

A number of publications concerning optimal design of experiments when controls belong to some functional space were published in the late 1970's. Now it is evident that the basic ideas behind these theoretical approaches are the same as in traditional experimental design theory (e.g., Fedorov, Uspensky 1977; Kozlov 1981; Pazman 1986). The differences become tangible in the application of general theoretical results to specific experimental problems.

The simplest statistical model describing at least a part of the above mentioned experiments is the following one:

$$y_i = \vartheta^T x_i + \varepsilon_i , \quad i = \overline{1,N} \tag{1}$$

where $\vartheta \in R^m$ is a vector of unknown parameters; ε_i are independent random values with zero means and finite variances σ_i^2; T stands for transposing. Variables x_i are defined by the integral

$$x_i = \int_V f(v) h_i(v) dv \quad , \tag{2}$$

$f(v)$ is a vector of given basic functions; $h(v)$ are some functions which can be chosen (controlled) by an experimenter, $0 \le h \le 1$. In some cases integral (2) must be a Lebesque one. If the least squares estimators are used to analyze experiments described by (1), then the quality of these estimators is defined by their dispersion (variance-covariance) matrices $D = E\{(\hat{\vartheta} - \vartheta_t)(\hat{\vartheta} - \vartheta_t)^T\}$, where the subscript τ stands for true values. It is well-known that in regular cases

$$D^{-1} = \bar{M} = \sum_{i=1}^{N} \sigma^{-2}(h_i)x(h_i)x^T(h_i). \tag{3}$$

Matrix M is usually called "information matrix." The objective of optimal experimental design is the search for controls $h_i^*(v)$ providing better dispersion matrices or (more accurately) some functions of them. The examples of epxeriments described by (1), (2) can be found in Fedorov, Uspensky (1977), Kozlov (1981), Twomey (1966).

OPTIMIZATION PROBLEMS IN EXPERIMENTAL DESIGN PROBLEM

As with traditional design theory (see above cited publications), the set of values $\xi_x = \{p_i, x_i\}$, $\Sigma p_i = 1$, where the weights p_i are the shares n_i/N of total number of measurements (or the shares t_i/T of total time available), which have to be done under the conditions x_i (or at the supporting points x_i), will be called a design (of an experiment). In the traditional case, the set $X \in R^m$ of feasible controls (operability region) is explicitly given. In the considered case, X is the mapping of a feasible set H of controls $h_i(v)$, and usually the construction of X is a problem of great difficulty. Therefore, it could be useful to consider designs in the original space also: $\xi_h = \{p_i, h_i(v)\}$.

From (3) it is evident that for model (1), (2) the information matrix depends upon a design ξ_x (or ξ_h), but does not depend upon the results of measurements. Due to this fact, the design problem can be formulated as the following minimization problem

$$\xi^* = Arg \min_{\xi} \Phi\{\bar{M}(\xi)\} \tag{4}$$

where ξ can have both possible subscripts. The function Φ (optimality criteria) describes the objectives of an experimenter.

It is reasonable to distinguish between two types of designs: continuous and discrete ones. In the first case, weights p_i can vary continuously between 0 and 1. This takes place when the weight is proportional to the time of measurement. We can go further and assume that any probabilistic measure $\xi_x = \xi(dx)$ or $\xi_H = \xi(dh)$ describes some design. In these cases:

$$\bar{M}(\xi_x) = N\int_X xx^T\xi(dx) = NM(\xi_x) \text{ or } \bar{M}(\xi_h) = N\int_H x(h)x^T(h)\xi(dh) = NM(\xi_h).$$

In what follows below, the subscript x or h will be omitted without any comments if it will not lead to confusion.

Assuming that $\Phi(NM) = \alpha(N)\Psi(M)$ (and it is true for the majority of optimality criteria used in practice) minimization problem (11) can be replaced by

$$\xi^* = Arg \min_{\xi} \Psi\{M(\xi)\}$$

where no values depend upon the total time or the total number of available measurements. This means that a continuous optimal design does not depend upon them also. This useful property is not valid in the discrete case.

CONTINUOUS OPTIMAL DESIGNS

For the sake of simplicity in this section and all subsequent sections only the case when

$$\Psi\{M\} = |M^{-1}| \quad \text{and} \quad \xi^* = Arg \max_{\xi} |M|$$

will be considered. Other criteria can be handled in a similar way.

Theorem 1 (Kiefer-Wolfowitz). If X is compact then

(1) *there exists an optimal design ξ_x^* containing no more than $m(m+1)/2$ supporting points.*

(2) *The following problems are equivalent: (a) maximization of $|M(\xi_x)|$, (b) minimization of $\max_{x \in X} \lambda(x) d(x,\xi_x)$, (c) $\max_{x \in X} \lambda(x) d(x,\xi_x) = m'$, where $\lambda(x_i) = \sigma_i^{-2}$ and $d(x,\xi_x) = x^T M^{-1}(\xi_x) x$,*

(3) *at the supporting points of an optimal design ξ_x^* the function $d(x,\xi_x)$ approaches its maximum.*

(4) *the set of optimal designs is convex.*

In a number of comparatively simple situations Theorem 1 gives a chance to construct optimal design analytically. For more complicated models it helps to develop numerical procedures and to understand some general features of optimal designs.

Example. Let now $f^T(v) = (\sin \pi v, \cos \pi v)$. It can be proved that for any slit function $h(v)$ the vector (Krein, Nudelman 1973, VII:3)

$$x^T = \{\int_{-\pi}^{\pi} h(v) \sin \pi v dv \; , \; \int_{-\pi}^{\pi} h(v) \cos \pi v dv\}$$

must belong to the circle $\{X: x_1^2 + x_2^2 \leq 2\}$.

The optimal designs for this operability region and response function $\vartheta^T x = \vartheta_1 x_1 + v_2 x_2$ can be easily constructed. For instance, optimal design can consist of the supporting points coinciding with all vertexes of any regular polygon refined to the circle X and their weights must be equal. One of the simplest optimal designs is

$$\xi_x^* = \begin{Bmatrix} (0;2) & (2;0) \\ 0.5 & 0.5 \end{Bmatrix}$$

and the corresponding optimal design in slit function space can (the solution of integral equation (2) is not unique) have the following supporting points:

$$h_1^*(v) = \begin{cases} 1, & -0.5 \le v \le 0.5, \\ 0, & v < -0.5,\ v > 0.5 \end{cases}, \text{ and } h_2^*(v) = -h_1^*(v) .$$

It is worthwhile to note that the widths of slit function "windows" have the same order as intervals of typical variations of basic functions.

NUMERICAL METHODS

If assumption (a) holds and there is a way to find X then the following numerical procedure can be used for optimal design construction:

$$\xi_{s+1} = (1-\alpha_s)\xi_s + \alpha_s \xi(x_s) \tag{5}$$

where $\xi(x_s)$ is the design concentrated at the single point

$$x_s = Arg \max_{x \in X} \lambda(x) d(x, \xi_s) \tag{6}$$

where $d(x,\xi_s) = x^T M^{-1}(\xi_s) x$.

The sequence $\{\alpha_s\}$ can be, for instance:

$$(a) \quad \sum_{s=0}^{s=\infty} \alpha_s = \infty , \quad \lim_{s \to \infty} \alpha_s = 0 \qquad (b) \quad \alpha_s = \frac{\lambda(x_s)d(x_s,\xi_s) - m}{\{\lambda(x_s)d(x_s,\xi_s) - 1\}m}$$

Both of them guarantee that

$$\lim_{s \to \infty} |M(\xi)| = \min_{\xi} |M(\xi)| \tag{7}$$

More sophisticated versions of this method are discussed in Ermakov (1983), Fedorov and Uspensky (1975).

In spite of the formal simplicity of iterative procedure, its practical usefullness is rather restricted: one must find out the way to construct X before using this procedure.

Procedures (5), (6) with $\lambda(x) \equiv const$ can be replaced by equivalent procedures in the space of functions $h(v)$. For that one has to replace the vector x with the vector $\int_V h(v) f(v) dv$:

$$\xi_{s+1} = (1-\alpha_s)\xi_s + \alpha_s \xi(h_s), \tag{8}$$

$$h_s = Arg \max_{h \in H} d(h, \xi_s), \tag{9}$$

Unfortunately, maximization problem (9) is more complicated than (6). One of the simplest approximation could be the following one:

- discretisize the set V (and therefore $V \times V'$ also), say with interval Δ;
- collect all points v_j on the corresponding grid which positively contribute to the sum

$$d(\xi_s) = \sum_{j,j'} f(v_j) M^{-1}(\xi_s) f(v_{j'}); \tag{10}$$

- put $h_s(v_j) = 1$ if v_j was chosen on the previous stage; otherwise $h_s(v_j) = 0$ some arguments for that can be found in the next section.

The sequences $\{j\}$ and $\{j'\}$ must be identical. Therefore if $j = j_0$ is included in sum (10), then $j' = j_0$ must be included also.

When approximation (10) is used then one can tell about Δ-optimal designs (ξ^*_Δ) which can be considered as some approximation of optimal designs. The iterative procedure (8), (9) and (10) guarantees that $\lim_{s \to \infty} |M(\xi_s)| = M(\xi^*_\Delta)$.

The idea of Δ-optimal designs can be advanced further. As it was observed before the widths of slit function windows were related to the intervals of variation of basic functions. Therefore, it is reasonable to decompose the set V into a comparatively moderate number of subsets Δ_j ($j = \overline{1,k}$), for instance, coinciding with the most typical fluctuations of basic functions. Assume that integrals

$$F_j = \int_{\Delta_j} f(v) dv$$

can be calculated. Then the operability region X can be approximated by X_Δ with elements $x = Fu$, where $F = (F_1, \ldots, F_k)$, $u^T = (u_1, \ldots, u_k) \in U_\Delta$ and $u_j = 1$, if $h(v) = 1$, $v \in \Delta_j$, and $u_j = 0$; otherwise, $j = \overline{1,k}$. Observing that $\vartheta^T x = \vartheta^T F u = \omega_T u$ and the information matrix equals

$$M = F \int_{U_\Delta} u u^T \xi(du) F^T = F M_\omega(\xi_u) F^T$$

where $\xi(du)$ describes a design ξ_u with supporting points in U_Δ , one can conclude that rank F must be equal to m (number of estimated parameters). Therefore, the decomposition of V should contain at least m subsets Δ_j ($k \geq m$).

When $k = m$ and of course $|F| \neq 0$, then $|M| = |F|^2 |M_\omega|$ and the design problem is reduced to the maximization of $|M_\omega|$. The latter problem coincides with the routine problem of "optimal weighting."

If $k > m$ then iterative procedure (5),(6) can be used with the replacement of the vector $\lambda^{1/2}(x) x$ with the vector Fu :

$$u_s = Arg \max_u u F^T M^{-1}(\xi_s) F u.$$

This maximization problem is a discrete one and at every s-th stage it demands no more than 2^k calculations of $uF^TM^{-1}(\xi_s)Fu$.

STRUCTURE OF SLIT FUNCTION

In the previous section, it was assumed that the slit function can equal 1 or 0. Some "physical" arguments were behind this assumption. The compactness of operability region X was also an essential assumption which was done to slimplify all final results. If one refuses this assumption, then instead of optimal designs, so-called optimal sequences must be considered and that leads to some technical difficulties. The following results (which are straightforward corollaries of well-known results from classical Markov's moment theory; see, for example, Karlin and Studden (1966) illuminate that both above mentioned assumptions are not very restrictive. For the sake of simplicity, we consider a one-dimensional case ($V\subset R^1$):

Assume now that:

(a) $0\le h(v)\le 1$, for any $v\in(a,b)$

(b) Functions $f(v)$ constitute a Tchebysheff system on the open interval (a,b), where a and b are possibly infinite.

Theorem 2. The operability region

$$X = \{x = \int_a^b f(v)h(v)dv:\ 0\le h(v)\le 1\}$$

is a compact convex set in R^m.

From Theorem 1 it is clear that all supporting points of any optimal design must be boundary points of X. Therefore, only these points had to be considered in the previous sections, and for them the following result takes place:

Theorem 3. The necessary and sufficient condition for x *to be a boundary point of* X *is the fulfillment of the condition*

$$h(v)\{1-h(v)\} = 0 \qquad (11)$$

almost everywhere in (a,b).

Let $h(v)$ be a function satisfying (11) and let $I(x)$ be the number of separate nondegenerate intervals where $h(v)=1$ with the special convention that an interval whose closure contains point a or b, is counted as 1/2. For any point $x\in X$, $I^*(x)$ stands for the least possible $I(x)$.

Theorem 4. A necessary and sufficient condition that x *belongs to the boundary of* X *is that* $I^*(x)\le(m-1)/2$. *Moreover, every boundary point corresponds to a unique* $h(v)$ *with* $I(x)=I^*(x)$.

Theorems 3 and 4 allow for the development of a comparatively simple algorithm of optimal design construction.

Let $v=(v_1,\ldots,v_{m-1})$, where $a \le v_1 \le \cdots \le v_{m-1} \le b$. According to Theorem 4, there exist optimal designs with all supporting points (in the operability region H) which have the following structures: $h_1(v) = \{1, v \in (a,v_1); 0, v \in (v_1,v_2); 1, v \in (v_2,v_3); \cdots \}$ and $h_2(v) = 1 - h_1(v)$.

That allows for the modification of the iterative procedure (5),(6) to the procedure with maximization in space which dimension is less or equal $(m-1)$:

$$\xi_{s+1} = (1-\alpha_s)\xi_s + \alpha_s \xi(h_s) \, , \; h_s = Arg \max_{\gamma, v} d(x_\gamma, \xi_s),$$

where $x_\gamma = \int_a^b f(v) h_\gamma(v) dv$.

REFERENCES

Ermakov, S.M., editor (1983): Mathematical Theory of Experimental Design. Moscow: Nauka, 391 (In Russian).

Fedorov, V.V. and A.B. Uspensky (1975): Numerical Aspects of the Least Squares Methods. Moscow: Moscow State University, 168.

Fedorov, V.V. and A.B. Uspensky (1977): On the Optimal Condition Choice of Spectroscopic Measurements. Moscow: Proc. of State Scient. and Research Center for the Study of Earth's Environment and Natural Resources, 4:42-53 (In Russian).

Karlin, S. and W.J. Studden (1966): Tchebycheff Systems: With Applications in Analysis and Statistics. New York: Wiley & Sons, 586.

Kozlov, V.P. (1981): Design of Regression Experiments in Functional Spaces, in "Mathematical Methods in the Design of Experiments", ed. Penenko, V.V. Moscow: Nauka, Novosibirsk, 74-101 (In Russian).

Krein, M.G. and A.A. Nudelman (1973): Markov Moment Problem and Extremal Problems. Moscow: Nauka, 552.

Pazman, A. (1986): Foundations of Optimum Experimental Design. Dordrecht: D. Reidel Publishing Company, 228.

Twomey, S. (1966): Indirect Measurements of Atmospheric Temperature Profiles from Satellites: Mathematical Aspects of the Inversion Problem. Monthly Weather Review, 99:363-366.

EXTREME ORDER STATISTICS
APPLIED FOR OPTIMUM ESTIMATION
IN "HARD" MP PROBLEMS

János C. Fodor, János Pintér

Budapest

Key words: mathematical programming, "hard" problems, optimum estimation, extreme order statistics.

ABSTRACT

In this paper the constrained optimization problem will be considered. We shall suppose that the global maximum z^* of our problem exists. Using a random sample from an appropriate distribution over the feasible set of the optimization problem we construct a confidence interval for z^* using asymptotic theory. Our approach is based on different generalizations of the concavity concept. Some of our results have direct relevance in the theory of extremes and in stochastic programming.

1. INTRODUCTION

In recent years, a growing interest concerning statistical optimum estimation methods could be observed. These estimation techniques are especially important e.g. when solving "hard" MP problems: the proper accomplishment of theoretically convergent optimization methods may encounter computational intractabilities. In such cases, relatively simple random sampling techniques may lead to statistical estimates which, in turn, can serve as useful addition to "exact" optimization procedures (e.g. leading to statistical termination or other decision rules). For illustrative examples, we refer to the works of Clough (1969), de Haan (1981) and Patel and Smith (1979) (further references see in C. Fodor and Pintér (1986)).

In the following the optimization problem

$$\max_{x \in D^m} g(x) \tag{1}$$

will be considered, where $D^m \subset \mathbb{R}^m$ is a non-empty subset of the real Euclidean m-space and $g:\mathbb{R}^m \to \mathbb{R}$ is a continuous real function. Below we shall suppose that the global optimum $z^* = g(x^*)$ of problem (1) exists (while x^* is not necessarily unique).

If D^m is bounded, then one can generate independent, uniformly distributed sampling points on it and calculate the respective function values of g. If the probability distribution function (p.d.f.) of the sample maximum is known to converge to one of the known extreme value p.d.f.'s, then the generated function values can be used for estimating z^*. The validity of the above convergence can be verified, under different assumptions concerning D^m and g (see references cited above). In the present paper similar results are given, based on different generalizations of the concavity concept. This provides a flexible framework for generating random samples on D^m and identifying the resulted extreme value p.d.f. Our results hold also for unbounded regions D^m. Besides, some of them have direct relevance in the theory of extremes and in stochastic programming: thus, it is hoped that they are potentially applicable not only in the outlined optimization context.

2. GENERALIZED CONCAVE MEASURES AND FUNCTIONS

The concept of r-concave measures was introduced by Borell (1975). Let P be a measure, defined on the Borel-measurable subsets of $\mathbb{R}^m$. Given a fixed $r \in \mathbb{R}$, $r \neq 0$, P is called r-concave, if for arbitrary convex subsets A, B $\mathbb{R}^m$ and $0 \le \lambda \le 1$ there holds

$$P(\lambda A + (1-\lambda)B) \ge \{\lambda \cdot [P(A)]^r + (1-\lambda)[P(B)]^r\}^{\frac{1}{r}} \quad .$$

Applying continuity argumentation, one can easily see that r=0 yields the known concept of logconcave measure cf. e.g. Prékopa (1971), (1973)).

The definition below is a generalization of r-concavity for functions. Let $G:\mathbb{R} \to \mathbb{R}$ be a continuous, strictly monotone function,

$h: \mathbb{R}^m \to \mathbb{R}$ be a continuous function. Assume that G is defined on the values of h. The function h is called G-concave, if for arbitrary $x_1, x_2 \in \mathbb{R}^m$ and $0 \leq \lambda \leq 1$ there holds

$$h(\lambda x_1 + (1-\lambda)x_2) \geq G^{-1}\{\lambda \cdot G[h(x_1)] + (1-\lambda) \cdot G[h(x_2)]\} .$$

As special cases, $G(x) = x$ (concavity), $G(x) = \log x$ (logconcavity), $G(x) = x^r$ (r-concavity) can be mentioned.
It is easy to see that if a measure P is r_0-concave for some $r_0 \in \mathbb{R}$, then P is r-concave for all $r \leq r_0$: a similar assertion holds with respect to r-concave functions. Some well-known results of Prékopa (1971), (1973) concerning logconcave measures and functions, can be generalized for the r-concave case, wehen $r \in \mathbb{R}$ is arbitrary.

THEOREM 2.1. a) Let P be an r-concave probability measure in $\mathbb{R}^m$. Let $A \subset \mathbb{R}^m$ be an arbitrary fixed convex set, $x \in \mathbb{R}^m$. Then $h(x) = P(A+x)$ is an r-concave function of x.
b) If P is an r-concave probability measure in $\mathbb{R}^m$, then the p.d.f. $F(x)$ and the function $1-F(x)$ are both r-concave.
c) Let $g_i(x,y)$ $i=1,\ldots,I$ x $\mathbb{R}^m$, y $\mathbb{R}^q$ respectively G_1-concave,..., G_I-concave functions in m+q variables and let $G_1(0)=\ldots=G_I(0)=0$. Suppose that the q-vector valued random variable ξ has an r-concave probability measure ($r \in \mathbb{R}$). Then

$$h(x) = P(g_i(x, \xi) \geq 0 \quad i=1,\ldots,I)$$

is an r-concave function of $x \in \mathbb{R}^m$.

PROOF. See C. Fodor and Pintér (1986).

3. GENERALIZED CONCAVITY IN THE THEORY OF EXTREMES

Let $X_1, \ldots, X_n, \ldots$ be independent, identically distributed (i.i.d.) random variables (r.v.'s). Define $F(x) = P(X_n < x)$, further, $Z_n = \max(X_1,\ldots,X_n)$. Then $P(Z_n < x) = [F(x)]^n$.
The p.d.f. F belongs to the domain of atraction of a non-degenerate p.d.f. H, if there exist real sequences $\{a_n\}$ and $\{b_n\}$ $b_n>0$ such that

$$(2) \quad \lim_{n\to\infty} [F(a_n + b_n x)]^n = H(x)$$

holds at all continuity points of H.
In the sequel the relation (2) will be denoted by $F \in D(H)$. We shall not go into details, concerning classical results on domains of attraction of the limiting p.d.f.'s, cf. e.g. Galambos (1978): instead, some new results applicable for our purposes will be given below. Define $\bar{F}(x) = 1-F(x)$ for $x \; \mathbb{R}$. Furtherly, as in C. Fodor, [1986], we shall define the following sets

$$V(\bar{F}) = \{r \in \mathbb{R} : \bar{F} \text{ is r-concave on some interval } (\underline{x}, \omega)\}$$
$$X(\bar{F}) = \{r \in \mathbb{R} : \bar{F} \text{ is r-convex on some interval } (\bar{x}, \omega)\}$$

where $\omega = \omega(F) = \sup\{x : F(x) < 1\}$ and $\underline{x}$, $\bar{x} < \omega$.
Define now a class of p.d.f.'s

$$K = \{F \text{ p.d.f.} : X(\bar{F}) \cup V(\bar{F}) \neq \emptyset \text{ and } |X(\bar{F}) \cap V(\bar{F})| \leq 1\} .$$

($|A|$ being the cardinality of the set A),

finally, let

$$r^*(\bar{F}) = \sup\{r : r \in V(\bar{F})\},$$

$$r_*(\bar{F}) = \inf\{r : r \in X(\bar{F})\}.$$

We proved in C. Fodor (1986) the following assertion.

THEOREM 3.1. $F \; K$ if and only if $r^*(\bar{F}) = r_*(\bar{F})$.

(Note that the possiblity of $r^*(\bar{F}) = r_*(\bar{F}) = +\infty$ or $-\infty$ is not excluded).

The parametric family $K(r)$ of p.d.f.'s is based on the above assertion. Let

$$K(r) = \{F \in K : r^*(\bar{F}) = r\} .$$

THEOREM 3.2. The following relations are valid:

(i) $D(H_{1,\gamma}) \cap K = K\left(-\frac{1}{\gamma}\right)$

(ii) $D(H_{2,\gamma}) \cap K = K\left(\frac{1}{\gamma}\right)$

(iii) $D(H_{3,0}) \cap K = K(0)$,

where

$$H_{1,\gamma}(x) = \begin{cases} \exp(-x^{-\gamma}) & \text{for} \quad x>0 \\ 0 & \text{for} \quad x\le 0 \end{cases}$$

$$H_{2,\gamma}(x) = \begin{cases} 1 & \text{for} \quad x\ge 0 \\ \exp(-(-x)^{\gamma}) & \text{for} \quad x<0 \end{cases}$$

$$H_{3,0}(x) = \exp(-e^{-x})$$

and γ is a positive constant.

As it is known from classical results of extreme value theory (see e.g. Galambos (1978)), the p.d.f.'s $H_{1,\gamma}$, $H_{2,\gamma}$, $H_{3,0}$ are the only possible limit distributions in (2), whose domains of attraction are not empty.

We can define for a probability distribution Q (as for a p.d.f. above) the quantity $r^*(Q) = \sup\{r \in \mathbb{R} : Q \text{ is } r\text{-concave measure}\}$ and let

$$\mathcal{M}(r) = \{Q \text{ prob. distr. } : r^*(Q) = r\} .$$

THEOREM 3.3. Let ζ be uniformly distributed on A, where A is a convex, bounded subset of $\mathbb{R}^m$. Then $Q_\zeta \in \mathcal{M}(\frac{1}{m})$, where Q_ζ is the probability distribution of ζ.

4. CONFIDENCE-INTERVAL ESTIMATES FOR THE OPTIMUM

Assume now that for estimating the optimum value z^* of (1), the sample of i.i.d.r.v.'s $X_i=g(\zeta_i)$ $\zeta_i \in D^m$ $i=1,2,\dots,n$ is taken. Obviously, for the p.d.f. $F(x)=P(X_i<x)$ we have $\omega(F)=z^*$. We shall assume that the feasible set D^m is convex; besides, let G be an (appropriately choosen) strictly monotone continuous real function for which there holds $G(0)=0$.

THEOREM 4.1. Assume that the (m+1) variate function

$$g_1(x,y) = -x + g(y) \qquad (x \in \mathbb{R}, \ y \in D^m)$$

is G-concave. Let ζ have a probability distribution Q_ζ on D^m with $Q_\zeta \in M(r)$, $r \geq 0$.

(i) If $r>0$ then $F \in D(H_{2,\frac{1}{r}})$.

(ii) If $r=0$ then $F \in D(H_{3,0})$.

PROOF. cf. C. Fodor and Pintér (1986).

As it is seen from the above Theorem, the resulted limit distribution depends mainly on the p.d.f. of the sample ζ.

Below we present two evident consequences of Theorem 4.1., which may have relevance with respect to "hard-to-solve" e.g. large dimensional, many constraints type convex programming problems.

COROLLARY 4.1. Let $g(y)$ be a concave function. If $Q_\zeta \in M(r)$ with $r>0$, then $F \in D(H_{2,\frac{1}{r}})$ and if $r=0$, then $F \in D(H_{3,0})$.

COROLLARY 4.2. If $g(y)$ is convex or concave and D^m is convex, bounded in $\mathbb{R}^m$ and ζ is uniformly distributed on D^m, then $F \in D(H_{2,m})$.

THEOREM 4.2. If for a p.d.f. F the relation (2) holds and $\omega=\omega(F)<+\infty$, then

$$(3) \qquad \lim_{n\to\infty} P\left(\frac{(1-b_n)\cdot Z_n - a_n + \omega\cdot b_n}{b_n} < x\right) = H(x)$$

is also valid with the same a_n, b_n as in (2).

PROOF. See C. Fodor and Pintér (1986).

The relationship (3) permits to define asymptotically valid confidence intervals for the unknown optimum z^*. As $Z_n \leq z^*$ $n=1,2,\dots$, therefore let u_p be the $(1-p)$-quantile of the p.d.f. H : $H(u_p)=1-p$ (the value $p>0$ is to be selected near to zero). Then, by Theorem 4.2, for sufficiently large n, one can write the approximate relation

$$P\left(Z_n \leq z^* \leq u_p + \frac{a_n + (b_n - 1)\cdot Z_n}{b_n}\right) \approx 1-p \quad .$$

The yielded approximate confidence interval obviously depends on the parameters a_n, b_n (for fixed n): these values are, of course, generally unknown and therefore have to be estimated from the sample (more details can be found e.g. in Gumbel (1958) or Pintér (1978)).

REFERENCES

Borell C. (1975): Convex set functions in d-spaces. Period. Math. Hungar. 6, No 2, 111-136.

Clough D. J. (1969): An asymptotic extreme-value sampling theory for estimation of a global maximum. Canad. Op. Res. Soc. J. 7, 102-115.

C. Fodor J. (1986): On domains of attraction of extreme value distributions (submitted for publication).

C. Fodor J. and Pintér J. (1986): Optimum estimation methods based on extreme order statistics (submitted to: Optimization).

Galambos J. (1978): The asymptotic theory of extreme order statistics. Wiley, New York.

Gumbel E. J. (1958): Statistics of extremes. Columbia Univ. Press, New York.

Haan L. de (1981): Estimation of the minimum of a function using order statistics. J. of the American Statistical Association. 76, No 374, 467-469.

Patel N. R. and Smith R. L. (1979): A statistical approach to solving concave minimization problems under linear constraints. Working Paper No 337, Graduate School of Business, University of Pittsburgh.

Pintér J. (1978): On the maximal distance between two series of empirical distribution functions, with application to an inventory problem. Methods of Operations Research, XXIX, 623-636.

Prékopa A. (1971): Logarithmic concave measures with application to stochastic programming. Acta Sci. Math. 32, 301-316.

Prékopa A. (1973): On logarithmic concave measures and functions. Acta Sci. Math. 34, 335-343.

J. C. Fodor
Computing Center, Eötvös Loránd University,
H-1502. Budapest 112, P.O. Box 157,
Hungary

J. Pintér
Research Centre for Water Resources
Development (VITUKI)
H-1453. Budapest, P.O. Box 27, Hungary

SEQUENTIAL ESTIMATION IN AN EXPONENTIAL CLASS OF MARKOV PROCESSES

Jürgen Franz, Wolfgang Winkler

Dresden

Key words: Markov processes, random time transformation, likelihood function, sufficiency, sequential estimation procedures, inequality of Cramér-Rao-Wolfowitz type, branching processes.

ABSTRACT

Sequential estimation problems have been considered for some special classes of Markov processes in the last ten years. In this note, a general exponential class of Markov processes is studied. Many widely used models for stochastic processes are of this type. The exponential class under consideration is closely related to homogeneous processes with independent increments by random time transformations. In the sequential case, i. e. given a Markov stopping time, an inequality of Cramér-Rao-Wolfowitz type is stated and properties of efficient sequential estimation procedures are derived. As an application, more detailed results are given for a special branching process. In particular, the distribution of the estimate can be obtained and the problem of constructing minimum variance unbiased estimates is considered.

1. EXPONENTIAL CLASS OF MARKOV PROCESSES

Let $X = \{X(t),\ t \geq 0\}$ be a time-continuous Markov process on $[\Omega, \mathfrak{A}, \mathfrak{P}]$ with paths belonging to $D^m[0,\infty)$, i. e. X is m-dimensional with values in $[R^m, \mathfrak{B}^m]$ and almost all paths are right-continuous functions having limits from the left. A filtration in $\mathfrak{A}$ is denoted by $\mathfrak{F} = \{\mathfrak{F}_t,\ t \geq 0\}$. We assume $\mathfrak{F}_t = \sigma\{X(u),\ u \leq t\}$ and $\mathfrak{A} = \mathfrak{F}_\infty = \{\mathfrak{F}_t,\ t \geq 0\}$.

The family of probability distribution is given by $\mathcal{P} = \{P_\theta, \theta \in \Theta\}$ with unknown parameter $\theta \in \Theta$, where Θ is a k-dimensional open interval in R^k. The measures restricted to $\mathcal{F}_t$ are denoted by $P_{\theta,t}$. We suppose that $P_{\theta_1,t}$ and $P_{\theta_2,t}$ are equivalent for every $t \geq 0$ and $\theta_1, \theta_2 \in \Theta$. Hence, Radon-Nikodym-derivatives exist and for fixed $\theta_0 \in \Theta$ the likelihood function of the process X observed up to t is given by

(1) $$L_\theta(t) = \frac{dP_{\theta,t}}{dP_{\theta_0,t}}, \quad \theta \in \Theta, \quad 0 \leq t < \infty.$$

In the following we put $\theta_0 = 0$. Now we define the exponential class of Markov processes.

<u>Definition:</u> a) Let $M = \{M(t), t \geq 0\}$ be a Markov submartingale on $[\Omega, \mathcal{F}_\infty, P_\theta]$ with paths in $D^k[0,\infty)$ and let $S = \{S(t), t \geq 0\}$ be a non-decreasing process on $[\Omega, \mathcal{F}_\infty, P_\theta]$ with paths in $D^1[0,\infty)$ and $S(0) = 0$ P_θ-a.s. Moreover, let γ denote a real-valued function on Θ, twice continuously differentiable and with $\gamma(0) = 0$. Then, the process X belongs to the (M,S)-exponential family if the likelihood function possesses the representation

(2) $$L_\theta(t) = \exp\{\theta^T M(t) - \gamma(\theta)\, S(t)\}, \quad \theta \in \Theta, \quad 0 \leq t < \infty.$$

b) The set of all processes X belonging to one of the (M,S)-exponential families is called the exponential class of Markov processes.

We shall give examples of exponential families:

1. Let X be a 1-dimensional Poisson process with parameter $\lambda \in (0,\infty)$. We have $L_\theta(t) = \exp\{(\ln\lambda)\, X(t) - (\lambda - 1)t\}$. Then, all $X = X_\theta$ with $\theta = \ln\lambda$ $(-\infty < \theta < \infty)$ form the (X,t)-exponential family of Poisson processes.

2. Assume that X is a 1-dimensional Wiener process with linear trend $\theta \cdot t$; the likelihood function is $L_\theta(t) = \exp\{\theta\, X(t) - \frac{\theta^2}{2} t\}$, $\theta \in R^1$. The family of all processes $X = X_\theta$ $(\theta \in R^1)$ is the (X,t)-exponential family of Wiener processes.

3. Let X be a linear time-homogeneous birth-and-death- process with a birth rate $\lambda X(t)$, a death rate $\mu X(t)$, $P(X(0) = x_0) = 1$, and an immigration in the state zero with a known rate ρ (see Franz (1982)). In this case, we have $L_\theta(t) = \exp\{(\ln\lambda)B(t) + (\ln\mu)D(t) - (\lambda + \mu - 2)S(t)\}$, where $\theta = (\ln\lambda, \ln\mu)^T \in R^2$, $M(t) = (B(t), D(t))^T$ and $S(t) = \int_0^t X(u)\, du$. The number of birth cases (without immigration) and the number of death cases are denoted by $B(t)$ and $D(t)$, respectively. Let $N(t)$ be the total number of jumps and let $N_0(t)$ be the number of immigration cases. Then, $N(t) = B(t) + D(t) + N_0(t)$ and $X(t) = x_0 + B(t) + N_0(t) - D(t)$. The family of all processes $X = X_\theta$, $\theta \in R^2$, forms the (M,S)-exponential family of linear birth-and-death processes with known zero-state immigration.

4. Let X be a 1-dimensional time-homogeneous Markov branching process with the state space $E = \{0,1,2,\ldots\}$, $P(X(0) = x_0) = 1$, and with split intensity $\lambda \in (0,\infty)$. We suppose that the offspring distribution is given by

$$(3) \qquad p_j := P(X_i = x_i | X_{i-1} = x_{i-1}) = \begin{cases} \mu^j (j!)^{-1} e^{-\mu} , & j = 1,2,\ldots, \\ e^{-\mu} , & j = -1 , \end{cases} \qquad (\mu \in (0,\infty))$$

where $j = x_i - x_{i-1}$, $X_i = X(t_i)$ and t_i $(i = 1,2,\ldots)$ are the splitting points of X. Because of $\sum_{i=0,2}^{\infty} jp_j < \infty$ the process X cannot have an explosion in a finite time interval. According to Athreya and Keiding (1975) the likelihood function is given by

$$(4) \qquad L_\theta(t) = \exp\{(\ln\lambda - \mu+1)\, N(t) + (\ln\mu)\, M_+(t) - (\lambda - 1)\, S(t)\}$$

where $\theta = (\ln\lambda - \mu+1, \ln\mu)^T \in R^2$, N(t) ist the total number of splits, S(t) is the total life time (see example 3), $M_+(t) = \sum_{i=1}^{N(t)} \max(0, X_i - X_{i-1})$ and $M(t) = (N(t), M_+(t))^T$. It holds $X(t) = x_0 + M_+(t) - N^-(t)$, where $N^-(t)$ denotes the number of death cases. We say that all $X = X_\theta$, $\theta \in R^2$, represent the (M,S)-exponential family of 1-dimensional linear Poisson branching processes.

5. Assume that X is a 1-dimensional diffusion process satisfying $d\,X(t) = \theta\, X(t) + (X(t))^{1/2}\, d\,W(t)$, $X(0) = x_0 > 0$ P_θ-a.s. $(\theta \in R^1)$, where W denotes a standard Wiener process. According to Brown and Hewitt (1975) the likelihood function is $L_\theta(t) = \exp\{\theta(X(t) - x_0) - \frac{\theta^2}{2} S(t)\}$ where $S(t) = \int_0^t X(u)\,du$. The familiy of all $X = X_\theta$, $\theta \in R^1$, is the (X,S)-exponential family of diffusion branching processes.

Based on the likelihood function (2) we introduce the score function

$$(5) \qquad U_\theta(t) = \frac{d}{d\theta}(\ln L_\theta(t)) = M(t) - \frac{d}{d\theta}\gamma(\theta)\, S(t).$$

It is known that the process $L_\theta = \{L_\theta(t), t \geq 0\}$ is a one-mean-martingale w.r.t. P_0. Moreover, assuming $E_\theta\, S(t) < \infty$ the score process $U_\theta = \{U_\theta(t), t \geq 0\}$ is a zero-mean-martingale w.r.t. P_θ $(\theta \in \Theta)$ (see, for instance, Feigin (1976)). This fact leads to the equation

$$(6) \qquad E_\theta\, M(t) = \frac{d}{d\theta}\gamma(\theta) \cdot E_\theta\, S(t).$$

Under the condition $E_\theta\, S^2(t) < \infty$ one can prove

$$(7) \qquad E_\theta\, U_\theta(t)\, U_\theta^T(t) = \left(\frac{d}{d\theta}\right)^2 \gamma(\theta) \cdot E_\theta\, S(t) .$$

We remark that the righthand term of (7) is called the expected Fisher information.

In this sense, $(\frac{d}{d\theta})^2 \gamma(\theta)\cdot S(t)$ is also called the observed Fisher information. We suppose that the inverse function $\dot{\gamma}^{-1}$ of $\dot{\gamma}(\theta) = \frac{d}{d\theta}\gamma(\theta)$ exists. Using (5) the ML-estimate of θ is

$$(8) \qquad \hat{\theta}(t) = \dot{\gamma}^{-1}\left(\frac{M(t)}{S(t)}\right).$$

From (2) it is easy to see that $W(t) = (S(t), M(t))$ is a sufficient statistic w.r.t. $P_{\theta,t}$.

We now give some characterizations of the exponential class of Markov processes. For this purpose we need the notion of a random transformation time.

The family $T = \{T(t),\ t \geq 0\}$ is said to be a process of random transformation times w.r.t. X if T is $\mathcal{F}$-adapted and if the following conditions hold P_θ-a.s.: (i) T is nondecreasing w.r.t. t, (ii) $0 \leq T(t) < \infty$ for $0 \leq t < \infty$, (iii) $T(0) = 0$ and $T(t) \to \infty$ as $t \to \infty$.

Theorem 1: Let an (M,S)-exponential family be given. Then, there exists a homogeneous process with independent increments $\tilde{M}$ such that M is generated from $\tilde{M}$ by the process S of random transformation times, i. e. $M(t) = \tilde{M}(S(t))$. The likelihood function of $\tilde{M}(s)$ is

$$(9) \qquad \tilde{L}_\theta(s) = \exp\{\theta^T\tilde{M}(s) - \gamma(\theta)s\} \qquad (0 \leq s < \infty,\ \theta \in \Theta),$$

and the corresponding cumulant function $\psi(u,\theta) = \frac{1}{s}\ln(E_\theta e^{u\tilde{M}(s)})$ has the form

$$(10) \qquad \psi(u,\theta) = \gamma(u+\theta) - \gamma(\theta) \qquad (u,\ u+\theta \in \Theta).$$

Proof: Let $\tilde{\mathcal{F}}_s = \sigma\{\tilde{M}(u),\ u \leq s\}$, $\tilde{P}_{\theta,s} = P_\theta | \tilde{\mathcal{F}}_s$ and $\tilde{S}$ denotes the inverse process related to S, i. e. $\tilde{S} = \{\tilde{S}(s),\ s \geq 0\}$, defined by

$$(11) \qquad \tilde{S}(s) = \inf\{t : S(t) \geq s\}.$$

Then, $\tilde{L}_\theta(s) = \frac{d\tilde{P}_{\theta,s}}{d\tilde{P}_{0,s}} = \frac{dP_\theta | \mathcal{F}_{\tilde{S}(s)}}{dP_0 | \mathcal{F}_{\tilde{S}(s)}}$. Applying a result of Döhler (1981) we obtain $\tilde{L}_\theta(s) = \exp\{\theta^T M(\tilde{S}(s)) - \gamma(\theta)\, S(\tilde{S}(s))\} = \exp\{\theta^T\tilde{M}(s) - \gamma(\theta)s\}$ and hence, $\tilde{M}(s)$ is a homogeneous process with independent increments. Moreover, we get

$$E_\theta e^{u\tilde{M}(s)} = \int e^{u\tilde{M}(s)} d\tilde{P}_{\theta,s} = \int e^{(u+\theta)\tilde{M}(s) - \gamma(\theta)s} d\tilde{P}_{0,s} = \int \exp\{(\gamma(u+\theta) - \gamma(\theta))s\}\, d\tilde{P}_{u+\theta,s}$$

$$= \exp\{(\gamma(u+\theta) - \gamma(\theta))s\}. \ \#$$

Theorem 2: Let X be a homogeneous Markov process with paths in $D^k[0,\infty)$ and let S be a process of random transformation times. Then, the following two statements are equivalent:

(i) X belongs to the (X,S)-exponential family.

(ii) For every $t \geq 0$ the statistic $W(t) = (S(t), X(t))$ is sufficient w.r.t. $P_{\theta,t}$ and, for arbitrary $\theta_1, \theta_2 \in \Theta$ $(\theta_1 \neq \theta_2)$ the function $p_{\theta_1\theta_2}(w) := = \frac{dP_{\theta_1,t}}{dP_{\theta_2,t}}(w)$ is continuous in $w = (s,x)$.

Sketch of proof: Obviously the second statement follows from (i). Applying the random transformation times $\tilde{S}$ the function $p_{\theta_1\theta_2}(W(t))$ goes over to $\tilde{p}_{\theta_1,\theta_2}(s, \tilde{X}(s))$ which has an exponential form of type (2), according to Winkler and Franz (1979). The random transformation times S shows that $p_{\theta_1\theta_2}(W(t))$ is also of form (2). #

2. EFFICIENT SEQUENTIAL ESTIMATION

If the observation of the underlying process X is finished at a Markov stopping time τ the likelihood function is given by (see Döhler (1981))

$$L_\theta(\tau) = \exp\left\{\theta^T M(\tau) - \gamma(\theta)\cdot S(\tau)\right\}. \tag{12}$$

Let $\delta = (\tau, Y(\tau))$ be a sequential procedure in order to estimate a parameter function $h(\theta)$, where $Y(\tau)$ is an estimate for $h(\theta)$. We assume

(H) $h(\theta)$ is nonconstant and differentiable w.r.t. θ, $\theta \in \Theta$.

(Y) $Y(\tau)$ satisfies the properties $E_\theta Y(\tau) = h(\theta)$ and $E_\theta Y^2(\tau) < \infty$, $\theta \in \Theta$.

Theorem 3: Suppose X belongs to the exponential class of Markov processes and let $\delta = (\tau, Y(\tau))$ be a sequential procedure for estimating $h(\theta)$. If the conditions (H) and (Y) are fulfilled the following inequality of Cramér-Rao-Wolfowitz type holds

$$\mathrm{var}_\theta\,(Y(\tau)) \geq (E_\theta S(\tau))^{-1}\,(\dot{h}(\theta))^T\,(\ddot{\gamma}(\theta))^{-1}\,\dot{h}(\theta) \quad (\dot{h}(\theta) = \tfrac{d}{d\theta}h(\theta)). \tag{13}$$

The equality sign holds in (13) at $\theta = \theta^* \in \Theta$ iff

$$Y(\tau) = h(\theta^*) + (E_{\theta^*}S(\tau))^{-1}\,(\dot{h}(\theta^*))^T\,(\ddot{\gamma}(\theta^*))^{-1}\,U_{\theta^*}(\tau). \tag{14}$$

The proof can be given in analogy to a theorem in Winkler and Franz (1979).

Efficiency is defined as usual: The sequential procedure δ is said to be efficient at $\theta = \theta^*$ if equality holds in (13) for $\theta = \theta^*$. In this case, the function $h(\theta)$ is efficiently estimable at $\theta = \theta^*$ and the estimator $Y(\tau)$ is efficient at $\theta = \theta^*$ too. A sequential procedure δ is called Θ'-efficiently if δ is efficient for all $\theta \in \Theta' \subseteq \Theta$.

Θ'-efficient sequential procedures are characterized by the following

Theorem 4: If the sequential procedure δ is Θ'-efficient and the conditions of theorem 3 are fulfilled, then there exist real numbers $c_0, c_1, \ldots, c_k$ and z with $c_0^2 + c^T c > 0$, $z \neq 0$ such that

(15) $$c_0 S(\tau) + c^T M(\tau) = z \qquad P_\theta\text{-a.s.},$$

where $c^T = (c_1, \ldots, c_k)$. Moreover, it holds

(16) $$E_\theta S(\tau) = (c_0 + c^T \dot{\gamma}(\theta))^{-1} z.$$

(15) can be proved by similar arguments used in Winkler and Franz (1979); (16) is a consequence of (15) and the moment relation $E_\theta M(\tau) = \dot{\gamma}(\theta) \cdot E_\theta S(\tau)$.

3. SEQUENTIAL PROCEDURES IN POISSON BRANCHING PROCESSES

Finally, we consider the family of linear Poisson branching processes (see example 4). We mention that $W(\tau) = (S(\tau), N(\tau), M_+(\tau))$ is a sufficient statistic w.r.t. $P_{\theta,\tau}$. In order to estimate a certain function $h(\theta)$ Θ'-efficient sequential procedures δ are characterized by the equation

(17) $$c_0 S(\tau) + c_1 N(\tau) + c_2 M_+(\tau) = z \qquad P_\theta\text{-a.s.}, \quad \theta \in \Theta',$$

where $c_0^2 + c_1^2 + c_2^2 > 0$, $z \neq 0$.

In such procedures, the efficiently estimable functions $h(\theta)$ and the corresponding efficient sequential estimates $Y(\tau)$ can be derived in a similar way proved for birth-and-death-processes in Franz (1982 a):

(18) $$h(\theta) = \frac{a_0 + a_1\gamma(\theta) + a_2 e^{\theta_2}\gamma(\theta)}{c_0 + c_1\gamma(\theta) + c_2 e^{\theta_2}\gamma(\theta)}, \qquad \gamma(\theta) = \exp(\theta_1 + e^{\theta_2}),$$

(19) $$Y(\tau) = \begin{cases} b_0 + b_1 N(\tau) + b_2 M_+(\tau) & \text{if } c_0 \neq 0, \\ \tilde{b}_0 + \tilde{b}_1 S(\tau) + \tilde{b}_2 M_+(\tau) & \text{if } c_0 = 0, \end{cases}$$

where $c_1 b_0 = a_0$, $z b_1 = a_1 - (c_2/c_1) a_0$, $z b_2 = a_2$, $c_1 \tilde{b}_0 = a_1$, $z\tilde{b}_1 = a_0 - (c_2/c_1) a_1$, $z\tilde{b}_2 = a_2$ and a_0, a_1, a_2 are real numbers.

Now, we turn to the problem of finiteness of procedures $\delta = (\tau, Y(\tau))$, i. e. $P(\tau < \infty) = 1$.

Theorem 5: Suppose that δ is a Θ'-efficient sequential procedure and the conditions of theorem 3 are fulfilled. If the additional conditions

(i) $c_0 + c_1\gamma(\theta) + c_2 e^{\theta_2}\gamma(\theta) > 0$,

(ii) $c_0 > 0$, $c_1 \leq 0$, $c_2 \leq 0$ $(z \in R_+^1)$ or $c_0 = 0$, $c_1 > 0$, $c_2 = 0$ $(\frac{z}{c_1} \in \mathbb{N})$

are valid, then the procedure δ is finite.

To prove this random transformation times are applied and the assertion is derived for a compound Poisson process (for the last assertion, see Stefanov (1982)). Now, we are interested in the common distribution of the sufficient statistic $W(\tau)$ under the condition (17).

In the case $c_1 \neq 0$, we obtain from (17)

$$N(\tau) = \frac{1}{c_1}(z - c_0\, S(\tau) - c_2\, M_+(\tau)).$$

The process $\tilde{M} = (\tilde{N}, \tilde{M}_+)$ (see theorem 1) is a compound Poisson process with probabilities

$$(20) \qquad p_{\tilde{N},\tilde{M}_+}(n,m,s;\theta) = \frac{n^m s^n}{n!\, m!} \exp\left\{(\ln\lambda - \mu)n + (\ln \mu)m - \lambda s\right\},$$

where $n, m = 0, 1, 2, \ldots;\ \theta = (\ln\lambda - \mu + 1, \ln \mu)$. The variable $\sigma = S(\tau)$ is a stopping time related to $\tilde{M}$. According to Franz (1982 b) the common distribution of $(\sigma, \tilde{M}_+(\sigma)) = (S(\tau), M_+(\tau))$ can be obtained from (20). In the special case $c_1 = 1$, $c_0 = c_2 = 0$, i. e. $N(\tau) = z$ P_θ-a.s., we have

$$(21) \qquad p^*_{\sigma,M_+}(s,m,z;\theta) = \frac{\lambda^z}{(z-1)!} s^{z-1} e^{-\lambda s} \frac{(\mu z)^m}{m!} e^{-\mu z},$$

where $0 \leq s < \infty$, $m = 0, 1, \ldots;\ z = 1, 2, \ldots$. Therefore, the variable σ is gamma distributed and independent of the Poisson distributed variable $M_+(\tau)$. Note, that for other values of c_i $(i = 0, 1, 2)$ the random variables σ and $M_+(\tau)$ can be dependent.

The knowledge of the distribution of $W(\tau)$ can be used to construct minimum variance (in sence of a Rao-Blackwell theorem) unbiased estimates $Y(W(\tau))$. In order to illustrate this we put

$$\int \sum_m y(s,m) p^*_{\sigma,M_+}(s,m,z;\theta)\, ds = h(\theta),$$

and the solution $Y(\tau) = Y(S(\tau), M_+(\tau))$ is a minimum variance unbiased estimate for $h(\theta)$. Especially, for $h(\theta) = \lambda^p$ we obtain $Y(\tau) = (z-1)!\ (S(\tau)^p \Gamma(z-p))^{-1}$ $(p < z)$, for $h(\theta) = \mu$: $Y(\tau) = \frac{M_+(\tau)}{z}$ and for $h(\theta) = \exp(-r\lambda)$: $Y(\tau) = $ $= \left(\frac{S(\tau) - r}{S(\tau)}\right)^{z-1}$ $(r \geq 0)$.

REFERENCES

Athreya, K.S.; Keiding, V. (1975): Estimation theory for continuous time branching processes. Preprint No. 6, Inst. of. Math. Statist. Univ. Copenhagen.

Brown, B. M.; Hewitt, J. I. (1975): Inference for the Diffusion Branching Process. J. Appl. Prob. 12, 588 - 594.

Döhler, R. (1981): Dominierbarkeit und Suffizienz in der Sequentialanalyse. Math. Operationsforsch. u. Statistik, Ser. Statistics 12,1; 101 - 134.

Fe gin, P. D. (1976): Maximum likelihood estimation for continuous time stochastic processes. J. Appl. Prob. 13, 712 - 736.

Franz, J. (1982 a): Sequential estimation and asymptotic properties in birth-and-death-processes. Math. Operationsforsch. Statist., Ser. Statistics 13,2; 231 - 244.

Franz, J. (1982 b): Minimax estimation related to efficient sequential procedures in special Markov processes. Preprint 07-11-82 Sekt. Math., Techn. Univ. Dresden.

Stefanov, V. T. (1982): Sequential estimation for compound Poisson process I + II. SERDICA Bulg. math. publ. 8, 183 - 189 + 255 - 261.

Winkler, W.; Franz, J. (1979): Sequential estimation problems for the exponential class of processes with independent increments. Scand. J. Statist. 6, 129 - 139.

Technische Universität Dresden
Sektion Mathematik
Mommsenstraße 13
DDR - 8027 Dresden

ANALYSIS OF STOCHASTIC PETRI NETS

BY THE CONCEPT OF NEAR-COMPLETE DECOMPOSABILITY

Josef Giglmayr

Heinrich-Hertz-Institut fuer Nachrichtentechnik Berlin GmbH

Key words: Generalized stochastic Petri nets, decomposition, transition rate matrix

ABSTRACT

The crucial problem when modelling real systems by stochastic Petri nets is the state space explosion and consequent the increase of the size of the transition rate matrix representing the stochastic Petri net. Within the present paper a stochastic Petri net is analysed by decomposing the transition rate matrix. In particular, the stationary marking probabilities are determined by solving several smaller matrix equations instead of solving the large equation system made up of the transition rate matrix. The applicability of the approach is shown by the simplified Petri net model of a broadband switching control. For this example rules for the decomposition providing exact results for the stationary marking probabilities are presented.

INTRODUCTION

The analysis of stochastic Petri nets (SPNs) starts with the computation of the stationary marking probabilities from which certain performance measures can be derived. Within the present paper we focuse on the computation of these steady-state probabilities. In particular, the steady-state marking probabilities are determined by solving N smaller matrix equations (N is the number of appropriate choosen subsystems) and the matrix equation describing the interaction between the subsystems instead of solving one large equation system made up of the transition rate matrix corresponding to the entire net. This procedure is based on the concept of near-complete decomposability developed by Courtois (1977) etc.

The variables introduced for the reason of the decomposition are the eigenvalues and the eigenvectors representing the subsystems, the transition probabilities describing the interaction between the subsystems and the steady-state probabilities for being in one of the subsystems (called the macro aggregation variables). Finally, the steady-state marking probabilities are approached by the micro aggregation probabilities which are obtained from the macro aggregate probabilities and the eigenvectors of the subsystems.

The application of this decomposition method in order to solve queueing network problems is presented in Courtois (1977), Kameda (1984) etc. Decomposition is also applied in Bobbio and Trivedi (1986) for the evaluation of the performability of systems (i.e. performance and availability). However, in the case presented the elements of the transition rate matrix are of the same order of magnitude and no rules for the decomposition can be derived from the location of these elements. Therefore, throughout this paper an attempt is made to find the appropriate decomposition directly from the transition rate matrix. In particular, the decomposition providing exact results for the stationary marking probabilities is presented which is an interesting result.

BACKGROUND-STOCHASTIC PETRI NETS

We assume the reader is familiar with PNs. SPNs are obtained by assigning Poisson rates to each transition. A generalized class of SPNs (GSPNs) was introduced by Marsan et al (1984) by allowing timed as well as immediate transitions where immediate transitions are assumed to fire in zero time. In GSPNs the concept of a random switch is included consisting of probabilities for the selection of simultaneously enabled immediate transitions. When one or more immediate transitions are enabled the associated marking is a vanishing state of the GSPN. Contrary, markings which enable only timed transitions are called tangible states of the GSPN. For the evaluation of the GSPN only tangible states must be considered.

The timed transitions in a SPN or a GSPN are assumed to fire with Poisson rates. By the firing of these transitions a sequence of tangible marking states M_k $(k \geq 1)$ is obtained. The steady-state probabilities for these marking states satisfy the following matrix equation

$$AP = 0 \tag{1}$$

where A is a square matrix (its size is determined by the number of marking states) and P is the marking probability vector. Throughout this paper eq(1) will be solved by the concept of near-complete decomposability.

THE EXAMPLE

The GSPN in Figure 1 (where thick bars represent timed transitions and thin bars indicate immediate transitions) models the flow of signaling data in a broadband switching control at a very abstract level. In particular, the signaling processor is analysing signaling data (t_1) and upon the result the token (representing a request) flows through the path on the left hand side of the GSPN (telephony with picture phone option) or through the path on the right hand side (distribution of TV programs). The simultaneously enabled immediate transitions t_2 and t_3 together with the probabilities $p(t_2)$ and $p(t_3)$ model the separation of signaling data according to the requested service. Switching of telephone and picture phone lines is modelled by transition t_4 and the distribution of TV programs by t_5. For a more refined model t_4 may be replaced by the subnets presented in Figure 2. There, signaling data are analysed by the group processor (t_4). Then, during a telephone conversation (established in t_6) the picture phone is connected (t_7) and disconnected (t_8), respectively. Finally, by the firing of t_9 the telephone line and the picture phone line are released simultaneously (Figure 2a) or the telephone line is released only when the picture phone is switched off (Figure 2b without t_{10}) or the release of the telephone line implies the release of the picture phone (Figure 2b with t_{10}).

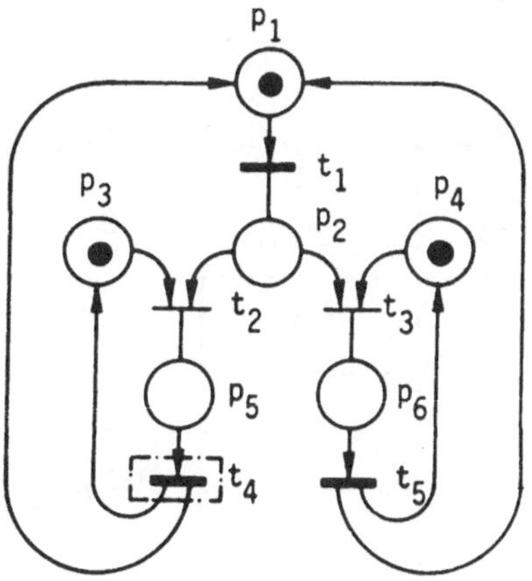

Figure 1. GSPN with one random switch

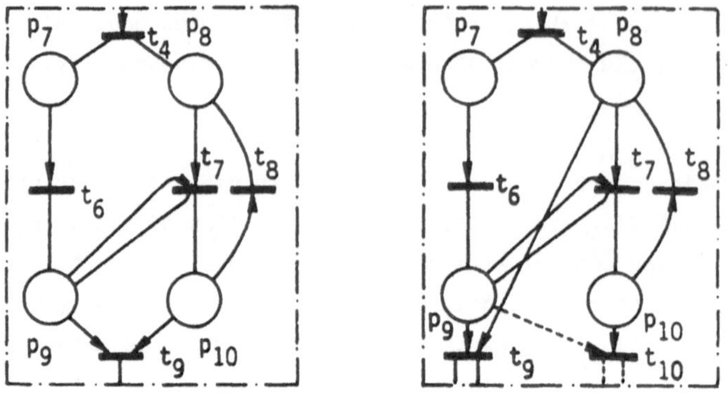

a) b)

Figure 2. Subnets for the refinement of Figure 1

COMPUTATION OF THE MARKING PROBABILITIES

For this reason we consider the adjoint to eq(1)

$$P^T(Q - E) = 0 \qquad (2)$$

which is obtained from $(AP)=(AP)^T=P^TA^T=0$ and where $Q=A^T+E$ is a stochastic matrix (E is an identity matrix and T means the transpose).

The GSPN in Figure 1 together with the subnet in Figure 2a is assigned as Example 1 and together with the subnet in Figure 2b as Example 2. The transition rate matrix for Example 1 is

$$Q_I(\overset{*}{Q}_{II}) = \begin{bmatrix} 1-\lambda_1 & p(t_2)\lambda_1 & 0 & 0 & 0 & p(t_3)\lambda_1 \\ 0 & 1-\lambda_4 & \lambda_4 & 0 & 0 & 0 \\ 0 & 0 & 1-\lambda_6 & \lambda_6 & 0 & 0 \\ 0\ (\lambda_9) & 0 & 0 & \begin{matrix}1-\lambda_7 \\ (1-\lambda_7-\lambda_9)\end{matrix} & \lambda_7 & 0 \\ \lambda_9\ (\lambda_{10}) & 0 & 0 & \lambda_8 & \begin{matrix}1-\lambda_8-\lambda_9 \\ (1-\lambda_8-\lambda_{10})\end{matrix} & 0 \\ \lambda_5 & 0 & 0 & 0 & 0 & 1-\lambda_5 \end{bmatrix}$$

where two possible decompositions are shown (the solid lines indicate decomposition 1 and the dashed lines decomposition 2). Similar, the transition rate matrix of Example 2 is

$$Q_{II} = \begin{bmatrix} 1-\lambda_1 & p(t_3)\lambda_1 & p(t_2)\lambda_1 & 0 & 0 & 0 \\ \lambda_5 & 1-\lambda_5 & 0 & 0 & 0 & 0 \\ 0 & 0 & 1-\lambda_4 & \lambda_4 & 0 & 0 \\ 0 & 0 & 0 & 1-\lambda_6 & \lambda_6 & 0 \\ \lambda_9 & 0 & 0 & 0 & 1-\lambda_7-\lambda_9 & \lambda_7 \\ \lambda_{10} & 0 & 0 & 0 & \lambda_8 & 1-\lambda_8-\lambda_{10} \end{bmatrix}$$

Please notice that the transition rate matrix $\overset{*}{Q}_{II}$ above (which is obtained from Q_I by replacing four elements by the elements in the parenthesis and the decomposition follows the dashed lines there) represents the same system as Q_{II} does.

Following the decomposition algorithm (cf. the appendix) the transition rate matrix is decomposed into Q=V+W (V is made up of the diagonal submatrices and W contains the residual part) and an appropriate matrix X has to be choosen which makes $\overset{*}{Q}=V+X$ stochastic. Then, from the stochastic submatrices we obtain eigenvalues and eigenvectors where the latter approaches approximately the steady-probabilities of being in one of the states of the subsystems. Please notice, a crucial problem is the procedure making the submatrices stochastic.

For the computation of the transition rate matrix $R=(r_{IJ})$ which describes the interaction between the subsystems, the transition rate matrices of the examples have to be divided into 16 (decomposition 1 of Example 1) or 9 (decomposition 2 of Example 1 and 2) submatrices which easily are obtained by continuing all solid or dashed lines of the transition rate matrix.

The solution of the matrix equation X(R-E)=0 together with the normalizing condition $\sum_I X_I=1$ gives the macro aggregate probabilities X_I where the latter approaches approximately the steady-state probability for being in one of the subsystems. Then the steady-state marking probability vector of the GSPN is obtained from the X_I's and the eigenvectors.

For decomposition 1 applied to Example 1 (the solid lines there) the exact results for the steady-state marking probability vector is obtained (even the decomposition condition is not satisfied). All other results are approximate and the decomposition condition may be hard to satisfy.

Please notice that instead of one 6 by 6 matrix equation we have solved two 2 by 2 equation systems and one of size 4 by 4 (decomposition 1 of Example 1) or three 2 by 2 equation systems and one of size 3 by 3 (decomposition 2 of Example 1 and Example 2).

RESULTS AND CONCLUSIONS

In the foregoing two examples were presented. In particular, decomposition 1 of Example 1 was found to provide exact results (which easily can be checked by comparing the explicit formulae) and all other results are approximate. The following discussion in order to obtain rules for the appropriate decomposition of the transition rate matrix.

Decomposing the transition rate matrix of a PN is equivalent to dividing the entire net into subnets. On the PN level each subnet interact with the remaining part of the net via the transitions which must be fired in order to depart from the net (inputs) and the transitions which must be fired in order to depart from the subnets (outputs). On the Markov level the inputs to a subsystem are represented by the elements of the transition rate matrix above the diagonal block representing the subsystem, and the outputs are represented by the elements to the left or to the right of the corresponding diagonal block (in the latter case the outputs act as inputs to the next subsystem).

Applying decomposition 1 to Example 1 (solid lines) each subsystem has only one input and one output (e.g. $p(t_2)\lambda_1$ is the input to subsystem 2 and λ_6 the corresponding output, the latter is the input to subsystem 3 and λ_9 is the output) whereas with decomposition 2 subsystem 3 has two inputs (λ_7 and $p(t_3)\lambda_1$) and two outputs (λ_9 and λ_5). In Example 2 described by Q_{II} subsystem 3 has one input (λ_6) and two outputs (λ_9 and λ_{10}). In the same example described by $\overset{*}{Q}_{II}$ subsystem 3 has two inputs (λ_7 and $p(t_3)\lambda_1$) and two outputs (λ_5 and λ_{10}).

Summarizing the results, decomposition 1 applied to Example 1 provides each subsystem with one input and one output and as shown in Giglmayr (1986) the approach presented provides exact results for the steady-state marking probabilities. Applying the decomposition 2 to Example 1 subsystem 3 has two inputs and two outputs and the results are approximate.

Now the question arises wether the exact results are destroyed by the multiple inputs to the subsystem, or by the multiple outputs or by both. Therefore, Example 2 was choosen where subsystem 3 has one input and two outputs and the result is still approximate (setting $\lambda_9=0$ and replacing λ_{10} by λ_9 provides the exact result from the foregoing). Hence, the exact result is destroyed by the multiple output. Now the transition rate matrix of Example 2 is rearranged ($\overset{*}{Q}_{II}$) and the decomposition is choosen in order to provide subsystem 3 with two inputs and two outputs. Setting $\lambda_{10}=0$ there, we have two inputs and one output and the result is still approximate which obviously is caused by the multiple input.

Our conclusions from the foregoing discussion is

- the concept of near-complete decomposability is able to provide exact results
- for this reason the subsystems must have only one input and one output
- multiple inputs as well as multiple outputs of a system give approximate results.

APPENDIX

The main steps of the decomposition algorithm are as follows:

1. Find appropriate subsystems (with one input and one output) by decomposing the transition rate matrix
2. Decompose the matrix Q in eq(2) into
   $Q=V+W$
   where V contains the (diagonal) submatrices Q_{IJ} and W contains the residual part of Q (the non-diagonal submatrices)
3. Find a matrix X such that
   $\overset{*}{Q}=V+X$
   where $\overset{*}{Q}$ is a completely decomposable transition rate matrix (i.e. a stochastic matrix which only contains stochastic submatrices in the diagonal)
4. Compute all eigenvalues of each submatrix $\overset{*}{Q}_I$ by
   $\det(\overset{*}{Q}_I - \lambda E) = 0$
   (we use $\overset{*}{\lambda}(i_I)$ to denote the eigenvalues of $\overset{*}{Q}_I$)
5. In the case approximate results are expected, check wether the Markov chain is nearly-complete decomposable (<) or not ($\geq$) by the condition
   $\varepsilon < [1 - \max_I \overset{*}{\lambda}(2_I)]/2$
   where ε represents the maximum degree of coupling between the subsystems Q_I which is expressed by the maximum sum of the matrix rows of the non-diagonal submatrices of Q

6. Compute the left eigenvector associated with the eigenvalue $\overset{*}{\lambda}(i_I)$
   $\overset{*}{e}^T(i_I)[\overset{*}{Q}_I - \overset{*}{\lambda}(i_I)E] = 0$
7. Obtain the transition rate matrix $R=(r_{IJ})$ such that
   $r_{IJ} = \sum_i \overset{*}{e}_{i_I}(1_I) \sum_j q_{i_I j_J}$
8. Obtain the stationary probability vector X of R such that
   $X(R - E) = 0$
9. Finally,
   $x_{i_I} = X_I \overset{*}{e}_{i_I}(1_I)$
   where the x_{i_I}´s approaches the probability vector in eq(1) and eq(2) (the X_I´s and x_{i_I}´s are the macro and micro aggregate probabilities, respectively).

REFERENCES

Bobbio A. and Trivedi K. (1986): An aggregation technique for the transient analysis of stiff Markov chains. To appear in IEEE Trans. on Computers

Courtois P.J. (1977): Decomposability - Queueing and Computer System Applications. 1st ed., Academic Press, New York

Giglmayr J. (1986): Analysis of stochastic Petri nets by the decomposition of the transition rate matrix. Submitted to IEEE Trans. on Computers

Kameda H. (1984): Optimality of central processor scheduling policy for processing a job stream. ACM Trans. Comp. Syst. 2, 78-90

Ajome-Marsan M. et al (1984): A class of generalized stochastic Petri nets for the performance evaluation of multiprocessor systems. ACM Trans. Comp. Syst. 2, 93-122

Heinrich-Hertz-Institut fuer Nachrichtentechnik Berlin GmbH
Einsteinufer 37
D-1000 Berlin 10
FRG

ON PROBABILISTIC INTERPRETATION OF CONSULTING SYSTEMS

Jiří Grim
Prague

1. INTRODUCTION

Consulting systems work with propositional variables which may be uncertain. To express the uncertainty the value of a propositional variable is characterized by a weight called degree of belief or certainty degree. A domain dependent knowledge-base of a consulting system is formulated by means of rules of the form:

$$\text{(1.1)} \qquad IF\ A\ \ THEN\ \ W(v_i) = w_{i|A}$$

in a way similar to that of human reasoning. In words, the propositional variable v_i (the consequent) is true with the weight $w_{i|A}$ if the antecedent A is true. The aim of the consulting systems is to derive the weights of some practically important goal variables given a knowledge base and the weights of a set of input variables called questions. For this purpose some inference rules are necessary, usually specified in a heuristic way. The information processing in the consulting systems is generally very complex and uneasy to analyse. This circumstance may become a serious difficulty when designing a practical consulting system.

In this paper we consider a probabilistic view of consulting systems (cf./4/,/5/). It is based essentially on the fact that the uncertainty of propositional variables may be well expressed by probabilities of their two possible values. A rule of the form (1.1) defines in this sense a conditional probability.

$$(2.5)\quad R \in \mathcal{R}; \quad R: W(v_i^{x_i}|A) = w_{i|A}^{x_i}(1-w_{i|A})^{1-x_i}; \quad x_i \in \{0,1\}$$

To obtain an acyclic structure of dependences we shall assume that $\mathcal{R}$ is a loop-free set of rules (cf./3/).

In view of the set $\mathcal{R}$ a variable $v \in \mathcal{V}$ is called a question if it occurs only in antecedents of rules and it is called a goal if it occurs only as a consequent. The variables which occur both in antecedents and consequents are intermediate. We denote $\mathcal{V}_0$ the set of all questions and $\mathcal{V}^*$ the set of all goals of $\mathcal{R}$:

$$(2.6)\quad \mathcal{V}_0 = \{v_1, v_2, \ldots, v_r\} \subset \mathcal{V}; \quad \mathcal{V}^* = \{v_{s+1}, v_{s+2}, \ldots, v_N\} \subset \mathcal{V}.$$

To enable a probabilistic interpretation of a consulting system we make the following assumptions:

Assumption I. A propositional variable $v_i \in \mathcal{V}$ having a weight $W(v_i) = w_i$ is a binary random variable taking the values 1 (true) or 0 (false) with the probabilities w_i or $(1 - w_i)$ respectively:

$$(2.7)\quad P\{v_i = x_i\} = p_i(x_i) = W(v_i^{x_i}) = w_i^{x_i}(1-w_i)^{1-x_i} = \begin{cases} w_i; & x_i = 1 \\ 1-w_i; & x_i = 0 \end{cases}$$

Assumption II. For each rule $R \in \mathcal{R}$

$$(2.8)\quad R: W(v_i|A) = w_{i|A}; \quad A = v_{i_1}^{x_{i_1}} \& \ldots \& v_{i_l}^{x_{i_l}}; \quad x_{i_\alpha} \in \{0,1\}$$

the weight $w_{i|A}$ defines a conditional distribution of the variable $v_i \in \mathcal{V}$:

$$(2.9)\quad \begin{aligned} &P\{v_i = x_i | v_{i_1} = x_{i_1}, \ldots, v_{i_l} = x_{i_l}\} = p_i(x_i|A) = \\ &= W(v_i^{x_i}|A) = w_{i|A}^{x_i}(1-w_{i|A})^{1-x_i} = \begin{cases} w_{i|A}; & x_i = 1 \\ 1-w_{i|A}; & x_i = 0 \end{cases} \end{aligned}$$

In view of these assumptions we are given a vector $\boldsymbol{v}$ of binary random variables taking the values from the corresponding binary space

$$(2.10)\qquad \boldsymbol{v} = (v_1, v_2, \ldots, v_N) \in \mathcal{X}; \quad v_i \in \mathcal{V}; \quad \mathcal{X} = \{0,1\}^N$$

The knowledge base is given in form of a system of conditional distributions defined by the set of rules $\mathcal{R}$ for all goal- and intermediate variables:

$$(2.11)\qquad \tilde{\mathcal{R}} = \{p_i(x_i|A_{ij}),\ j = 1,2,\ldots,m_i;\ i = r+1,\ldots,N\}$$

The input information supplied by the user is available as a set of marginal distributions of questions, s.c. questionnaire:

$$(2.12)\qquad Q = \{p_1(x_1), p_2(x_2), \ldots, p_r(x_r)\}$$

Thus the aim of a probabilistic consulting system is to derive a joint distribution of the goal variables $\mathcal{V}^*$ given a knowledge base $\tilde{\mathcal{R}}$ and a questionnaire Q.

3. A PROBABILISTIC CONSULTING SYSTEM

It can be seen that a probabilistic consulting system is fully determined by the distribution of the vector $\boldsymbol{v}$:

$$(3.1)\qquad \mathcal{P}\{v_1 = x_1, v_2 = x_2, \ldots, v_N = x_N\} = p(\boldsymbol{x}); \quad \boldsymbol{x} = (x_1, \ldots, x_N) \in \mathcal{X}.$$

Introducing notation

$$(3.2)\qquad \begin{aligned} \boldsymbol{x}^{(i)} &= (x_1, x_2, \ldots, x_i) \in \mathcal{X}^{(i)}; & \mathcal{X}^{(i)} &= \{0,1\}^i \\ \boldsymbol{x}^* &= (x_{s+1}, \ldots, x_N) \in \mathcal{X}^*; & \mathcal{X}^* &= \{0,1\}^N \end{aligned}$$

we can express the joint distribution of goals as a marginal of $p(\boldsymbol{x})$:

$$(3.3)\quad P\{v_{\Delta+1}=x_{\Delta+1},\ldots,v_N=x_N\}=p(\boldsymbol{x}^*)=\sum_{\boldsymbol{x}^{(\Delta)}\in\mathcal{X}^{(\Delta)}} p(\boldsymbol{x}^*|\boldsymbol{x}^{(\Delta)})\,p(\boldsymbol{x}^{(\Delta)})$$

Now, the application of probabilistic consulting systems reduces to evaluation of the sum (3.3). The corresponding evaluation method represents a natural probabilistic inference machine.

Let us recall that the set $\mathcal{R}$ is loop-free and therefore the dependence structure of rules corresponds to an acyclic oriented graph:

$$(3.4)\quad \vec{\mathcal{G}}=\{\mathcal{V},\mathcal{E}\};\quad \mathcal{E}=\{(\overrightarrow{v_k,v_i}):\ v_k\in\mathcal{V}(A_{ij});\ 1\le j\le m_i;\ i=\Delta+1,\ldots,N\}$$

Here the variables of the set $\mathcal{V}$ are the nodes and the set of edges $\mathcal{E}$ is defined by means of the input sets of rules. Using a well known property of the acyclic oriented graphs we can order the nodes in accordance with the orientation of all edges. Consequently, we may assume without any loss of generality that it holds:

$$(3.5)\quad (\overrightarrow{v_k,v_i})\in\mathcal{E}\ \Rightarrow\ k<i$$

or equivalently

$$(3.6)\quad v_k\in\mathcal{V}(A_{ij})\ \Rightarrow\ k<i$$

Thus, by means of rules, a consequent depends only on variables having a smaller index and simultaneously the notation (2.6) is applicable. Using this facts and a well known expansion formula, we can write

$$(3.7)\quad p(\boldsymbol{x}^*)=\sum_{\boldsymbol{x}^{(\Delta)}\in\mathcal{X}^{(\Delta)}} p(\boldsymbol{x}^*|\boldsymbol{x}^{(\Delta)})\,p_\Delta(x_\Delta|\boldsymbol{x}^{(\Delta-1)})\ldots p_3(x_3|\boldsymbol{x}^{(2)})\,p_2(x_2|x_1)\,p(x_1)$$

whereby the occuring conditional probabilities can be deduced from the knowledge base $\tilde{\mathcal{R}}$.

To simplify the formula (3.7) we omit any dependence not implied by the set of conditional distributions $\tilde{\mathcal{R}}$. First, as the goals are not interrelated by rules, we can write

$$(3.8)\quad p_i(x_i|\mathbf{x}^{(i-1)}) = p_i(x_i|\mathbf{x}^{(s)}); \qquad i = s+1,\dots,N$$

and, applying an analogous argument to questions, we obtain

$$(3.9)\quad p_i(x_i|\mathbf{x}^{(i-1)}) = p_i(x_i); \qquad i = 2,3,\dots,r.$$

After substitutions (3.8),(3.9) the formula (3.7) can be rewritten as follows:

$$(3.10)\quad p(\mathbf{x}^*) = \sum_{\mathbf{x}^{(s)}\in\mathcal{X}^{(s)}} \Big[\prod_{i=1}^{r} p_i(x_i)\Big]\Big[\prod_{i=r+1}^{s} p_i(x_i|\mathbf{x}^{(i-1)})\Big]\Big[\prod_{i=s+1}^{N} p_i(x_i|\mathbf{x}^{(s)})\Big]$$

Let us note that for any elementary conjunction

$$(3.11)\quad A = v_{i_1}^{\xi_{i_1}} \& \dots \& v_{i_\ell}^{\xi_{i_\ell}}, \qquad \xi_{i_\alpha} \in \{0,1\};$$

there is a unique subset $\mathcal{A}$ of $\mathcal{X}$:

$$(3.12)\quad \mathcal{A} = \{\mathbf{x}\in\mathcal{X} : x_{i_\alpha} = \xi_{i_\alpha}, \quad \alpha = 1,2,\dots,\ell\} \subset \mathcal{X}$$

such that A is true iff $\mathbf{v}\in\mathcal{A}$. We make use of this equivalence without introducing a different symbol. In this sense an antecedent A_{ij}, $(\mathcal{V}(A_{ij})\subset\mathcal{V})$ will be viewed in the following as a subset of $\mathcal{X}$, and we can use the notation:

$$(3.13)\quad P\{v_i = x_i | \mathbf{v}\in A_{ij}\} = p_i(x_i|A_{ij}) = w_{i|j}^{x_i}(1-w_{i|j})^{1-x_i}$$

Simultaneously we denote $A^{(i)}$ the restriction of the set A, $(A\equiv\mathcal{A})$ on the subspace $\mathcal{X}^{(i)}$:

$$(3.14)\quad A^{(i)} = \{\mathbf{x}^{(i)}\in\mathcal{X}^{(i)} : x_{i_\alpha} = \xi_{i_\alpha}, \ i_\alpha \le i, \ \alpha = 1,2,\dots,\ell\}$$

Again, to reduce the general dependences in (3.10) we use the following easily verified implication (cf. (3.14), Assumption II.):

$$(3.15)\quad \mathbf{x}^{(i-1)} \in A_{ij}^{(i-1)} \Rightarrow p_i(x_i|\mathbf{x}^{(i-1)}) = p_i(x_i|A_{ij}) = p_i(x_i|A_{ij}^{(s)}).$$

To enable a simple application of this relation we make first the following assumptions :

Assumption III. (exclusiveness of antecedents)

$$(3.16)\quad j \neq l \Rightarrow A_{ij} \cap A_{il} = \emptyset; \quad i = r+1,\dots,N.$$

Assumption IV. (completness of antecedents)

$$(3.17)\quad \bigcup_{j=1}^{m_i} A_{ij} = \mathcal{X}; \qquad i = r+1,\dots,N.$$

From the relations (3.16),(3.17) it follows that the intersections

$$(3.18)\quad B(j_{r+1},\dots,j_N) = \bigcap_{i=r+1}^{N} A_{ij_i}^{(s)}; \qquad 1 \le j_i \le m_i;$$

form a partition of the subspace $\mathcal{X}^{(s)}$ and therefore we can rearrange the sum (3.10):

$$(3.19)\quad \sum_{\mathbf{x}^{(s)} \in \mathcal{X}^{(s)}} \approx \sum_{j_{r+1}=1}^{m_{r+1}} \dots \sum_{j_N=1}^{m_N} \sum_{\mathbf{x}^{(s)} \in B(j_{r+1},\dots,j_N)}$$

Now, using the relation (3.15), we can write (cf.(3.10)):

$$(3.20)\quad p(\mathbf{x}^*) = \sum_{j_{r+1}=1}^{m_{r+1}} \dots \sum_{j_N=1}^{m_N} \Big[\prod_{i=s+1}^{N} p_i(x_i|A_{ij_i})\Big] W(j_{r+1},\dots,j_N)$$

where

$$(3.21)\quad W(j_{r+1},\dots,j_N) = \sum_{\mathbf{x}^{(s)} \in B(j_{r+1},\dots,j_N)} \Big[\prod_{i=1}^{r} p_i(x_i)\Big]\Big[\prod_{i=r+1}^{s} p_i(x_i|\mathbf{x}^{(i-1)})\Big] =$$

$$= \sum_{\mathbf{x}^{(s)} \in B(j_{r+1},\dots,j_N)} \Big[\prod_{i=1}^{r} p_i(x_i)\Big]\Big[\prod_{i=r+1}^{s} p_i(x_i|A_{ij_i})\Big]$$

A successful probabilistic interpretation of the consulting systems could clarify the underlying information processing and establish a useful link to the existing methods of statistical decision-making (/1/,/2/).

2. A PROBABILISTIC VIEW OF THE CONSULTING SYSTEMS

Let $\mathcal{V}=\{v_1, v_2, \ldots, v_N\}$ be a finite set of propositional variables. For any variable $v \in \mathcal{V}$ let $W(v)$ denote a certainty degree of v (degree of belief, weight). We define

(2.1) $$W(v) = w \in \langle 0,1 \rangle, \quad W(\neg v) = 1 - w; \quad v \in \mathcal{V},$$

whereby $w=1$ means "v is certainly true", $w=0$ means "v is certainly false" and $w=\frac{1}{2}$ means "v is unknown". We introduce notation:

(2.2) $$v^x = \begin{cases} v, & x=1 \\ \neg v, & x=0 \end{cases}, \quad W(v^x) = w^x(1-w)^{1-x} = \begin{cases} w, & x=1 \\ 1-w, & x=0 \end{cases}$$

Further let $\mathcal{R}$ be a nonempty set of rules of the form (1.1) representing a problem dependent knowledge base. The antecedent A of a rule R is assumed to be an elementary conjunction

(2.3) $$A = v_{i_1}^{x_{i_1}} \& \ldots \& v_{i_\ell}^{x_{i_\ell}}, \quad x_{i_\alpha} \in \{0,1\}, \quad \mathcal{V}(A) = \{v_{i_1}, \ldots, v_{i_\ell}\} \subset \mathcal{V},$$

whereby $\mathcal{V}(A)$ is a set of input variables of R and $v \in \mathcal{V}, (v \notin \mathcal{V}(A))$ is an output variable - the consequent of R. To simplify notation we shall write the rules equivalently in the form

(2.4) $$R \in \mathcal{R}, \qquad R: W(v_i | A) = w_{i|A},$$

or more generally in the form

are nonnegative weights. Consequently (cf.(3.13)), the resulting distribution $p(\boldsymbol{x}^*)$ is a finite mixture of multivariate Bernoulli distributions:

$$(3.22)\quad p(\boldsymbol{x}^*) = \sum_{j_{r+1}=1}^{m_{r+1}} \cdots \sum_{j_N=1}^{m_N} W(j_{r+1},\dots,j_N) \prod_{i=s+1}^{N} w_{i|j_i}^{x_i} (1-w_{i|j_i})^{1-x_i}$$

4. COMPUTATIONAL ASPECTS

As it follows from the formula (3.22) to determine the joint distribution of goals $p(\boldsymbol{x}^*)$ we have to compute the weights (3.21). This can be easily done as long as the antecedents have the form of elementary conjunctions. The following obvious lemma suggest a practically applicable procedure.

Lemma 4.1. Let $B(j_{r+1},\dots,j_N) \subset \mathcal{X}^{(s)}$ be an intersection of the form (3.18) defined by

$$(4.1)\quad B(j_{r+1},\dots,j_N) = \{\boldsymbol{x}^{(s)} \in \mathcal{X}^{(s)} : x_{i_\alpha} = \xi_{i_\alpha},\ \alpha = 1,2,\dots,m\}$$

where

$$(4.2)\quad i_1 < i_2 < \dots < i_k \le r < i_{k+1} < \dots < i_m \le s$$

Then it holds

$$(4.3)\quad W(j_{r+1},\dots,j_N) = \sum_{\boldsymbol{x}^{(s)} \in B(j_{r+1},\dots,j_N)} \left[\prod_{i=1}^{r} p_i(x_i)\right]\left[\prod_{i=r+1}^{s} p_i(x_i|A_{ij_i})\right] =$$

$$= \left[\prod_{\alpha=1}^{k} p_i(\xi_{i_\alpha})\right]\left[\prod_{\alpha=k+1}^{m} p_{i_\alpha}(\xi_{i_\alpha}|A_{i_\alpha j_{i_\alpha}})\right]$$

Let us recall now that the formula (3.22) is based on the Assumptions III,IV which are usually not satisfied by a practical knowledge base. We shall consider first the possibility that the condition (3.17) does not hold. In such a case we can write

$$(4.4) \qquad \mathcal{X} - \bigcup_{j=1}^{m_i} A_{ij} = A_{i0} \neq \emptyset$$

for some consequents $v_i \in \mathcal{V}$. One possibility to avoid this difficulty is a formal definition of additional rules for the nonempty antecedents (4.4):

$$(4.5) \qquad p_i(x_i|A_{i0}) = \frac{1}{2}; \qquad (A_{i0} \neq \emptyset)$$

However, this approach introduces "nonstandard" antecedents and does not distinguish between missing information and a "truly" uniform distribution. Another and even simpler way is to confine the sum in (3.22) to the available antecedents only. Denoting

$$(4.6) \qquad \sigma_W = \sum_{j_{s+1}=1}^{m_{s+1}} \cdots \sum_{j_N=1}^{m_N} W(j_{s+1}, \ldots, j_N); \qquad (0 \leq \sigma_W \leq 1)$$

we can express the distribution of goals again as a finite mixture:

$$(4.7) \quad p(x^*) = \frac{1}{\sigma_W} \sum_{j_{s+1}=1}^{m_{s+1}} \cdots \sum_{j_N=1}^{m_N} W(j_{s+1}, \ldots, j_N) \prod_{i=s+1}^{N} w_{i|j_i}^{x_i} (1 - w_{i|j_i})^{1-x_i}$$

whereby the number $(1-\sigma_W)$ may be viewed as a measure of missing information.

The Assumption III. is reasonable from a logical point of view but rarely satisfied. Again the condition (3.16) can be restored formally if we replace the pair of conditional distributions with overlapping antecedents

$$(4.8) \qquad p_i(x_i|A_{i1}),\ p_i(x_i|A_{i2}) \in \tilde{\mathcal{R}}; \qquad A_{i1} \cap A_{i2} \neq \emptyset$$

by the modified distributions

$$\bar{p}_i(x_i|A_{i_1} - A_{i_1} \cap A_{i_2}) = p_i(x_i|A_{i1}); \qquad (0 \le \vartheta \le 1);$$

$$\bar{p}_i(x_i|A_{i2} - A_{i1} \cap A_{i2}) = p_i(x_i|A_{i2}); \tag{4.9}$$

$$\bar{p}_i(x_i|A_{i1} \cap A_{i2}) = \vartheta \cdot p_i(x_i|A_{i1}) + (1-\vartheta)\, p_i(x_i|A_{i2});$$

However, the formulas (4.9) become complicated in case of several overlapping antecedents and the Lemma 4.1 cannot be applied directly to the new regularized antecedents. For this reason a dialog with experts could enable a more efficient regularization of the knowledge base.

REFERENCES

/1/ GRIM J.: Multivariate statistical pattern recognition with nonreduced dimensionality.
Kybernetika (1986) 22, 142-157

/2/ GRIM J.: Sequential decision-making in pattern recognition based on the method of independent subspaces.
Proc. DIANA II. Conference on discriminant analysis, Liblice near Prague, 26.5.-30.5.1986

/3/ HÁJEK P.: Combining functions for certainty degrees in consulting systems.
Int. J. Man-Machine Studies (1985) 22, 59-76

/4/ HAVRÁNEK T., HÁJEK P.: On connections between model search methods for contingency tables and intensional expert systems.
Tenth Prague Conference on Information Theory, Statistical Decision Functions and Random Processes. Prague 1986

/5/ PEREZ A., JIROUŠEK R.: Constructing an intensional expert system (INES).
Medical Decision Making: Diagnostic Strategies and Expert Systems. J.H. van Bemel, F. Grémy and J.Zvárová (eds.), North-Holland, Amsterdam 1985, 307-315

Czechoslovak Academy of Sciences
Institute of Information Theory and Automation
182 08 Prague 8
Pod vodárenskou věží 4
Czechoslovakia

FELLER'S ONE-DIMENSIONAL DIFFUSIONS AS UNIQUE WEAK SOLUTIONS TO STOCHASTIC DIFFERENTIAL EQUATIONS

Jürgen Groh

Jena

Key words: One-dimensional diffusions, stochastic analysis

ABSTRACT

Continuous strong Markov processes X on the line generated by Feller's generalized second order differential operator $D_m D_p^+$ are considered. In the case of inaccessible boundaries of the state space R and an identical natural scale $p(x) = x$ these diffusion processes are local martingales. Further, it is supposed that the speed measure m contains a strictly positive absolutely continuous component. Then the diffusion X is characterized as weak solution to a stochastic differential system involving local time. This representation is unique in law. Further, Itô's change of variable formula and a theorem concerning absolutely continuous change of law are given. Finally, the case of a general natural scale is considered.

1. INTRODUCTION

We are concerned with Feller's one-dimensional diffusions X generated by the generalized second order differential operators $D_m D_p^+$. To omit difficulties arising from boundary conditions or explosions we assume that both boundaries $-\infty$, $+\infty$ of the state space R are inaccessible in the terminlogy of Feller (1952). At first we assume an identical natural scale $p(x) = x$, then according to Arbib (1965), Lai (1973) the diffusion X is even a martingale.

Furthermore, we suppose that the speed measure m has the structure

$$dm(x) = \sigma(x)^{-2} \cdot 2dx + d\mu(x), \qquad x \in R \tag{1}$$

with a measurable function σ, bounded and bounded away from zero on compact intervals. With μ we denote a measure on R which is singular with respect to the Lebesgue measure, i.e. there exist two disjoint Borel sets Λ and Γ whose union is R and $\mu(\Lambda) = \int 1_\Gamma(x)dx = 0$. Then we can give a representation of the diffusion X as unique weak solution to some stochastic differential system. Essentially, we use the fact that a diffusion X with generator $D_m D_x^+$ can be constructed via stochastic time change from some underlying Brownian motion process.

2. PRELIMINARIES

Let $X = (X_t, F_t, P_x^o)$ be a diffusion process in the sense of Dynkin (1965) generated by the differential operator $D_m D_x^+$. We regard this diffusion as constructed via a time change T from some Brownian motion process B^o. This is not a serious restriction because under the present conditions on the speed measure m the time change process is invertible. Thus, let us agree that $B^o = (B_t^o, F_t^o, P_x^o)$ is the "right" Brownian motion over the sample space (Ω^o, F^o) starting P_x^o-a.s. from $x \in R$ and with the standard local times $\{l^o(t,y);\ y \in R\}$, compare Trotter (1958), Itô and McKean (1965). The time change T, defined for all $t \geq 0$ by the relation

$$t = \int_R l^o(T_t, y)\, dm(y) \tag{2}$$

leads by means of the substitutions

$$X_t = B^o_{T(t)}, \qquad F_t^X = F^o_{T(t)}, \qquad t \geq 0$$

to the diffusion $X = (X_t, F_t^X, P_x^o)$. As it is well-known, even the process X has local times, which can be expressed by

$$l(t,y) = l^o(T_t, y), \qquad t \geq 0,\ y \in R.$$

Moreover, for each locally integrable function f it holds

$$\int_0^t f(X_u)du = \int_R l(t,y)f(y)dm(y), \qquad t \geq 0$$

3. THE STOCHASTIC EQUATION

In Groh (1982b) we have shown an explicit expression for the natural increasing process, or quadratic variation of the diffusion $X = (X_t, F_t^X, P_x^o)$, it holds

$$\langle X\rangle_t = \int_0^t \sigma^2(X_s)I_\Lambda(X_s)ds, \qquad t \geq 0.$$

Observe that the quadratic variation $\langle X\rangle$ depends essentially on the behaviour of the diffusion X within the "nonsingular" set Λ. Furthermore, $\langle X\rangle$ is absolutely continuous with respect to the Lebesgue measure. Following Doob (1953), the diffusion X can be represented as a stochastic integral with respect to some Brownian motion process, where it may be necessary to enlarge the underlying probability space by adjunction of another Brownian motion. This procedure will be crucial in constructing a stochastic differential equation for X, for details compare Groh (1984a).

THEOREM 1. Given a diffusion X generated by the infinitesimal operator $D_m D_x^+$. Then one can find a filtered probability space (Ω, F, F_t, P_x) carrying a Brownian motion $B = \{B_t, t \geq 0\}$ and the diffusion X as well, such that the relations

$$(3) \qquad \begin{cases} X_t = X_0 + \int_0^t \sigma(X_s)1_\Lambda(X_s)dB_s, & t \geq 0, \\ \int_0^t 1_\Gamma(X_s)ds = \int_R l(t,y)d\mu(y), & t \geq 0, \\ X_0 = x \end{cases}$$

holds P_x-a.s. for all $x \in R$. In other words, the pair (X,B) forms a weak solution to the stochastic differential system (3).

4. WEAK UNIQUENESS

Now we can ask for which class of processes is determined by the system of equations (3).

By a solution to the differential system (3) we understand a triplet of stochastic processes $\mathfrak{X} = (X_t, B_t, L_t)$ defined on a filtered probability space $(\Omega, F, F_t, P_x;\ x \in R)$ satisfying P_x-a.s. the following conditions:

(i) B_t is a Brownian motion starting from zero.

(ii) $L_t = \{l(t,y);\ y \in R\}$ is a family of local times for the continuous martingale X. The processes $l(t,y)$ are P_x-a.s. non-negative, (t,y)-continuous and increasing in t with

$$\int_0^t 1_{\{u:\ X_u = y\}}(s)\, l(ds,y) = l(t,y).$$

(iii) The processes X_t, B_t, and L_t satisfy the equations (3).

Let $\mathfrak{X} = (X_t, B_t, L_t)$ be a solution to (3). Applying a generalization of the Lévy-Doob theorem, given by Arbib (1965), one can conclude that the process X is a diffusion with infinitesimal generator $D_m D_p^+$. Consequently, the law of the process X is uniquely determined by (3).

THEOREM 2. The solution to the stochastic differential system (3) is unique in law.

5. ITO'S CHANGE OF VARIABLE FORMULA

The knowledge of the natural increasing process $\langle X \rangle$ allows us to deduce an explicit version of Itô's formula for the process X.

THEOREM 3. Let f be a twice continuously differentiable function on R. Then, for every $t \geq 0$ and $x \in R$, it holds the equation

$$f(X_t) = f(X_0) + \tfrac{1}{2}\int_0^t f''(X_u)\sigma^2(X_u) I_\Lambda(X_u)\,du +$$

$$+ \int_0^t f'(X_u)\sigma(X_u) I_\Lambda(X_u)\,dB_u, \qquad P_x\text{-a.s.}$$

This assertion is an immediate consequence of the change of variable formula given in Kunita and Watanabe (1967), which contains the equation $f(X_t) - f(X_0) = \int_0^t f'(X_u)dX_u + \frac{1}{2}\int_0^t f''(X_u)d\langle X\rangle_u$ $t \geq 0$, as a special case.

The theorem can be used to construct a stochastic differential equation for a diffusion with a non-identical but twice differentiable natural scale p, compare Groh (1982b).

6. ABSOLUTELY CONTINUOUS CHANGE OF LAW

Now we are concerned with diffusions, absolutely continuous with respect to the given process $X = (X_t, F_t, P_x)$, determined by the generator $D_m D_x^+$. An analytical condition for the absolute continuity was given by S. Orey (1974) in terms of the corresponding speed measures and scale functions. Let (X_t, F_t, Q_x) be another diffusion process with generator $D_n D_s^+$ and inaccessible boundaries. The probabilities Q_x are absolutely continuous with respect to P_x for all $x \in R$ if and only if $(dn/dm)\cdot(ds/dx) = 1$ and the second derivative s" exists almost everywhere. Using the stochastic differential for X one obtains an explicit formula for the corresponding Radon-Nikodym derivatives, compare Groh (1984b).

THEOREM 4. For every $x \in R$ and $t \geq 0$ it holds

$$\frac{dQ_x \mid F_t}{dP_x \mid F_t} = \exp\left\{ \int_0^t b(X_u)\sigma(X_u)I_\Lambda(X_u)dB_u - \right.$$
$$\left. - \tfrac{1}{2}\int_0^t b^2(X_u)\sigma^2(X_u)I_\Lambda(X_u)du \right\},$$

where $b = - s''/2s'$. The process

$$C_t = B_t - \int_0^t b(X_u)\sigma(X_u)I_\Lambda(X_u)du, \qquad t \geq 0$$

is a Brownian motion over the filtered probability space (Ω, F, F_t, Q_x). Finally, X_t forms a solution to the stochastic differential equation under the measures Q_x, $x \in R$

$$X_t = X_0 + \int_0^t b(X_u)\sigma^2(X_u)I_\Lambda(X_u)du + \int_0^t \sigma(X_u)I_\Lambda(X_u)dC_u, \quad t \geq 0.$$

7. DIFFUSIONS WITH GENERAL NATURAL SCALE

In the last two sections we have derived the stochastic differentials for diffusions generated by $D_m D_p^+$ with twice differentiable scale function p. Now we extend our results to diffusions with not necessarily smooth natural scales. Because we are interested in the representation of X as solution to some stochastic differential equation, the process X should be at least a semimartingale. This implies that the natural scale p is locally the difference of two bounded convex functions, compare Cinlar, Jacod, Protter and Sharpe (1980). Concerning the speed measure m we assume as in the foregoing sections that it contains a strictly positive absolutely continuous component. Both boundaries are assumed to be inaccessible. Under these conditions one can construct a diffusion coefficient σ, a singular measure μ concentrated on Γ, $\Lambda = R \setminus \Gamma$, and a measure generating function β such that the following theorem is true, compare Groh (1985). Here $L_t^a(X)$ stands for the left continuous version of local times for the semimartingale X, see Yor (1978)

$$(X_t - a)^+ = (X_0 - a)^+ + \int_0^t 1_{\{X_s > a\}} dX_s + \tfrac{1}{2} L_t^a(X), \qquad t \geq 0.$$

THEOREM 5. Given a diffusion X generated by the operator $D_m D_p^+$ with scale functions m and p as described above. Then one can find a filtered probability space (Ω, F, F_t, P_x) carrying a Brownian motion $B = \{B_t,\ t \geq 0\}$ and the diffusion X, such that the relations

$$(4) \quad \begin{cases} X_t = X_0 + \int_0^t \sigma(X_s) 1_\Lambda(X_s) dB_s + \int_R L_t^a(X) \sigma(a)^{-2} d\beta(a), & t \geq 0, \\ \int_0^t 1_\Gamma(X_s) ds = \tfrac{1}{2} \int_R L_t^a(X) d\mu(a), & t \geq 0, \\ X_0 = x \end{cases}$$

holds P_x-a.s. for all $x \in R$. Consequently, (X,B) forms a weak solution to the stochastic differential system (4). This solution is unique in law.

We conclude our report with a modified version of Itô's formula which is valid for all functions from the domain of definition of the differential generator $D_m D_p^+$ (which are not necessarily twice differentiable in the ordinary sense). The only assumption here is that both boundaries $-\infty$, $+\infty$ are inaccessible. Observe that in this case the process p(X) is a local martingale.

THEOREM 6. Given a generalized differentiable function f from $\mathfrak{D}(D_m D_p^+)$. Then for all $x \in R$ and $t \geq 0$ it holds P_x-a.s.

$$f(X_t) - f(X_0) = \int_0^t (D_m D_p^+ f)(X_s) ds + \int_0^t (D_p^+ f)(X_s) dp(X_s).$$

REFERENCES

Arbib M.A. (1965): Hitting and martingale characterizations of one-dimensional diffusions, Zeitschrift Wahrscheinlichkeitstheorie Verw. Gebiete 4, 232-247.

Cinlar E., J. Jacod, P. Protter and M.J. Sharpe (1980): Semimartingales and Markov processes, Zeitschrift Wahrscheinlichkeitstheorie Verw. Gebiete 54, 161-219.

Doob J.L. (1953): Stochastic Processes, John Wiley & Sons, New York.

Dynkin E.B. (1965): Markov Processes, Vols. 1-2, Springer-Verlag, Berlin.

Feller W. (1952): The parabolic differential equations and the associated semi-groups of transformations, Ann. Math. 55, 468-519.

Feller W. (1958): On the intrinsic form for second order differential operators, Illinois J. Math. 2, 1-18.

Fisk D.L. (1966): Sample quadratic variation of sample continuous second order martingales, Zeitschrift Wahrscheinlichkeitstheorie Verw. Gebiete 6, 273-278.

Groh J. (1982a): A stochastic differential equation for a class of Feller's one-dimensional diffusions, Math. Nachr. 107, 267-271.

Groh J. (1982b): On a stochastic calculus for Feller's one-dimensional diffusions, Preprint N/82/10, Friedrich-Schiller-Universität Jena.

Groh J. (1984a): Stochastic calculus for Feller's one-dimensional diffusions, Mededelingen uit het Wiskundig Instituut, Katholieke Universiteit Leuven, No. 168.

Groh J. (1984b): On absolute continuity of Feller's one-dimensional diffusion processes, Math. Nachr. 116, 337-348.

Groh J. (1985): Feller's one-dimensional diffusions as weak solutions to stochastic differential equations, Math. Nachr. 122, 157-165.

Groh J. (1986): On Brownian motion with irregular drift, Illinois J. Math.

Ito K. and H.P. McKean, Jr. (1965): Diffusion processes and their sample paths, Springer-Verlag, Berlin.

Kunita H. and S. Watanabe (1967): On square integrable martingales, Nagoya Math. J. 30, 209-245.

Lai T.L. (1973): Space-time processes, parabolic functions and one-dimensional diffusions, Trans. Amer. Math. Soc. 175, 409-438.

Orey S. (1974): Conditions for the absolute continuity of two diffusions, Trans. Amer. Math. Soc. 193, 413-426.

Trotter H.F. (1958): A property of Brownian motion path, Illinois J. Math. 2, 425-433.

Wang A. (1977): Generalized Itô's formula and additive functionals of Brownian motion, Zeitschrift Wahrscheinlichkeitstheorie Verw. Gebiete 41, 153-159.

Wong E. (1971): Representations of martingales, quadratic variation and applications, SIAM J. Control 9, 621-633.

Yor M. (1978): Sur la continuité des temps locaux associés à certaines semi-martingales, Astérisque 52-53, 23-35.

Friedrich-Schiller-Universität
Sektion Mathematik
DDR-6900 Jena, UHH

DUAL VARIANT OF SOME STRENGHTENING

OF THE MAXIMAL ERGODIC THEOREM

Blahoslav Harman

Liptovský Mikuláš

Key words: dynamical system, measure preserving transformation

ABSTRACT

Dual variant of some special generalization of the classical maximal ergodic theorem is surveyed .

The aim of the paper is to complete some special generalization of the maximal ergodic theorem. The idea of the proof of the Theorem 2 is due to Garsia [1] . For detail studying of the classical case see e.g. [2],[3],[4] .

Let R be the set of the real numbers. Let $\mathcal{R}: R^2 \to \langle 0,1\rangle$ be the map which satisfied the following conditions:

i/ $\forall x,y \in R$, $x \neq y$: $\mathcal{R}(x,y) > 0 \Longleftrightarrow \mathcal{R}(y,x) = 0$, $\mathcal{R}(x,x) = 0$

ii/ $\forall x,y \in R$: $\mathcal{R}(x,y) \geqq \sup_{u \in R} \mathcal{R}(x,u)\, \mathcal{R}(u,y)$

iii/ $\forall x \in R$: $\mathcal{R}(0,x) > 0 \Longleftrightarrow x > 0$

iv/ $\forall x,y,z \in R$: $\mathcal{R}(x,y) > 0$, $\mathcal{R}(y,z) > 0 \Rightarrow \mathcal{R}(x,y) \leqq \mathcal{R}(x,z)$.

A nontrivial example of such a map is as follows:

$$\mathcal{R}_{\alpha}(x,y) = \left(\frac{1}{2} + \frac{1}{\pi}\operatorname{arctg}(y)\right)\left(1 - e^{\alpha(x-y)}\right) \chi_{(0;\infty)}(y-x) \text{ for all } \alpha>0.$$

From above mentioned conditions it is easy to see that the map is a nondecreasing function of the second variable and a nonincreasing of the first one.

Let $(X, \mathcal{S}, \lambda, \tau)$ be a dynamical system, i.e. X is a non-empty set, $\mathcal{S}$ be a σ-algebra on X, λ be a measure on $\mathcal{S}$ and

$\tau: X \to X$ be a measure λ preserving transformation. Let a be a real number , $f \in L_1(\lambda)$. For our purposes let us introduce the following notations.

$T: R^X \to R^X$, $f \mapsto Tf$ where $(Tf)(x) = f(\tau x)$

$S_0^{(a)} f: X \to R$, $x \mapsto 0$ for a R

$S_k^{(a)} f = f + Tf + \dots + T^{k-1} f - ka$ for $k = 1,2,3,\dots$

${}^+S_n^{(a)} f = \max_{0 \le k \le n} S_k^{(a)} f$ ${}^-S_n^{(a)} f = \min_{0 \le k \le n} S_k^{(a)} f$

$$\mu_n^{(a)} f = \max_{0 \le k \le n} \mathcal{R}(0, S_k^{(a)} f)$$

$$\varrho_n^{(a)} f = \max_{0 \le k \le n} \mathcal{R}(S_k^{(a)} f, 0)$$

It is easy to see that ${}^+S_n^{(a)} f \geq 0$ and ${}^-S_n^{(a)} f \leq 0$. If moreover X is a set of finite measure then all above mentioned functions are integrable.

Lemma 1 . Let $(X, \mathcal{S}, \lambda, \tau)$ have the above mentioned meaning. Let $f \in L_1(\lambda)$. Then $\int f \, d\lambda = \int Tf \, d\lambda$.

Proof : See e.g. [3] .

The following theorem was formulated and proved in [5] :

Theorem 1 . Let $(X, \mathcal{S}, \lambda, \tau)$ be a dynamical system . Let $f \in L_1(\lambda)$. Then

$$\int f(\mu_n^{(a)} f) \, d\lambda + \int (1 - \mu_n^{(a)} f)\, {}^+S_n^{(a)} f \, d\lambda \geq a \int \mu_n^{(a)} f \, d\lambda .$$

In the dual variant of this theorem we shall replace the functions $\mu_n^{(a)} f$ and ${}^+S_n^{(a)} f$ by the functions $\varrho_n^{(a)} f$ and ${}^-S_n^{(a)} f$ respectively. Since the map $\mathcal{R}$ does not satisfy necessarily the condition $\mathcal{R}(x,y) = \mathcal{R}(-y,-x)$, then $\mu_n^{(-a)}(-f)$ might not be equal to the $\varrho_n^{(a)} f$. Hence the following theorem is not a straightforward consequence of the preceding one.

Theorem 2 . Let $(X, \mathcal{S}, \lambda, \tau)$ be a dynamical system. Let $f \in L_1(\lambda)$, $\lambda(X) < \infty$, $a \in R$. Let $\varrho_n^{(a)} f$ and ${}^-S_n^{(a)} f$ have the above mentioned meaning. Then

$$\int f(\varrho_n^{(a)} f)\, d\lambda + \int (1 - \varrho_n^{(a)} f)\, {}^{-}S_n^{(a)} f\, d\lambda \leq a \int \varrho_n^{(a)} f\, d\lambda .$$

Proof : From the definitions of $S_k^{(a)}$ and ${}^{-}S_n^{(a)}$ it follows:

$${}^{-}S_n^{(a)} f \leq S_k^{(a)} f \qquad \text{for } k = 0,1,2,\dots,n .$$

Due to elementary properties of the operator T we have
$T\,{}^{-}S_n^{(a)} f \leq TS_k^{(a)} f = Tf + T^2 f + \dots + T^k f - ka$ and then
$T\,{}^{-}S_n^{(a)} f + f - a \leq f + Tf + \dots + T^k f - (k+1)a = S_{k+1}^{(a)} f$. From the last inequality we obtain
$T\,{}^{-}S_n^{(a)} f + f - a \leq \min_{0 \leq k \leq n} S_{k+1}^{(a)} f \leq \min_{1 \leq k \leq n} S_k^{(a)} f$. Because of $\min_{1 \leq k \leq n} S_k^{(a)} f$ is negative if and only if $\varrho_n^{(a)} f > 0$ and $S_0^{(a)} f \equiv 0$, it follows $(T\,{}^{-}S_n^{(a)} f + f - a)\, \varrho_n^{(a)} f \leq (\min_{0 \leq k \leq n} S_k^{(a)} f)\, \varrho_n^{(a)} f =$

$$= (\varrho_n^{(a)} f)({}^{-}S_n^{(a)} f) .$$

After a short arrangement and integrating of the last inequality we have

$$\int f(\varrho_n^{(a)} f)\, d\lambda + \int (1 - \varrho_n^{(a)} f)\, {}^{-}S_n^{(a)} f\, d\lambda + \int (T\,{}^{-}S_n^{(a)} f - {}^{-}S_n^{(a)} f)\, d\lambda$$
$$- \int (1 - \varrho_n^{(a)} f)\, T\,{}^{-}S_n^{(a)} f\, d\lambda \leq a \int \varrho_n^{(a)} f\, d\lambda .$$

By using Lemma 1 $\int (T\,{}^{-}S_n^{(a)} f - {}^{-}S_n^{(a)} f)\, d\lambda = 0$.

Moreover $\int (1 - \varrho_n^{(a)} f)\, T\,{}^{-}S_n^{(a)} f\, d\lambda \leq 0$. The theorem is proved .

Remark. Let $E = \{(x,y) \in R^2 ;\ x < y\}$. Let $\mathcal{R}(x,y) = \chi_E$. It is possible to show in this case that

$$\mu_n^{(a)} f = \chi_{A_n} , \quad \varrho_n^{(a)} f = \chi_{B_n} \qquad \text{where}$$

$$A_n = \{x \in X ;\ \exists k : f(x) + f(\tau x) + \dots + f(\tau^{k-1} x) > ka\}$$
$$B_n = \{x \in X ;\ \exists k : f(x) + f(\tau x) + \dots + f(\tau^{k-1} x) < ka\}.$$

It is easy to see that

$({}^{+}s_n^{(a)}f)(x) = 0$ iff $(\mu_n^{(a)}f)(x) = 0$

$({}^{-}s_n^{(a)}f)(x) = 0$ iff $(\varrho_n^{(a)}f)(x) = 0$.

The assertions of the Theorem 1 and 2 will have the forms

$$\int_{A_n} f \, d\lambda \geq a\,\lambda(A_n) \quad \text{and} \quad \int_{B_n} f \, d\lambda \leq a\,\lambda(B_n) \; .$$

They make the generalization of the classical maximal ergodic theorem obvious.

REFERENCES

1 Garsia, A.M.: A simple proof of Eberhard Hopf s maximal ergodic theorem. J.Math and Mech. 14(1965), 381-382.

2 Halmos, P.: Lectures on ergodic theory. Tokyo 1953.

3 Friedman, N.A.: Introduction to ergodic theory. Van Nostrand, New York 1970.

4 Kornfeld, I.P., Sinai, Ja.G., Fomin, S.V.: Ergodičeskaja teorija. Nauka, Moskva 1980 .

5 Harman, B.: Some strenghtening of the maximal ergodic theorem. Proceedings of the conference on Ergodic Theory and Related Topics II. Martin Luther Universität - Halle Wittenberg 1986. To appear.

THE LOGARITHMIC GAMMA DISTRIBUTION - A USEFUL TOOL IN RELIABILITY STATISTICS

Gisela Härtler

Berlin

Key words: Logarithmic gamma distribution, accelerated life tests, Bayesian reliability prediction

ABSTRACT

Frequently estimation, testing, and reliability prediction is based on random variables following gamma type distributions. In the case of life distributions of exponential and Weibull-type it is well known that total life and generalized total life, respectively, is a random variable of gamma type. Consequently, their logarithms are distributed according to the logarithmic gamma distribution. This distribution function is characterized by some very useful properties, one of them of extreme practical relevance: the logarithmic gamma distribution can be approximated very well by the normal distribution if the shape parameter belongs to a certain subset of the parameter space. This subset is equivalent to the most common parameter values of practice. Therefore some complicated techniques of reliability statistics can be simplified considerably, e.g. accelerated life testing and Bayesian reliability prediction.

THE PROPERTIES OF THE LOGARITHMIC GAMMA DISTRIBUTION GENERALLY

Let be x a random variable defined on the whole real axis $-\infty < x < +\infty$. The probability density function of x is given by the expression

(1) $$f(x) = (a/\theta)^a[\Gamma(a)]^{-1}\exp[ax]\exp[-a/\theta)\exp x],$$

with parameters $a>0$, $-\infty < \theta < +\infty$. The probability density $f(x)$ is called to be of logarithmic gamma type since for $x = \ln z$ the random variable z follows a gamma distribution with scale parameter θ and shape parameter a.
The characteristic function of x is given by

(2) $$\varphi_x(t) = (\theta/a)^{it}\Gamma(a+it)/\Gamma(a).$$

The moments of x are obtained by differentiating $\varphi_x(t)$ and the setting $t = 0$. As result follows:
(i) The expectation of x is given by

(3) $$E[x] = \ln\theta + \psi(a) - \ln a,$$

with $\psi(a)$ denoting the digamma function of a.
(ii) The moments of order m, $m>1$, are simply polygamma functions of the order $(m-1)$,

(4) $$\mu_m[x] = \psi^{(m-1)}(a).$$

In many applications the parameter a denotes number of failures in a sample and is consequently an integer. In that case the polygamma function can be calculated according to the expression [2]

(5) $$\psi^{(m-1)}(a) = (-1)^{m-1}(m-1)![-\zeta(m)+1+1/2^m+ \dots + 1/(a-1)^m],$$

where $\zeta(m)$ denotes the Riemann zeta function.
(iii) The variance of x is simply the trigamma function of a

(6) $$\mathrm{Var}[x] = \psi'(a).$$

It is important to notice that the variance of x is independent on θ.
Calculations of higher moments, skewness and excess show that for not too small values of a $(a>3)$ the probability density of the logarithmic gamma distribution is very close to the probability density function of a normal distribution.

The application of the logarithmic gamma distribution and its approximation by the normal distribution was investigated first by BARTLETT and KENDALL [1] in 1946. It seems that this paper didn't find the adequate resonance for a long time. Recently several authors independently noticed the facilities of the logarithmic gamma distribution, especially with respect to some applications in reliability theory [5,6,7,8,11,12,15].

ACCELERATED LIFE TESTING

Let be given a stress dependent life model consisting in life distribution of Weibull type with linear acceleration function. The Weibull distribution

(7) $$F(t) = 1-\exp[-(t/\eta)^b], \quad t \geqq 0,$$

depends on the shape parameter $b>0$ assumed to be known and the scale parameter η depending on stress. The stress U is assumed to be multicomponent with $u_1, u_2, \ldots, u_\mu$, $\mu \geqq 1$, denoting the individual stress components. The stress dependence is given by the following model

(8) $$\ln(\eta^b) = \sum_{j=0}^{\mu} c_j u_j, \quad u_0=1,$$

with c_j denoting parameters which have to be estimated or tested against a hypothetical set of parameter values.

Let us assume that experiments are performed at $k>\mu$ different stress levels $U_i = (u_{i1}, u_{i2}, \ldots, u_{i\mu})$, $i=1,2,\ldots,k$. Every experiment consists in the observation of n items up to the rth failure (n may be different for different stress levels). The statistic called generalized total life is given by the formula [4]

(9) $$S_r = \sum_{l=1}^{r} t_{(l)}^b + (n-r)t_{(r)}^b,$$

where $t_{(l)}$ are the observed failure times $t_{(1)} \leqq t_{(2)} \leqq \ldots \leqq t_{(r)}$. It is well known [3] that the statistic

(10) $$T = (2S_r/\eta^b)$$

follows a $\chi^2(2r)$ distribution. If the parameter $\ln(\eta^b)$ is estimated by

$$Y = \ln S_r - \psi(r) \tag{11}$$

the distribution of Y is of logarithmic gamma type with expectation 0 and variance $\psi'(r)$. Further, if r is not too small (r>3) one may even assume that the distribution of Y is normal. Therefore equation (8) is a linear regression model

$$Y_i = \sum_{j=0}^{m} c_j u_{ij} + \varepsilon_i , \quad i=1,2,\dots,k, \tag{12}$$

with $\varepsilon_i \sim N(0,\psi'(r))$. If the experiments are designed so that r becames equal for every stress level U_i, i=1,2,...,k, the standard linear regression can be applied. For different values of r at different stress levels the weighted linear regression becames applicable. Using these approaches one may even apply optimum design of experiments in order to determine the appropriate stress levels.

BAYESIAN RELIABILITY PREDICTION

Usually reliability prediction techniques are based on the exponential distribution

$$F(t) = 1 - \exp[-\lambda t], \quad t \geqq 0, \tag{13}$$

with hazard rate $\lambda > 0$. Prediction consists in the product of a basic hazard rate and some factors expressing additional influences [9,10,14],

$$\lambda = \lambda_0 \pi_1 \pi_2 \cdots \pi_n . \tag{14}$$

Both basic model and influencing factors can be interpreted as prior information in the Bayesian sense. The following assumptions are used:

(i) The probability distribution of the predicted hazard rate is of gamma type corresponding with the conjugated prior of the total life statistic.

(ii) The logarithm of the predicted hazard rate depends by a li-

near model on the logarithms of influencing factors corresponding with the structure of known prediction models.
Now reliability prediction can be put into the Bayesian framework. Using the facilities of logarithmic gamma distribution one can even calculate the posteriori density of the predicted hazard rate.
The main property of the logarithmic gamma distribution in this connection is that expectation and variance of the sum of independent logarithmic gamma variables is equal to the sum of individual expectations and variances, respectively.
The predicting equation (14) will be transformed into

$$\ln\lambda = \ln\lambda_0 + \sum_{i=1}^{n} \ln\pi_i . \tag{15}$$

The individual terms are considered as random variables with logarithmic gamma distributions. Inserting into (15) the coefficients known e.g. for reliability prediction of integrated circuits is corresponding with the usual approach. Prediction is given in the form of a point estimate of the hazard rate.
The Bayesian approach is based additionally on an idea about the precision of the individual terms. The determination of precision is a rather subjective matter (the usual objection against the Bayesian paradigma!). Following LINDLEY [13] "...the subjective approach has the great strength of encouraging cooperation and discussion among scientists and of suggesting new experiments" the subjective valuation of the variances of the random variables seems to be advantageous in any case. Fortunately, for $a>3$ the logarithmic gamma distribution is rather close to the shape of a normal distribution. Therefore it is possible to express the subjective feeling about the precision of the terms $\ln\lambda_0$, $\ln\pi_i$, $i=1,2,\ldots,n$, in form of symmetric intervals. Assuming the bounds of those intervals to be the lower and upper quantiles of a normal distribution one gets values for their variances. Since the sum of expectations and variances is equal to expectation and variance of the posterior distribution the posteriori density will be given as approximate normal density.

REFERENCES

[1] Bartlett,M.S.,Kendall,D.G. (1946): The statistical analysis of variance heterogeneity and the logarithmic transformation. Journal of the Royal Statistical Society, vol.8,B,128-138

[2] Danos,M.,Rafelski,J. (1984): Pocketbook of mathematical functions. Verlag Harry Deutsch, Thum-Frankfurt/Main.

[3] Dubey,S.D. (1966): Some test functions for the parameters of the Weibull distribution. Naval Research Logistics Quarterly, 13, 113-128

[4] Härtler,G.(1983): Statistische Methoden für die Zuverlässigkeitsanalyse. Verlag Technik Berlin.

[5] Härtler,G. (1984) Schätzung und Test der Parameter der Arrheniusbeziehung bei exponentialverteilten Ausfällen. Preprint 84-8, Akademie der Wissenschaften der DDR/Zentralinstitut für Elektronenphysik.

[6] Härtler,G. (1985a): Ein Downstep-Stress Test für Arrheniusmodell und Exponentialverteilung. Preprint 85-6, Akademie der Wissenschaften der DDR/Zentralinstitut für Elektronenphysik.

[7] Härtler,G. (1985b): A Bayesian approach to reliability prediction. Proceedings of the Sixth Symposium on Reliability in Electronics, Budapest, I, 66-73.

[8] Härtler, G. (1985c): Parameter estimation for the Arrhenius model. To appear in IEEE Transactions on Reliability.

[9] Jääskeläinen,P. (1980): LSI reliability prediction based on time. Microelectronics & Reliability 20, 351-356.

[10] Jääskeläinen,P. (1982): LSI reliability in a medium size company. Proceedings of the Fifth Symposium on Reliability in Electronics, Budapest, I, 333-344.

[11] Jones,R.A.,Scholz,F.W.,Ossiander,M.,Shorack,G.R. (1985): Tolerance bounds for log gamma regression models. Technometrics 27, 109-118

[12] Lawless,J.F., Singhal,K. (1980): Analysis of data from life-test experiments under an exponential model. Naval Research Logistics Quart erly 27, 323-334

[13] Lindley,D.V. (1980): The Bayesian approach to statistics. University of California, Berkeley 1980

[14] Palo,S. (1983): Reliability prediction of microcircuits. Microelectronics & Reliability 23, 283-294

[15] Singpurwalla,N.D., Castellino,V.C., Goldschen,D.Y. (1975): Inference from accelerated life tests using Eyring type re-parametrization. Naval Research Logistics Quarterly, 22, 289-296

Akademie der Wissenschaften der DDR
Zentralinstitut für Elektronenphysik
Hausvogteiplatz 5 - 7
1086 Berlin
DDR

MODEL SEARCH METHODS FOR CONTINGENCY TABLES

AND INTENSIONAL EXPERT SYSTEMS

Tomáš Havránek

Prague

Key words: model search, categorial data, expert systems

ABSTRACT

The aim of the present paper is to introduce some possibilities of construction a knowledge base using some statistical data. In the first part a hypothetical simple expert system using results of the data analytic GUHA system is described. In the second part possibilities for a design of data analytic methods generating knowledge for a given intensional expert system INES are presented.

INTRODUCTION

The paper contains no final solutions; it appears that in the field of knowledge base construction more aspects are still completely unclear than clear. We hope that the present text can serve as a source of some ideas for further research, but on the other hand it contains some (heuristic) suggestions for practical steps towards knowledge base construction from statistical data. The main obstacle is that in such a case we are forced to analyse contingency tables having such dimensions for that the statistical theory does not suggests any reasonable analytic methods. The approach presented here is strongly affected by data analytic methods called GUHA methods; as a source reference the book of Hájek and Havránek (1978a) can serve. The present author would like to thank Petr Hájek, whose criticism improved substantially the content and readibility of the paper.

As a data we consider a matrix consisting of rows corresponding to some objects (situations) and columns that are records of values of some categorial variables corresponding to some propositions or nodes of an intended knowledge base. For simplicity we shall assume that variables are two-valued (yes, no or 1,0), possibly with some special value X for "unknown".

MAIN RESTRICTION FOR THE GUHA PRODUCED KNOWLEDGE BASE

In MYCIN-like expert systems, e.g. EQUANT by Hájek and Hájková (1984), the knowledge base consits of rules of the form

IF (antecedent) THEN (consequent) WITH WEIGHT (weight)

or in shortened EQUANT notation $A \Rightarrow C(w)$. The weight is a number between -1 and 1, but in the sequel we shal trasform it into the interval 0,1 with 1/2 corresponding to 0. The antecedent is usually an elementary conjunction of propositions, consequent an atomic proposition.

The program ASSOC of the GUHA package generates and evaluates (given a data matrix) hypotheses of the form $A \Rightarrow C$ and produces (in a solution file) a set of such hypotheses supported by data together with estimated probability $w = \hat{P}(C/A)$ of C conditioned by A. See Hájek and Havránek (1978a,b). In the simpliest case, A is an elementary conjuction and C is an atomic proposition. Due to the formal analogy it is a strong temptation to use this program for generating a knowledge base for EQUANT system. But the direct way, i.e. to use ASSOC produced hypotheses as knowledge base rules is not possible due to the extensional character of EQUANT. This question will be discussed in more detail elsewhere.

We shall turn our attention now to another question, namely how to use knowledge generated by ASSOC in an expert system. In fact, we shall suggest such a hypothetical expert system.

As a design choice, we made two main restrictions for such an expert system:

a) In the knowledge base we consider only questions and goals (and no intermediate nodes - propositions). Each rule leads from questions directly to goals.

b) Answers to questions are only yes, no or unknown. The source of quantified uncertainty is only the link between questins and goals.

Both these restrictions are natural if we are generating rules from data, where each proposition is realy evaluated 1,0 or X, and in that we have information concerning questions and goals together.

The present system should be considered as a part of a larger system. The larger system can use both the knowledge generated form data as well as knowledge obtained from experts. Results of a consultation of the present system, i.e. goals with some weights conditioned by concrete answers to questions can serve as answers to some questions (as some evaluated propositions) of the expert knowledge oriented part of the larger system (think EQUANT). The two parts of the system are linked not by merging knowledge bases and inference machines, but by some evaluated propositions as a result of the work of the first part. The need of evaluating these propositions - some goals of the first system - can be invoked by the consultation run of the second one.

STRUCTURE OF A KNOWLEDGE BASE

ASSOC produced knowledge base KB consists of rules of the form A=> C(w), where A is an elementary conjunction of some questions (input propositions) and C is a goal such that w is a reliably estimated conditional probability P(C/A) in the given data set M. In more detail, if q1,...,qn are questions, then A is an elementary conjuction $ai_1 \& \ldots \& ai_k$, where ai_j is qi_j or $\neg qi_j$.

A1: In the sequel, we shall suppose that if P(C/A) is reliably estimated, then P(C/A′) is reliably estimated for each subconjunction A′of A /A′ $\preceq$ A/.

Two points are to be noted here: First, if we have a data matrix, we have an estimate for the joint probability $P_{q1,\ldots qn,C} = \{P(a1 \& \ldots \& an \& C)\}$, but practically, we have reliable estimates only for some conditional probabilities P(C/A). Second, in the prgram ASSOC the notion of conservative improvement is used: Together with each chosen A=> C(w), we obtain information describing for which A′, A $\preceq$ A′ , P (C/A)= P (C/A′). This information should be included into KB additionally to A=> C(w) and then used by the inference machine. For the sake of simplicity we shall omit this aspect now in the description of the inference machine.

INFERENCE MACHINE

Suppose that a knowledge base KB and a list of goals that are actually to be evaluted is given. We try now to evaluate a goal from this list as follows:

(i) Find all questions ql,...,qk that occur in antecedents of the rules of KB in that C is the consequent (let this set be KB(C)).

(ii) If a question was asked during the consultation in evaluating a previous goal, use the answer obtained, else ask this question (possible answers are yes, no or unknown). Given these answers, evaluate antecedents of rules in KB(C) using the three-valued logic as in Hájek and Havránek (1978a). Let the list of rules with true antecedents be KB(C,true).

(iii) The set KB(C,true) is partially ordered by $\preceq$ between antecedents. Find the list of maximal elements KB(C,true,max). Rationale: if $A_1 \preceq A_2$ and both A_1 and A_2 are true, then the rule $A_2 \Rightarrow C(w_2)$ should be used as giving the better prediction for C. Note: In many cases, the fair way is now to communicate to the user the list of actual knowledge AK=KB(C,true,max) possibly linearly ordered by some reliability measure (say the estimated probability P (A) for antecedents). If we plan to communicate with some other expert system, it is desirable to proceed further.

(iv) Consider the linearly ordered list AK (note that con juction of all antecedents form AK is true). Apply a combining function to first two rules in AK[1]: $A_1 = C(w_1)$ and [2]: $A_2 = C(w_2)$ and exclude these rules from AK. Include the obtained new rule [12]: $A_1 \& A_2 \Rightarrow C(w_{12})$ into AK. Reduce AK to maximal elements w.r.t. $\preceq$ as in (iii). If AK is one-element set, then stop and communicate to the user [12], else repeat (iv).

COMBINING FUNCTION

Consider the case in which A_1 and A_2 have no common question. Due the point (iii) we have no reliable estimate for $P(C/A_1 \& A_2)$ and hence for $P(C \& A_1 \& A_2)$ (supposing that if we have a reliable estimate for $P(C \& A)$ and $P(A)$ we have a reliable estimate for $P(C/A)$). In such a situation we can "heuristically" suppose that A_1 and A_2 are conditional independent given C. Note that at this moment we can suppose that $P(A_1 \& A_2) > 0$ since $A_1 \& A_2$ is true in a particular instance (consultation). In general, note that it is not possible that qi is in A_1 and ¬qi in A_2 and vice versa, hence there is a maximal subconjuction B such that $A_1 \& A_2 = A_1^- \& A_2^- \& B$, $A_1 = A_2^- \& B$, $A_2 = A_2^- \& B$.Due to (iii) A_1^- and A_2^- are nonempty. Using the same rationale as above, we shall suppose that A_1^- and A_2^- are conditionally independent given C&B ($P(A_1 \& A_2 \& C) = P(A_1 \& C) \cdot P(A_2 \& C) / P(B \& C)$, $P(A_1 \& A_2 \& \neg C) = P(A_1 \& \neg C) \cdot P(A_2 \& \neg C) / P(B \& \neg C)$) Computing under these conditions $P(C/A_1 \& A_2)$ and using the estimated weights, we obtain the combined weight

$$w_{12} = \left(1 + \frac{(1-w_1)(1-w_2)}{w_1 \quad w_2} \quad \frac{w_0}{1-w_0} \right)^{-1} \qquad (x)$$

where w_0 is the weight in B = $C(w_0)$.

Note:(a) Due to the assumption A1, if $A_1 \Rightarrow C(w_1)$ and $A_2 \Rightarrow C(w_2)$ are in KB then $B \Rightarrow C(w_0)$ is in KB too. (b) In the particular case when A_1 and A_2 are disjoint, then w_0 is the weight in the dummy rule $\emptyset \Rightarrow C(w_0)$, i.e. the apriori (estimated) weight of C. In this case the expression (x) corresponds to the PROSPECTOR combining function used in EQUANT (with $w_0 = 1/2$). If we consider odds, defined as $o_i = w_i/(1-w_i)$, then (x) equals to $o_{12} = o_1 o_2 / o_0$.

CHOICE OF RELIABLE RULES

For constructing a knowledge base using ASSOC we suppose that a data matrix M is given. Let now for a given A nad C, r be the frequency of A and a the frequency of A&C. The most straightforward way is to suppose that the estimate w=a/r is reliable if r=B, where B is given number (say B=25 for standard error less or equal 0.1). If now q1,...,qn is a given set of questions for a goal C, we can construct KB(C) specifying the control parameters of ASSOC as follows: (a) antecedent variables are q1,...,qn in both forms (qi as well as ¬qi), (b) maximal length of antecedent is n, (c) succedent variable is C in both forms, (d) quantifier is FIMPL (founded implication) with CPROB=0.5 and with BASE expressing our demand of reliability (perhaps in dependence on an apriori estimate of P(C) and (c) improvement is consevartive.Note that if we choose a number B above then as a necessary condition for $r \geq B$ we have to use BASE = $\lceil B.CPROB \rceil + 1$; results from the obtained solution file are to be then selected w.r.t $r \geq B$.

This way of generating the knowledge base can be computationally expensive; the use of improvement is substantial. The obtained list of rules is segmented in accordance to the lenght of antecedents (number of questions in antecedents). If we obtain a message that for a given length l no rule was found (no hypothésis was found), then it is clear that the same is true for each $l' > l$ and the construction of the knowledge base is finished.

Clearly further refinement can be considered e.g. based on a use of confidence intervals for P(C/A) for choice of reliable estimates. If $\langle \underline{w}^{\alpha}, \overline{w}^{\alpha} \rangle$ is the confidence interval based on a and k, then the rule is chosen if $w_C \notin \langle \underline{w}^{\alpha}, \overline{w}^{\alpha} \rangle$, where w_C is an (apriori) estimate of P(C). Then only rules with weight distinguishble from w_C are chosen. This way can be realized again by ASSOC but it leads to some efects in the inference mechanism and will be discussed elsewhere as well as other possibilities of defining reliable estimates.

DISCUSSION

There is a great number of open questions concerning the above suggested expert system. First of them is the question of the contex of classification /discrimination methods. Perhaps the present method has the advantage that not all goals are automatically eveluated and if a goal is evaluated (on some demand) then "all available" information is used. Clearly we can consider theoretically more adequate ways of estimating the final weight in (iv) above. Our suggestion has two advantages: it is simple and it is, in a degree, intensional. On the other hand we do not take into account possible dependences between goals. Perhaps in some case, a more elegant solution in the spirit of Perez and Jiroušek(1985) INES intesional expert system is possible.

In Perez and Jioušek (1985) it is supposed that the formal structure of the knowledge base correspond to a set of N nodes,able to take different states (say two possible states) according to the value of the stochastic variable assigned to each node. As an input for a construction of the knowledge base ~~over~~ such nodes, a set of marginal distributions is used. There is a computer program constructing then an intensional knowledge base (in fact an approximation of the joint probability over all nodes).

If we have a data set represented by configurations of states $\underline{i}$ and their frequencies $n_{\underline{i}}$ we can principially estimate all marginal probabilities including the joint probabilities $p_{\underline{i}}$ by $n_{\underline{i}}/n$, where $n = \sum_{\underline{i}} n_{\underline{i}}$. But clearly such estimates can be highly unreliable to be used in some decisions . The statistical question is to find acceptable smaller set of marginal probabilities to be considered e.g. as an input to INES.

If $v \subseteq V = \{1,\ldots,N\}$ is a set of nodes (variables), let $n^v_{\underline{i}}$ be $n_{\underline{i} \wedge v}$ and $n^v = \{n^v_{\underline{i}} : \underline{i} \wedge v, \underline{i} \in \{0,1\}^N\}$. The most straightforward way is to use non-zero marginals, i.e. such n^v in which each $n^v_{\underline{i}} > 0$ and then to put set of all such n^v (or n^v/n) into the INES system. This approach is really simple, but it has two flaws: (i) some $n^v_{\underline{i}} = 0$ can represent a relevant information (are not only a consequence of small frequencies with respect to the dimensionality of the table), and (ii) we still can use much more marginals (and parameters) than necessary.

UNRESTRICTED SEARCH FOR MINIMAL MODELS

Consider now that we do not necessarily construct the whole knowledge base with say N = 100 nodes, but sometimes its fragment with N = 10 to 30 nodes. We can consider only graphical model for contingency tables. These models for dependency structure between nodes (variables) are uniquelly expressible by graphs over the set of nodes V. There is here a simple ordering of models: $m_1 < m_2$ if the set of edges of m_1 is a proper subset of the set of edges of m_2. Let now call a model accepted if it is not rejected by a goodness-of-fit test in the given data. If S is a set of accepted models, we can seek for a set of minimal acceptable models A = S with respect to the ordering $<$. An effective procedure for such a search is described in Edwards and Havránek (1985). Let us denote $\hat{p}$ (m) the joint probability estimated under the model m. The resulting models can be e.g. ordered further by $I(\hat{p}(m))$, cf.Perez and Jiroušek (1985). Utilizing the whole set A has the advantage, that we can choose from some extramathematical reasons a suboptimal model.

For sparse tables (such we consider here) it is necessary to use the upward search starting with the model of total independence

$(V,\emptyset)$ and then considering models with the edge etc.(the decision in the point 2 of the algorithm of Edwards and Havránek (1985) will be to use $D_a(R)$, A at any case). In this procedure, a model m is not evaluated, if there was found a model $m_0 \leq m$, m_0 accepted. If the table is sparse and multidimensional, some more complex models cannot be evaluated due to computational (numerical) problems with IPF estimating algorithm. Then we have to add a rule that if a model m_1 cannot be evaluated, then all models m such that $m_1 \leq m$ are not considered. Then the procedure can stop with $A = \emptyset$.

It seems to be reasonble to avoid the use of the IPF algorithm and consider not the graphical models but their proper subclass, namely decomposable models. Edwards (1984) suggested a stepwise procedure for this case based on nonasyptotic tests; his procedure is a backward (downward) elimination procedure starting with the complete graph and deleting edges. In our context it seems to be more useful to consider an upward (forward) procedure starting with $(V,\emptyset)$ and adding edges. We can consider a branching procedure: if decomposable model m_0 is tested, then (a) if it is accepted, then do not consider all models m, $m_0 \leq m$ and put m_0 into A, (b) if it is not accepted, test all its immediate decomposable succesors obtained by adding one edge. We can use simple asymptotic likelihood ratio chi-square test. No IPF algorithm is necessary, but still the procedure can be hardly tractable. On the other hand Lemma 1 in Edwards (1984) shows that principially each decomposable model can be assesed by this way.

There is an interesting modification. We can use zero partial association test (ZPA tests; cf. Lemma 2 in Edwards, 1984) to test whether the edge in question (e.g. between nodes A, B) corresponds to zero partial association (i.e. independence conditioned by other connected nodes). This can be done in a marginal table n^V corresponding to a component of the model m_0 to which the edge should be added. We shall not consider a new model $m_1 = m_0 \cup (A,B)$ and its successors if this hypothesis is not rejected. The test used can be non asymptotic exact test or a Monte-Carlo test (but this can lead to a great complexity). Maximal models that cannot be improved can be used as a base for further considerations.

SIMPLIFIED PROCEDURES

The above procedures are computationally expensive. Some nonexhaustive but more effective procedures are needed for cases that are untractable.

First we can modify the edge adding procedure from the above paragraph to be a stepwise upward procedure: M:=(V,∅) and do L1: consider edges not in m; create succesively all models by adding one edge to m; if such a model m_1 is decomposable, compute $\chi^2_{LR}(m_1/m) = \chi^2_{LR}(m) - \chi^2_{LR}(m_1)$; find such an m_1 with maximal $\chi^2_{LR}(m\ /m)$; test ZPA the edge m_1-m; if significant go to L1, else stop.

Consider now models containing cliques at most of the cardinality 2. Then $\chi^2_{LR}(m_1/m)$ corresponds to I(i,j) with the estimated probability $\hat{p}^{\,ij} = n^{ij}/n$ of the edge in question. Hence we can use firstly the greedy algorithm as proposed by Perez and Jiroušek (1985) to find a starting decomposable model based on two dimensional marginals. Such model m_0 is the best one in the sense of maximalizing ML-fit (and I(m)) among all decomposable models considering only two dimensional marginals. Note that maximalizing I(m) is equivalent to minimilizing the χ^2_{LR} statistics (corresponsding to $H(m_s,m)$ where m_s is the saturated model). Now one can start the procedure with $m:=m_0$ instead of m:=(V,∅).

REFERENCES

Edwards D.(1984): A computer intensive approach to the analysis of sparse multidimensional contingency tables. In:COMPSTAT 84, Physica-Verlag, 355-359.

Edwards D., Havránek T.(1985): A fast procedure for model search in multidimensional contigency tables, Biometrika 72,339-351.

Hájek P., Hájková M.(1984): The expert system EQUANT-brief description and user s manual. In: New enhancements in GUHA software, Mathematical Institute of Czech.Acad.Sci.,21-36.

Hájek P., Havránek T.(1984a): Mechanizing hypothesis formation-mathematical foundations for a general theory, Springer-Verlag.

Hájek, P. , Havránek T.(1984b): The GUHA method - its aims and techniques. Int.J.Man-Machine Studies 10, 3-22.

Perez A., Jiroušek R.(1985): Constructing an intesional expert system (INES). In: Medical Decision Making - Diagnostic Strategies and Expert Systems, North-Holland, 307-315.

General Computing Center
Czechoslovak Academy of Science
Pod vodárenskou věží 2
182 07 Prague
Czechoslovakia

THE HÁJEK-RÉNYI TYPE INEQUALITY FOR TRACIAL STATES IN A VON NEUMANN ALGEBRA

Ewa Hensz

Łódź

Hájek-Rényi inequality, von Neumann algebra, tracial state, laws of large numbers

ABSTRACT

A non-commutative version of the Hájek-Rényi inequality is given. This result plays a similar role in the theory of limit theorems in von Neumann algebras as the Hájek-Rényi inequality does in the classical case, namely, it simplifies the proofs of some strong laws of large numbers.

INTRODUCTION

In the classical probability theory an inequality which has been found by Hájek and Rényi (1955) is very useful in proving strong laws of large numbers. The inequality in question states that if $X_1, X_2, \ldots,$ is a sequence of mutually independent random variables with mean values $E(X_k) = 0$ and finite variances $E(X_k^2) = d_k^2$ and $c_1, c_2, \ldots$ is a nonincreasing sequence of positive numbers, then for every $\varepsilon > 0$ and all integers n and m $(n < m)$ we have

$$(1) \qquad P\{\max_{n \le k \le m} c_k |X_1 + \ldots + X_k| \ge \varepsilon\} \le$$

$$\le \varepsilon^{-2}(c_n^2 \sum_{k=1}^{n} d_k^2 + \sum_{k=n+1}^{m} c_k^2 d_k^2).$$

Clearly, the inequality (1) is an extension of the famous inequality of Kolmogorov (the case when all $c_k = 1$). The use of the

Hájek-Rényi inequality (1) instead of the Kolmogorov one makes it possible to simplify the proofs mentioned.

Recently, various results concerning the strong limit theorems, in particular, the laws of large numbers have been extended to the non-commutative context. It is worthwhile to notice here the results of Batty (1979), Jajte (1982), (1985), Łuczak (1985); see also Jajte (1985a).

In this setting, the role of a random variable is played by an element of a von Neumann algebra $\mathcal{A}$ and a probability measure is replaced by a normal faithful state on $\mathcal{A}$. If this state is tracial, the von Neumann algebra can be replaced by an algebra $\tilde{\mathcal{A}}$ consisting of some unbounded operators.

The basic tool in studies of strong limit theorems by Batty (1979) has been an inequality of Kolmogorov type. The aim of this note is to give a non-commutative version of the Hájek-Rényi inequality which extremely abridges the proof of the non-classical version of Kolmogorov strong law of large numbers.

BASIC NOTIONS AND DEFINITIONS

For the general theory of von Neumann algebras, the reader is referred to Takesaki (1979) or Sakai (1971). Here we only establish the notation and recall some definitions.

Let $\mathcal{A}$ be a finite, countably decomposable von Neumann algebra of operators acting in a Hilbert space H with a faithful normal tracial state τ. Let $\tilde{\mathcal{A}}$ denote the $*$-algebra of all operators in H measurable in the sense of Segal-Nelson (see Nelson (1974)). For $p \geq 1$, let $\|x\|_p = \tau(|x|^p)^{1/p}$ for $x \in \mathcal{A}$. By $L^p(\mathcal{A},\tau)$ we mean the completion of $\mathcal{A}$ in the norm $\|\ \|_p$.

The identity mapping on $\mathcal{A}$ extends to a continuous injection of $L^p(\mathcal{A},\tau)$ in $\tilde{\mathcal{A}}$, and we shall regard $L^p(\mathcal{A},\tau)$ as a subspace of $\tilde{\mathcal{A}}$. Also τ extends by continuity to a positive linear functional of norm 1 on $L^1(\mathcal{A},\tau)$. For $x \in L^2(\mathcal{A},\tau)$ we have $x^*x \in L^1(\mathcal{A},\tau)$ and $\|x\|_2^2 = \tau(x^*x)$.

Let $\mathcal{B}_1$ and $\mathcal{B}_2$ be a von Neumann-subalgebras of $\mathcal{A}$. $\mathcal{B}_1$ is said to be *independent* of $\mathcal{B}_2$ (with respect to τ) if for $x \in \mathcal{B}_1$ and $y \in \mathcal{B}_2$

$$\tau(xy) = \tau(x)\tau(y).$$

A sequence $(\mathcal{B}_n)$ of von Neumann-subalgebras of $\mathcal{A}$ is *successively independent* if $\mathcal{B}_n$ is independent of the von Neumann algebra $\mathcal{R}\{\mathcal{B}_m : m < n\}$ generated by $\mathcal{B}_m$ for $m < n$. A sequence of operators (x_n) in $\mathcal{A}$ (or in $\tilde{\mathcal{A}}$) is *successively independent* if $\mathcal{R}(x_n)$ are successively independent.

A sequence (x_n) in $\tilde{\mathcal{A}}$ is said to be *convergent almost uniformly* (a.u.) to $x \in \tilde{\mathcal{A}}$ if for every $\varepsilon > 0$ there is a projection p in $\mathcal{A}$ with $\tau(p) < \varepsilon$ such that $(x_n - x)p^{\perp} \in \mathcal{A}$ and $\|(x_n - x)p^{\perp}\| \to 0$ as $n\to\infty$ (when p is a projection in $\mathcal{A}$, $p^{\perp}$ denotes $1 - p$ where 1 is the identity operator).

Finally, let $e.(x)$ denote the spectral measure of self-adjoint operator x, i.e. $x = \int \lambda e_{d\lambda}(x)$.

MAIN RESULTS

THEOREM. Let (x_k) be a successively independent sequence in $L^2(\mathcal{A},\tau)$ with $\tau(x_k) = 0$, $s_k = \sum_{r=1}^{k} x_r$, and let (c_k) be a non-increasing sequence of positive numbers. For $\varepsilon > 0$ and positive integers n and m $(n < m)$ there exists a projection q in $\mathcal{A}$ such that

$$(2) \qquad \tau(q) \le \varepsilon^{-2}\Big(c_n \sum_{r=1}^{n} \|x_r\|_2^2 + \sum_{r=n+1}^{m} c_r^2 \|x_r\|_2^2\Big)$$

$$(3) \qquad \|c_k s_k q^{\perp}\| \le \varepsilon, \qquad\qquad n \le k \le m.$$

P r o o f. We shall define inductively a finite sequence of projections p_k $(n \le k \le m)$ in $\mathcal{A}$. Let us put $p_n = e_{(\varepsilon^2,\infty)} \cdot (c_n^2 s_n^* s_n)$ and for given $p_n,\dots,p_{k-1}$ let us set $p_k = e_{(\varepsilon^2,\infty)}(c_k^2 (\sum_{r=n}^{k-1} p_r)^{\perp} s_k^* s_k (\sum_{r=n}^{k-1} p_r)^{\perp})$ for $k \le m$. It is clear that $p_n,\dots,p_m$ are mutually orthogonal projections in $\mathcal{A}$, p_k belongs to $\mathcal{R}(x_r : r \le k)$. Putting $q = \sum_{r=n}^{m} p_r$ we obtain for $n \le k \le m$

$$\|c_k s_k q^{\perp}\|^2 = \|q^{\perp} c_k^2 s_k^* s_k q^{\perp}\| \le$$

$$\leq \| p_k^{\perp} \Big(\sum_{r=n}^{k-1} p_r \Big)^{\perp} c_k s_k^* s_k \Big(\sum_{r=n}^{k-1} p_r \Big)^{\perp} p_k^{\perp} \| \leq \varepsilon^2$$

which gives the inequality (3).

To estimate $\tau(q)$ we shall follow a nice idea of Hájek and Rényi and consider the operator y in $\mathcal{A}$ of the form

$$y = \sum_{k=n}^{m-1} (c_k^2 - c_{k+1}^2) s_k^* s_k + c_m^2 s_m^* s_m .$$

By the independence of (x_n), we have

$$\tau(y) = \sum_{k=n}^{m-1} (c_k^2 - c_{k+1}^2) \tau(s_k^* s_k) + c_m^2 \tau(s_m^* s_m) =$$

$$= \sum_{k=n}^{m-1} (c_k^2 - c_{k+1}^2) \tau\Big(\sum_{r,s=1}^{k} x_r^* x_s \Big) + c_m^2 \tau\Big(\sum_{r,s=1}^{k} x_r^* x_s \Big) =$$

$$= \sum_{k=n}^{m-1} (c_k^2 - c_{k+1}^2) \sum_{r=1}^{k} \tau(x_r^* x_r) + c_m^2 \sum_{r=1}^{k} \tau(x_r^* x_r).$$

The change of the order of summation gives

$$\tau(y) = c_n^2 \sum_{r=1}^{n} \tau(x_r^* x_r) + \sum_{r=n+1}^{m} c_r^2 \tau(x_r^* x_r) = c_n^2 \sum_{r=1}^{n} \|x_r\|_2^2 + \tag{4}$$

$$+ \sum_{r=n+1}^{m} c_r^2 \|x_r\|_2^2 .$$

On the other hand, by properties of trace, we have

$$\tau(y) \geq \tau(yq) = \sum_{r=n}^{m} \tau(y p_r) =$$

$$= \sum_{r=n}^{m} \Big(\sum_{k=n}^{m-1} (c_k^2 - c_{k+1}^2) \tau(s_k^* s_k p_r) + c_m^2 \tau(s_m^* s_m p_r) \Big) \geq$$

$$\geq \sum_{r=n}^{m} \Big(\sum_{k=r}^{m-1} (c_k^2 - c_{k+1}^2) \tau(s_k^* s_k p_r) + c_m^2 \tau(s_m^* s_m p_r) \Big).$$

Now, by the independence of (x_n), we have for $n \le r \le k \le m-1$

$$\tau(s_k^* s_k p_r) = \tau(((s_k - s_r) + s_r)^*((s_k - s_r) + p_r)) =$$

$$= \tau(p_r(s_k - s_r)^*(s_k - s_r)p_r) + \tau((s_k - s_r)^* s_r p_r)$$

$$+ \tau(p_r s_r^*(s_k - s_r)) + \tau(p_r s_r^* s_r p_r) \ge$$

$$\ge \tau(s_k - s_r)^* \tau(s_r p_r) + \tau(p_r s_r^*)\tau(s_k - s_r) +$$

$$+ \tau(p_r s_r^* s_r p_r) = c_r^{-2}\tau(c_r p_r s_r^* s_r p_r) =$$

$$= c_r^{-2}\tau(c_r^2 p_r (\sum_{s=n}^{r-1} p_s) s_r^* s_r (\sum_{s=n}^{r-1} p_s) p_r) \ge$$

$$\ge c_r^{-2}\varepsilon^2 \tau(p_r)$$

Now, two above estimations yield

$$\tau(y) \ge \sum_{r=n}^{m} (\sum_{k=r}^{m-1} (c_k^2 - c_{k+1}^2)\varepsilon^2 c_r^{-2}\tau(p_r) + c_m^2 c_r^{-2}\varepsilon^2 \tau(p_r) = \tag{5}$$

$$= \varepsilon^2 \sum_{r=n}^{m} \tau(p_r) = \varepsilon^2 \tau(q).$$

Then, the inequality (2) follows from (4) and (5).

Remark. Putting $p = \sum_{r=n}^{\infty} p_r$ and passing in (2) to the limit when $m \to \infty$, we obtain the limit formulae

$$\tau(p) \le \varepsilon^{-2}(c_n^2 \sum_{r=1}^{n} \|x_r\|_2^2 + \sum_{r=n+1}^{\infty} c_r^2 \|x_r\|_2^2) \tag{2'}$$

and

$$\|c_k s_k p^{\perp}\| \le \varepsilon \quad \text{for} \quad k \ge n. \tag{3'}$$

Taking $c_k = \frac{1}{k}$ we obtain at once the strong law of large numbers of Kolgomorov-Batty, namely:

Let (x_n) be a successively independent sequence in $L^2(\mathcal{A},\tau)$

such that $\sum_{n=1}^{\infty} n^{-2}\|x_n - \tau(x_n)\|_2^2$ is convergent and let $s_n = \sum_{r=1}^{n} x_r$. Then $n^{-1}(s_n - \tau(s_n)) \to 0$ a.u. as $n\to\infty$.

REFERENCES

Batty C.J.K. (1979): The Strong Law of Large Numbers for States and Traces of a W^*-Algebra. Z. Wahrscheinlichkeitstheorie Verw. Gebiete, 48, 177-191.

Hájek J. Rényi A. (1955): Generalization of an inequality of Kolmogorov. Acta Math. Acad. Sci. Hungar. 6, 281-283.

Jajte R. (1982): A non-commutative extension of Hsu-Robbin's law of large numbers. Bull. Acad. Polon. Sci. Ser. Sci. Math. 30, No 11-12, 533-537.

Jajte R. (1985): Strong Limit Theorems for Orthogonal Sequences in von Neumann Algebras, PAMS, 94, No 2, 229--235.

Jajte R. (1985a): Strong Limit Theorems in Non-Commutative Probability. Lect. Notes in Math. No 1110, Springer Verlag, Berlin-Heidelberg-New York--Tokyo.

Łuczak A. (1985): Laws of large numbers in von Neumann algebras and related results. Studia Math. 81, No 3, 231-243.

Nelson E. (1974): Notes on Non-commutative Integration. J. Func. Anal. 15, 103-116.

Sakai S. (1971): C^*-algebras and W^*-algebras. Springer Verlag, Berlin-Heidelberg-New York-Tokyo.

Takesaki M. (1979): Theory of operator algebras I. ibidem.

University of Łódź
Institute of Mathematics
ul. Banacha 22
90-238 Łódź
POLAND

ON THE INCREASE OF CONDITIONAL ENTROPY IN MARKOV CHAINS

Yasuichi Horibe

Hamamatsu

Key words: Markov chain, conditional entropy, convex-increasing property, doubly stochastic transition matrix, latin square, permutation matrix.

ABSTRACT

Convex increase in n of the n-step conditional entropy $H(X_n|X_0)$ is shown for a stationary Markov chain $X_0,X_1,\ldots$. The convergence of the conditional entropy directly gives a simple variation of the Rényi's information-theoretic proof of the classical Markov's limit theorem for ergodic chains. The convex increase of the conditional entropy is then considered in the case of doubly stochastic transitions. Sufficient conditions are discussed for the entropy of the state distribution to present an identical convex-increasing behavior regardless of the initial state, using latin squares, permutation matrices, and groups.

INTRODUCTION

Let $X_0,X_1,\ldots$ be a stationary finite Markov chain with state space $\{1,2,\ldots,N\}$. Our interest will be in the behaviors of the n-step conditional entropy $H^{(n)} \equiv H(X_n|X_0) = \sum_j P\{X_0=j\}H_j^{(n)}$ and the "individual" n-step conditional entropy $H_j^{(n)} \equiv H(X_n|X_0=j)$. The latter is defined as the entropy of the conditional probability distribution of X_n, given $X_0 = j$. Generally, the entropy $H(\mathbf{p})$ of the probability distribution (or vector) $\mathbf{p} = (p_1,\ldots,p_N)$ is given, as usual, by the following formula:

$$H(\mathbf{p}) = -\sum_i p_i \log p_i.$$

Shaw(1984) calls $H(X_n) - H(X_n|X_0) = H(X_0) - H^{(n)}$ "stored information" of the chain after n units of time, and $H^{(n)}$ indicates the uncertainty or the unpredictability of X_n when X_0 is known. A natural property that the sequence $H^{(0)}$, $H^{(1)}$, ... should possess may be the following convex-increasing property:

Theorem 1. $0 \leq H^{(n+1)} - H^{(n)} \leq H^{(n)} - H^{(n-1)}$, $n = 1,2,\ldots$ $(H^{(0)} \equiv 0)$.

The second inequality here shows that $H^{(n)}$ is convex with respect to n, i.e., the speed of the increase of entropy decreases. This convexity is stated in Shaw(1984), based on the following beautiful intuitive reasoning alone [Shaw (1985)]: " ... sharper distributions representing greater stored information will spread faster than broad distributions." A proof of this theorem is now given in the following section.

PROOF OF THEOREM 1

Monotonicity of $H^{(n)}$: Let us use the following inequality:

$$H(X_{n+1}|X_0,X_1) \leq H(X_{n+1}|X_0),\ n \geq 1.$$

[For the well-known fundamental entropy equalities and inequalities, see Ash (1965).] The left side entropy becomes:

$$H(X_{n+1}|X_0,X_1) = H(X_{n+1}|X_1) = H(X_n|X_0) = H^{(n)},$$

where the first equality is due to the Markov property of the chain $X_0,X_1,\ldots$ and the second to the stationarity. Hence $H^{(n)} \leq H^{(n+1)}$.

Convexity of $H^{(n)}$: We have

$$0 \leq H(X_1|X_0,X_2) = H(X_0,X_1,X_2) - H(X_0,X_2)$$

$$= H(X_0) + H(X_1|X_0) + H(X_2|X_0,X_1) - (H(X_0) + H(X_2|X_0))$$

$$= H(X_1|X_0) + H(X_2|X_1) - H(X_2|X_0) = H^{(1)} + H^{(1)} - H^{(2)}.$$

Hence $H^{(2)} - H^{(1)} \leq H^{(1)}$.

In order to show the convexity inequality for $n \geq 2$, we use

$$H(X_n|X_0,X_1,X_{n+1}) \leq H(X_n|X_0,X_{n+1}).$$

The left side entropy here is equal to $H(X_n|X_1,X_{n+1})$ by the Markov property, and

$$H(X_n|X_1,X_{n+1}) = H(X_1,X_n,X_{n+1}) - H(X_1,X_{n+1})$$

$$= H(X_1) + H(X_n|X_1) + H(X_{n+1}|X_1,X_n) - (H(X_1) + H(X_{n+1}|X_1))$$

$$= H(X_n|X_1) + H(X_{n+1}|X_n) - H(X_{n+1}|X_1) = H^{(n-1)} + H^{(1)} - H^{(n)}.$$

Similarly we have $H(X_n|X_0,X_{n+1}) = H^{(n)} + H^{(1)} - H^{(n+1)}$ for the right side entropy. Hence $H^{(n-1)} + H^{(1)} - H^{(n)} \leq H^{(n)} + H^{(1)} - H^{(n+1)}$, so that $H^{(n+1)} - H^{(n)} \leq H^{(n)} - H^{(n-1)}$, completing the proof.

Note the following: The convexity inequality proved above can be rewritten as

$$H(X_0,X_{n+1}) + H(X_1,X_n) \leq H(X_0,X_n) + H(X_1,X_{n+1}),$$

and this inequality can be generalized (by a similar proof) to

$$H(X_i,X_m) + H(X_j,X_k) \leq H(X_i,X_k) + H(X_j,X_m),\ i<j<k<m.$$

A VARIATION OF THE RÉNYI'S INFORMATION-THEORETIC PROOF OF A CLASSICAL LIMIT THEOREM

Let us revisit the famous limit theorem due essentially to A.A.Markov (1856-1922). Denote the transition matrix of the chain by $P = (p_{jk})$. From the fact that $|P - I| = 0$ (I: identity matrix) it is easy to see that there exists a probability vector $\mathbf{q} = (q_1,\ldots,q_N)$ such that $\mathbf{q}P = \mathbf{q}$ [see Rényi(1970, p.294)]. We may therefore take $P\{X_0=j\} = q_j$.

<u>Theorem 2</u> (Markov): If the chain is ergodic, i.e., there exists n_0 such that all the elements $p_{jk}^{(n_0)}$ of P^{n_0} are strictly positive, then $\mathbf{q}$ is uniquely determined, with all $q_j > 0$ (since $\mathbf{q}P^{n_0} = \mathbf{q}$), and every row vector $\mathbf{p}_j^{(n)} = (p_{j1}^{(n)},\ldots,p_{jN}^{(n)})$ of P^n converges to the same $\mathbf{q}$.

When the chain is ergodic, we see $H_j^{(n)} = H(\mathbf{p}_j^{(n)}) \to H(\mathbf{q})$ for each j by the continuity of the function $H(\mathbf{p})$, hence $H^{(n)} \to H(\mathbf{q}) = H(X_0)$ ("the chain tends to completely forget the initial state"). The fact that $H^{(n)}$ converges, however, is a simple consequence of the bounded monotonicity: $H^{(n)} \leq H^{(n+1)} \leq H(X_0)$. We use this fact to offer a simple proof of Theorem 2, as a variation of the well-known information-theoretic proof due to Rényi(1961)(1970).

<u>Proof</u>: Let $P^{n_1}, P^{n_2}, \ldots$ $(n_1 < n_2 < \ldots)$ be any convergent subsequence of the sequence of points $P, P^2, P^3, \ldots$. They belong to the compact (and convex) set of all N by N stochastic matrices in the N^2-dimensional Euclidean space. Such a subsequence necessarily exists(Bolzano-Weierstrass). Suppose $\mathbf{p}_j^{(n_s)} \to \mathbf{p}_j'$, $j = 1,\ldots,N$. It then suffices to show $\mathbf{p}_1' = \ldots = \mathbf{p}_N' = \mathbf{q}$. Put

$$(1)\qquad d_j^{(n_s)} = H\Big(\sum_k p_{jk}^{(n_0)}\mathbf{p}_k^{(n_s)}\Big) - \sum_k p_{jk}^{(n_0)} H(\mathbf{p}_k^{(n_s)}).$$

Since $H(\mathbf{p})$ is convex in $\mathbf{p}$, the following is due to the Jensen inequality for this function.

$$d_j^{(n_s)} \geq 0.$$

Multiply (1) by q_j and sum over j, then we readily have

(2) $$\sum_j q_j d_j^{(n_s)} = H^{(n_0+n_s)} - H^{(n_s)}.$$

Let s go to infinity in (2), then the right hand side tends to zero by the convergence of $H^{(n)}$. Since $d_j^{(n_s)} \geq 0$ and $q_j > 0$, $d_j^{(n_s)}$ must approach zero. Hence we have

$$H(\sum_k p_{jk}^{(n_0)} \mathbf{p}_k') = \sum_k p_{jk}^{(n_0)} H(\mathbf{p}_k'),$$

by the continuity of $H(\mathbf{p})$. This means that the equality holds in Jensen inequality, which implies $\mathbf{p}_1' = \ldots = \mathbf{p}_N'$, since $p_{jk}^{(n_0)} > 0$ and H is strictly convex. Letting $s \to \infty$ in the relation $\mathbf{q}P^{n_s} = \mathbf{q}$, we immediately have $\mathbf{p}_1' = \ldots = \mathbf{p}_N' = \mathbf{q}$.

A DOUBLY STOCHASTIC CASE

When P is doubly stochastic (i.e., each column is also a probability vector), we have the monotonicity $H_j^{(n-1)} \leq H_j^{(n)}$, n = 1,2,..., for each j = 1,...,N. This is due to the well-known property of the entropy: $H(\mathbf{p}) \leq H(\mathbf{p}P)$ for any probability N-vector $\mathbf{p}$ [see Ash(1965)]. Here $H_j^{(0)} = H(\mathbf{p}_j^{(0)}) = 0$, $\mathbf{p}_j^{(0)} = (0,..,0,\underset{(j)}{1},0,...,0)$, and, therefore, $\mathbf{p}_j^{(n)} = \mathbf{p}_j^{(0)} P^n$.

Clearly, in the doubly stochastic case, the uniform distribution (1/N,...,1/N) can be taken as $\mathbf{q}$. Hence $H_j^{(n)}$ tends to the maximum possible entropy log N, provided the chain is ergodic.

We shall be interested in conditions on doubly stochastic P for the following to hold:

(*) $$H_1^{(n)} = H_2^{(n)} = \ldots = H_N^{(n)}, \quad n = 1,2,\ldots .$$

If this holds, then $H^{(n)} = \sum_j (1/N) H_j^{(n)} = H_j^{(n)}$, and $H_j^{(n)}$ is convex-increasing with respect to n, from Theorem 1, and thus the entropy of the state distribution of the chain starting at any state presents an identical convex-increasing behavior. This seems to correspond, as an ideal case, to the Shaw's words in Introduction.

Now let $\{\Pi_1,\ldots,\Pi_N\}$ be a set of N by N permutation matrices such that the elements of the matrix $\Pi_1 + \ldots + \Pi_N$ are all 1. A permutation matrix is a square 0,1-matrix every row and column of which has exactly one 1. Let us call such a set "latin square", because the linear combination $a_1\Pi_1 + \ldots + a_N\Pi_N$ represents a latin square [e.g., Biggs(1985)] of N letters $a_1,\ldots,a_N$. We shall consider such doubly stochastic matrices P that can be expressed as

$$P = p_1\Pi_1 + \dots + p_N\Pi_N,$$

where $\{\Pi_1,\dots,\Pi_N\}$ is a latin square and $(p_1,\dots,p_N)$ is a probability vector.

If, for each n, P^n can be expressed as above, i.e., as "a convex linear combination on a latin square", then (*) holds, for, then, the entropy of the probability distribution composed of the coefficients in the convex linear combination becomes the common value of the entropies in (*).

Theorem 3: Let $\{\Pi_1,\dots,\Pi_N\}$ be a "latin group" (i.e., a latin square and at the same time a group under the matrix multiplication). Then for any transition matrix of the form $P = p_1\Pi_1 + \dots + p_N\Pi_N$, (*) holds.

Proof: It is sufficient to show by induction on n that P^n can be expressed as a convex linear combination on the given latin group, for every n. Suppose P^n has already been in the form $p_1'\Pi_1 + \dots + p_N'\Pi_N$. Develop the following:

$$P^{n+1} = (p_1\Pi_1 + \dots + p_N\Pi_N)(p_1'\Pi_1 + \dots + p_N'\Pi_N).$$

Then, since $\{\Pi_i\Pi_1, \Pi_i\Pi_2, \dots, \Pi_i\Pi_N\} = \{\Pi_1, \dots, \Pi_N\}$ for each $i = 1,2,\dots,N$ by the group property, P^{n+1} is expressed as a similar form $p_1''\Pi_1 + \dots + p_N''\Pi_N$ with $p_1'' + \dots + p_N'' = \sum_{i,j} p_i p_j' = 1$, completing the proof.

The cyclic group $\{I,\Pi,\Pi^2,\dots,\Pi^{N-1}\}$ of order N is a latin square, where Π is the permutation matrix with 1 in the positions (1,2),(2,3),...,(N-1,N), (N,1).

When N = 4, it can be checked that any latin square of the form $\{I,\Pi_2,\Pi_3,\Pi_4\}$ becomes a group - there are four cases.

Conversely, given an arbitrary group $G = \{g_1,\dots,g_N\}$ of order N, we may construct $p_1\Pi_1 + \dots + p_N\Pi_N$, according to the following rule, such that $\{\Pi_1,\dots,\Pi_N\}$ becomes a latin group isomorphic to G.

Rule: If $g_k = g_j g_i^{-1}$, take p_k as the (i,j)-th element.

It is easy to verify that the resulting $\{\Pi_1,\dots,\Pi_N\}$ is in fact a latin group and is isomorphic to the following permutation group:

$$\left\{ \begin{pmatrix} g_1, & g_2, & \dots, & g_N \\ g_k g_1, & g_k g_2, & \dots, & g_k g_N \end{pmatrix}, \ k = 1,\dots,N \right\}.$$

This group is known to be isomorphic to G. [A.Cayley, see Biggs(1985)]

Consider one more example. Let $P_1 = p_1 I + p_2\Pi + p_3\Pi^2 + \dots + p_N\Pi^{N-1}$ be a transition matrix made on the cyclic group mentioned above. A matrix closely related to this is the following P_2: $P_2 = \Pi_0 P_1 = P_1^T \Pi_0$, where P^T is the transpose of P and Π_0 is the permutation matrix with 1 in the positions (1,1), (2,N),(3,N-1),(4,N-2),...,(N-1,3),(N,2). The latin square $L_0 = \{\Pi_0, \Pi_0\Pi, \Pi_0\Pi^2, \ldots, \Pi_0\Pi^{N-1}\}$ in this case does not have the identity I, hence is not a group. We can show, however, the following:

$$P_2^{2n-1} = \Pi_0 P_1 (P_1^T P_1)^{n-1} \text{ and } P_2^{2n} = (P_1^T P_1)^n. \text{ (note } \Pi_0^2 = I.)$$

Since $\Pi_1^T = \Pi_1^{-1}$ for any permutation matrix Π_1, and $L = \{I, \Pi, \ldots, \Pi^{N-1}\}$ is a cyclic group, we have:

$$(\sum p_i \Pi^i)^T = \sum p_i \Pi^{-i} = \sum p_i' \Pi^i,$$

$$(\sum p_i \Pi^i)(\sum p_i' \Pi^i) = \sum p_i'' \Pi^i.$$

From this it is seen that P_2^n can be expressed as a convex linear combination on latin square L_0 or L, for each n. Hence we have (*) for this example.
Problem: Characterize a class of latin squares (including all the latin groups) each of which makes (*) hold for any transition matrix that is a convex linear combination on that latin square.

REFERENCES

Ash R. (1965): Information theory, New York: Interscience.

Biggs N.L. (1985): Discrete mathematics, Clarendon Pr., Oxford.

Rényi A. (1961): On measures of entropy and information, Proc. of the Fourth Berkeley Symposium on Mathematical Statistics and Probability, Vol.1, Univ. of Calif. Pr., 547-561.

(1970): Foundations of probability, Holden-Day, Inc..

Shaw R. (1984): The dripping faucet as a model chaotic system, Aerial Pr..

(1985): personal communication.

Department of Information Sciences
Faculty of Engineering
Shizuoka University
Hamamatsu 432 Japan

ON THE ROBUSTNESS OF SEARCHING ALGORITHMS FOR MULTI-OBJECTIVE MARKOV DECISION PROBLEMS

Gerhard Hübner

Hamburg

Key words: Markov decision problems, multiple objectives, nearly optimal solutions, robustness

ABSTRACT

The paper considers Markov decision problems with multiple objectives, especially discounted problems with two objectives. Either one objective is minimized whilst the other has to obey a constraint, or a complete set of policies with non-dominated pairs of values is looked for. Three methods are considered: linear programming, a modified policy iteration and a Lagrangean searching procedure. When calculations are carried out only approximatively the second method may be not robust, i.e. the error may increase from one step to the next, as is shown by an example. Variants to avoid this difficulty are discussed.

I. INTRODUCTION

There are a lot of applications of Markov decision processes where more than one objective function has to be regarded when looking for good policies. The most famous seem to be inventory problems where it is difficult to compare numerically holding costs to the amount of shortage when demand exceeds the stock on hand (cp. e.g. Derman and Klein (1965), p. 276). Usually it is impossible to minimize both objectives at a time, so mostly only one is minimized (e.g. the holding costs) whereas the other (the amount of shortage) is restricted by a prescribed constant (guaranteeing a certain level of service). But the difficulty remains to fix this constant, so increasingly managers ask for a set of non-dominated solutions in order to choose ultimately one of the proposed policies by arguments hard to formulate in advance.

Both types of problems may be solved by using linear programming methods. These methods are well-known (cp. e.g. Derman (1970)) and shall not be discussed here in detail. But the software packages available usually do not exploit the special structure of Markov decision models. So two different types of algorithms are used here, the first being a variant of policy iteration, the second a Lagrangean search procedure. In both cases the consequences of approximative calculations are discussed.

2. THE MODEL

We consider a finite stationary Markov decision model with a finite set of states S and for each state $s \in S$ a finite set D_s of feasible actions. The transition law is given by the probabilities $p(s,a,s')$ for going to state s' when being in state s and choosing action $a \in D_s$. The one-step expected costs for the two objectives are $c(s,a)$ and $d(s,a)$ (resp.). All costs are discounted to the beginning by a discount factor $\beta \in (0,1)$. For definiteness we assume a fixed starting distribution $q = (q(s))$ where $q(s)$ is the probability of starting in state $s \in S$. For any stationary (possibly randomized) policy π ($\pi \in \Pi_s$) let V_π and W_π be the expected discounted total costs for the first and second objective (resp.). The restriction to stationary policies follows e.g. from Hartley (1979), cp. Derman (1970). Let $V_\pi(s)$ and $W_\pi(s)$ be the corresponding value functions if a fixed starting state $s \in S$ is assumed. Then $V_\pi = \sum_s q(s)\, V_\pi(s)$, $W_\pi = \sum_s q(s)\, W_\pi(s)$ and $V_\pi(\cdot)$, $W_\pi(\cdot)$ are the unique solutions of

$$V_\pi(s) = L_\pi V_\pi(s), \quad W_\pi(s) = M_\pi W_\pi(s)$$

where

$$L\,v(s,a) := c(s,a) + \beta \sum_{s'} p(s,a,s')v(s')$$

$$M\,w(s,a) := d(s,a) + \beta \sum_{s'} p(s,a,s')w(s')$$

and $L_\pi v(s)$, $M_\pi w(s)$ are the one-step expected values of $Lv(s,a)$,

$Mw(s,a)$ with respect to the distribution of actions prscribed by policy π. If π is deterministic ($\pi \in \Pi_{SD}$) with $\pi(s,a_s)=1$, $s \in S$, then

$$L_\pi v(s) = Lv(s,a_s), \quad M_\pi w(s) = Mw(s,a_s).$$

The set of all pairs $\{(V_\pi, W_\pi), \pi \in \Pi_S\}$ forms a convex polyhedron with vertices corresponding to deterministic policies (see Hartley (1979)). Two vertices play a special role: Let (V^o,W^o) be defined by $W^o := \min_\pi W_\pi$ and $V^o := \min\{V_\pi, W_\pi = W^o\}$. Similarly let $V^1 := \min_\pi V_\pi$ and $W^1 := \min\{W_\pi, V_\pi = V^1\}$. These vertices exist and correspond to policies π^o and π^1 (resp.). The edge of the polyhedron between (V^o,W^o) and (V^1,W^1) is the set of (Pareto-)efficient policies.

3. SEARCH ALONG THE EDGE

One possibility to obtain a complete set of non-dominated policies is to start at (V^o,W^o) with policy π^o (see section 2) and then going from vertex to vertex along the extreme edge. An algorithm for doing so goes back to Sladky (1967), p. 361, who considered the average cost case, but the discounted case works similarly.

LEMMA. If π_1 and π_2 are deterministic stationary policies which differ only by one action (a_1 and a_2) in one state s_o, then

$$\frac{L_{\pi_2} V_{\pi_1}(s_o) - V_{\pi_1}(s_o)}{M_{\pi_2} W_{\pi_1}(s_o) - W_{\pi_1}(s_o)} = \frac{V_{\pi_2} - V_{\pi_1}}{W_{\pi_2} - W_{\pi_1}}$$

if the first denominator does not vanish.

The proof is easily completed by observing $L_{\pi_2} V_{\pi_1}(s) = V_{\pi_1}(s)$, $s \neq s_o$, and

$$V_{\pi_2} - V_{\pi_1} = q(I-\beta P_{\pi_2})^{-1}(L_{\pi_2} V_{\pi_1}(\cdot) - V_{\pi_1}(\cdot)),$$

$$W_{\pi_2} - W_{\pi_1} = q(I-\beta P_{\pi_2})^{-1}(M_{\pi_2} W_{\pi_1}(\cdot) - W_{\pi_1}(\cdot)).$$

From this Lemma the following algorithm derives using the steepest descend:

ALGORITHM.

STEP 0: Use the starting policy π^{o} given above with values $(V^{o}=V_{\pi^{o}}, W^{o}=W_{\pi^{o}})$. Set $\pi_{o} := \pi^{o}$.

STEP 1: If π_k is given, define π_{k+1} by

$\pi_{k+1}(s_o) = a_o,\ \pi_{k+1}(s) = \pi_k(s),\ s \neq s_o,$ where

$$\frac{L V_{\pi_1}(s_o,a_o) - V_{\pi_1}(s_o)}{M W_{\pi_1}(s_o,a_o) - W_{\pi_1}(s_o)} = \min \frac{L V_{\pi_1}(s,a) - V_{\pi_1}(s)}{M W_{\pi_1}(s,a) - W_{\pi_1}(s)}$$

and the minimum is taken over all (s,a) with denominator > 0 and numerator < 0. If no such (s_o,a_o) exists then $(V^1 = V_{\pi_k}, W^1 = W_{\pi_k})$, stop.

Repeat Step 1.

What about this algorithm, if step 0 or step 1 result in approximate values only, e.g. because of rounding errors or of inaccurate model parameters?

One might suppose that an error for (V_{π_k}, W_{π_k}) should not increase when calculating $(V_{\pi_{k+1}}, W_{\pi_{k+1}})$. But this is not true as the following example shows:

EXAMPLE. Consider a model with $S=\{1,2\}$, $D_1=\{0,1,2\}$, $D_2=\{0,1\}$, $\beta=0.98$, $q(1)=1$ and $p(s,a,s')$, $c(s,a)$, $d(s,a)$ according to the following table:

s,a	p(s,a,1)	p(s,a,2)	c(s,a)	d(s,a)
1,0	.5	.5	0	0
1,1	.9	.1	-.835	.4175
1,2	.05	.95	-5.02	2.54
2,0	.05	.95	0	0
2,1	.9	.1	-1	1

There are six policies in Π_{SD} which shall be numbered according to the decisions chosen, e.g. π_{21} is defined by $\pi_{21}(1,2)=1$, $\pi_{21}(2,1)$

= 1. The pertinent values are numbered in the same way and are given in the following table:

	π_{00}	π_{10}	π_{20}	π_{11}	π_{21}	π_{01}
V..	0	-17.25	-17.32	-42.56	-148.91	-17.60
W..	0	8.63	8.76	23.73	87.89	17.60

The polyhedron of V- and W-values is shown in the following picture:

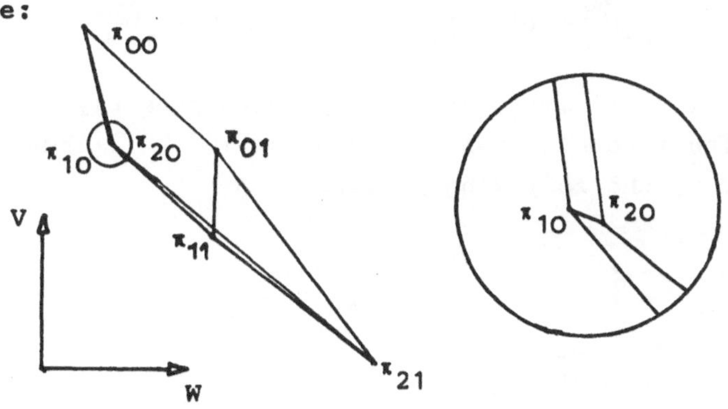

In this example a minimal error (about 1 %) in the first iteration of the algorithm leads to the wrong policy π_{20} instead of π_{10} with almost equal values. Then the next iteration results in policy π_{21} instead of π_{11} with values differing enormously.

To avoid this effect additional tests have to be included in the algorithm ensuring that no policy dominates the actual policy π_k by more than a prescribed distance.

4. A LAGRANGEAN SEARCH PROCEDURE

To obtain efficient policies we define

$$Z_\pi^\lambda := \lambda V_\pi + (1-\lambda)W_\pi, \quad \lambda \in [0,1], \quad \pi \in \Pi_{SD} .$$

For each fixed λ, a minimal Z_π^λ may be determined by policy iteration, by successive approximations or by similar methods (including linear programming).

We shall not try here to find all deterministic efficient policies by Lagrangean search. This is related to the methods of section 3, of Hartley (1979) and of White and Kim (1980).

Instead of we assume that for each λ we may obtain policies π_λ and bounds $\underline{V}^\lambda < V_{\pi_\lambda} \leq \overline{V}^\lambda$, $\underline{W}^\lambda \leq W_{\pi_\lambda} \leq \overline{W}^\lambda$, such that $\hat{z}^\lambda := \min_\pi z_\pi^\lambda \geq \lambda \underline{V}^\lambda + (1-\lambda)\underline{W}^\lambda$. If this is done for a finite set of parameters λ_k lower bounds for all (V_π, W_π) are obtained by

$$\lambda_k \underline{V}^{\lambda_k} + (1-\lambda_k)\underline{W}^{\lambda_k} \leq \lambda_k V_\pi + (1-\lambda_k) W_\pi$$

and upper bounds for all efficient (V_π, W_π) with possibly randomized policies π by linear combination of two pairs $(\overline{V}^\lambda, \overline{W}^\lambda)$. These bounds are demonstrated by the following picture.

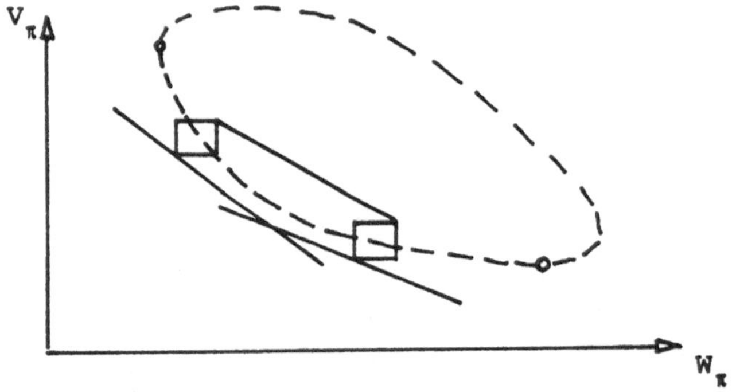

By this method the set of efficient pairs of values may be approximated as tight as needed, but no estimate on the number of parameters λ_k is known to me.

5. CONCLUSIONS

The algorithm of Sladky presented in section 3 is a good finite method to obtain all efficient solutions of a Markov decision problem with two objectives, but in extreme cases it may be numerically instable, whereas the Lagrangean method of section 4 is numerically stable.

Both methods also provide solutions to problems where one objective has to obey a constraint and the other is to be minimized.

When these algorithms are extended to more dimensions additional structural problems arise in addition to the enumerating and robustness problems of two dimensions. The resulting generalized methods have to be compared to other algorithms proposed in the literature, e.g. those of Hartley (1969), Furukawa (1982) and White/Kim (1980).

ACKNOWLEDGEMENT

I am grateful to Dr. P. Mandl and Dr. K. Sladký, Prague, for valuable discussions on the topic of this paper.

REFERENCES

Derman, C. (1970): Finite state Markovian decision processes. Academic Press, New York.

Derman, C., Klein, M. (1965): Some remarks on finite horizon Markovian decision models. Oper.Res.13, 272-278.

Furukawa, R. (1982): Recurrence set relations in stochastic multiobjective dynamic decision processes. Optimization 13, 113-122.

Hartley, R. (1979): Finite, discounted, vector Markov decision processes. Notes in Decision Theory Nr. 85, Univ. of Manchester.

Sladky, K. (1967): A problem concerning an optimal service policy for several facilities (Czech.). Kybernetika 4, 352-376.

White, C.C., Kim, K.W. (1980): Solution procedures for vector criterion Markov decision processes. Large scale Systems 1, 129-140.

Universität Hamburg
Institut für Math. Stochastik
Bundesstraße 55
D-2000 Hamburg 13

THEOREMS ON SELECTORS IN TOPOLOGICAL SPACES II

Adam Idzik

Warsaw

Key words: measurable multifunctions, Castaing s representation theorem, Filippov s implicit function theorem.

ABSTRACT

This paper is a second part of the author s paper "Theorems on selectors in topological spaces I" |in: Transactions of the Ninth Prague Conference on Information Theory,..., Academia, Prague 198̃|.

Theorems similar to the Castaing s representation theorem, measurability of functions defined on product spaces and implicit function theorems are presented in a new framework.

MEASURABILITY AND WEAK MEASURABILITY

In this section we shall consider some properties of measurable multifunctions, which we apply in the next sections.

The following properties of multifunctions we reformulate from Himmelberg (1975):

(4.1) PROPOSITION. Let X be a topological space with the property that every open set is F_σ, then a - measurability of a multifunction $\varphi : T \to P(X)$ implies weak a - measurability.

(4.2) PROPOSITION. Let J be at most countable set and let $\varphi_n : T \to P(X)$ be a multifunction for each $n \in J$. Then

(i) if J is finite and φ_n is measurable (weakly measurable), so is the mulfifunction $\bigcup_{n\in J}\varphi_n : T \to P(X)$ defined by $(\bigcup_{n\in J} \varphi_n)(t) = \bigcup_{n\in J} \varphi_n(t)$,

(ii) if J is infinite and φ_n is a - measurable (weakly a - measurable), so is the multifunction $\bigcup_{n\in J} \varphi_n$, and

(iii) if X is second countable and each φ_n is weakly a - measurable, then so is the multifunction $\prod_{n\in J} \varphi_n : T \to P(X^J)$ defined by $(\prod_{n\in J} \varphi_n)(t) = \prod_{n\in J} \varphi_n(t)$.

(4.3) PROPOSITION. If X is a subspace of Y, then $\varphi : T \to P(X)$ is measurable (weakly measurable, a - measurable, weakly a - measurable etc.) as a multifunction into X iff $\varphi : T \to P(X)$ is measurable (weakly measurable, a - measurable, weakly a - measurable, etc.) as a multifunction into Y.

(4.4) PROPOSITION. $\varphi : T \to P(X)$ is weakly measurable (weakly a - measurable, etc.) iff the multifunction $\overline{\varphi} : T \to P(X)$, defined by $\overline{\varphi}(t) = \overline{\varphi(t)}$, is weakly measurable (weakly a - measurable, etc.).

Now we give a generalization of Himmelberg s theorem:

(4.5) THEOREM (cf. Theorem 3.1 in Himmelberg (1975) and Theorem 3.2 in Himmelberg ,...(1981)). Let X be a perfectly normal space and $\varphi : T \to P(X)$ be a multifunction with countably compact values. If φ is weakly a - measurable (weakly m - measurable), then it is am - measurable (m - measurable).

Proof. Let F be a closed subset of X. Because X is perfectly normal, then there exists a family $\{G_n\}_{n\in N}$ of open sets such that $F \subset G_{n+1} \subset \overline{G}_{n+1} \subset G_n$ $(n\in N)$ and $F = \bigcap_{n\in N} G_n = \bigcap_{n\in N} \overline{G}_n$. Furthermore, for fixed $t\in T$, $\varphi(t)$ is countably compact and $\{t|\varphi(t) \subset X-F\} =$
$= \{t|\varphi(t) \subset \bigcup_{n\in N} (X - \overline{G}_n)\} = \bigcup_{n\in N} \{t|\varphi(t) \subset (X - \overline{G}_n)\} =$
$= \bigcup_{n\in N}\{t|\varphi(t) \subset X - G_n\}$. Thus $\varphi^{-1}(F) = T - \{t|\varphi(t) \subset X - F\} =$
$= T - \bigcup_{n\in N}\{t|\varphi(t) \subset X - G_n\} = T - \bigcup_{n\in N} (T - \varphi^{-1}(G_n))$. This ends the proof.

From Propositions 4.1 - 4.4 and Theorem 4.5, analogously as Theorem 4.1 in Himmelberg (1975) we can prove the following

(4.6) THEOREM. Let X be separable metrizable, and let $\varphi_n : T \to P(X)$ be a weakly a - measurable multifunction with closed values for each n in an at most countable set J. Also assume that for each $t\in T$, $\varphi_n(t)$ is compact for some $n\in J$. Then $\bigcap_{n\in J} \varphi_n : T \to P(X)$ defined by $(\bigcap_{n\in J} \varphi_n)(t) = \bigcap_{n\in J} \varphi_n(t)$ is am - measurable.

CASTAING'S REPRESENTATION THEOREM

The first theorems on a characterization of measurable multifunctions was done by Castaing (1967) (see Theorem 5.3 and Theorem 5.4). Himmelberg (1975) generalized Castaing's theorems (see Theorem 5.6). Here, applying theorems on selectors proved in the previous sections, we extend some results of Himmelberg.

(5.1) THEOREM. (cf. Theorem 5.6 in Himmelberg (1975). Let X be a topological space and let $\varphi : T \to \mathcal{P}(X)$ be a multifunction. If there exists a countable family $\{f_n\}_{n\in N}$ of weakly a - measurable selectors for φ such that $\varphi(t) = \overline{\{f_n(t)\}}_{n\in N}$ for all $t\in T$, then φ is weakly a - measurable.

Proof. By Proposition 4.2 (ii) the multifunction defined by $t\to\{f_n(t)\}_{n\in N}$ is weakly a - measurable and by Proposition 4.4 the multifunction $t \to \varphi(t) = \overline{\{f_n(t)\}}_{n\in N}$ is weakly a - measurable.

To formulate the next theorems we need the following

(5.2) DEFINITION. We say that a family $\mathcal{D} = \{D_n\}_{n\in N}$ of closed subsets of a topological space X is dense in X, if for every $x\in X$ and an open set $G\ni x$ there exists $n\in N$ such that $x\in D_n\subset G$.

In the terminology of Engelking (1977), see p. 170, $\mathcal{D}$ is a closed, countable network for X. For regular spaces the existence of a countable network implies the existence of a closed, countable network.

(5.3) THEOREM. Let $\varphi : T \to \mathcal{P}(X)$ be a measurable multifunction which has values complete with respect to a family $\mathcal{A} = \{\mathcal{A}_n\}_{n\in N}$ of covers of X ($\mathcal{A}_n \subset P(X)$; $n\in N$). If there exists a family $\mathcal{D} = \{D_n\}_{n\in N}$ dense in X, $\mathcal{A}_n$ has a countable closed refinement $\mathcal{B}_n$ ($n\in N$) and the family $\mathcal{B} = \{\mathcal{B}_n\}_{n\in N}$ separates points, then there exists a countable family $\{f_n\}_{n\in N}$ of am - measurable selectors for φ such that $\varphi(t) = \overline{\{f_n(t)\}}_{n\in N}$ for all $t\in T$.

Proof. For $n\in N$ we define a multifunction $\varphi_n : T \to \mathcal{P}(X)$ as follows:

$$\varphi_n(t) = \begin{cases} \varphi(t) \cap D_n & \text{on } \{t \mid \varphi(t) \cap D_n \neq \emptyset\} \\ \varphi(t) & \text{on } \{t \mid \varphi(t) \cap D_n = \emptyset\} \end{cases}$$

The multifunction φ_n is measurable and has values complete with respect to the family $\mathcal{A}$. Thus, by Theorem 2.3 in Idzik (1983) there exists an am - measurable selector f_n for φ_n. It is easy to check that the family $\{f_n\}_{n\in N}$ has required properties.

Similarly, applying Theorem 2.5 in Idzik (1983) we can prove

(5.4) THEOREM. Let $\varphi : T \to P(X)$ be a measurable multifunction which has values * - complete with respect to a family $\mathcal{A} = \{\mathcal{A}_n\}_{n\in N}$ ($\mathcal{A}_n \subset P(X)$; $n\in N$). If there exists a family $\mathcal{D} = \{D_n\}_{n\in N}$ dense in X, $\mathcal{A}_n$ has a countable star-finite closed refinement $\mathcal{B}_n$ ($n\in N$) and the family $\mathcal{B} = \{\mathcal{B}_n\}_{n\in N}$ * - separate points, then there exists a countable family $\{f_n\}_{n\in N}$ of m - measurable selectors for φ such that $\varphi(t) = \overline{\{f_n(t)\}_{n\in N}}$ for all $t\in T$.

(5.5) COROLLARY (cf. Corollary 1 in Idzik (1981)). Let X be a separable metrizable space and $\varphi : T \to P(X)$ be a measurable multifunction which has values complete with respect to a fixed metric in X. Then there exists a countable family $\{f_n\}_{n\in N}$ of m - measurable selectors for φ such that $\varphi(t) = \overline{\{f_n(t)\}_{n\in N}}$ for all $t\in T$.

Now, denote by $\mathcal{T}_\sigma$ the σ - field generated by $\mathcal{T}$ and say that a multifunction $\varphi : T \to P(X)$ is σ - measurable (weakly σ - measurable) if $\varphi^{-1}(D) \in \mathcal{T}_\sigma$ for every closed (open) set $D \subset X$. The families $\mathcal{T}$, $\mathcal{T}_a$, $\mathcal{T}_m$, $\mathcal{T}_{am}$ etc. are contained in $\mathcal{T}_\sigma$ and measurability, a - measurability etc. (weak measurability, weak a - measurability etc.) implies σ - measurability (weak σ - measurability). If $\mathcal{T} = \mathcal{T}_\sigma$, then $\mathcal{T} = \mathcal{T}_a = \mathcal{T}_m = \mathcal{T}_{am}$ = etc. and all concepts of the measurability (weak measurability) coincides.

(5.6) THEOREM. (Castaing's representation). Let X be a normal space with a separating family $\mathcal{B}$ and a family $\mathcal{D}$ dense in X. And let $\varphi : T \to P(X)$ be a multifunction with countable compact values. Then φ is weakly σ - measurable if and only if there exists a countable family $\{f_n\}$ of weakly σ - measurable selectors for φ such that $\varphi(t) = \overline{\{f_n(t)\}_{n\in N}}$ for all $t\in T$.

Proof. Let $\mathcal{T} = \mathcal{T}_\sigma$. If the multifunction φ with countably compact values is weakly σ - measurable, then by Theorem 4.5 φ is σ - measurable. Because every countable compact set is complete with respect to $\mathcal{B}$, then by Theorem 5.3 (putting $\mathcal{A} = \mathcal{B}$) there exists a countable family $\{f_n\}_{n\in N}$ of σ - measurable selectors (and by Proposition 4.1 weakly σ - measurable selectors) for φ such that $\varphi(t) = \overline{\{f_n(t)\}_{n\in N}}$ for all $t\in T$.

The converse theorem follows from Theorem 5.1.

MEASURABILITY OF PRODUCT FUNCTIONS

In the proof of Theorem 5.6 we used the fact that for a topological space X with the property that every open set is F_δ, then δ - measurability of a function $f : T \to X$ implies weak δ - measurability (see Proposition 4.1). More generally, it is true that for an arbitrary topological space X and a function $f : T \to X$ measurability (a - measurability, m - measurability, δ - measurability etc.) is equivalent to weak measurability (weak m - measurability, weak a - measurability, weak δ - measurability etc.). This follows from the equality $f^{-1}(D) = T - f^{-1}(X-D)$ for any open or closed subset D of X.

Now we shall generalize a theorem of Kuratowski on the measurability of functions of two variables (see Kuratowski (1966), § 31, V). Let us start from the following

(6.1) LEMMA. Let X be a second countable space and Y be a space. Denote by $\{D_m\}_{m\in N}$ a countable base in X and by $S=\{s_i\}_{i\in N}$ a countable dense subset of X. If a function $g : X \to Y$ is continuous and a closed set $F = \bigcap_{n\in N} G_n = \bigcap_{n\in N} \bar{G}_n$, where $\bar{G}_n$ is the closure of an open set G_n in Y $(n\in N)$, then

$$\{x \mid g(x) \in F\} = \bigcap_{n\in N} \bigcup_{m\in N} \bigcap_{s_i\in D_m} \{D_m \mid g(s_i) \subset G_n\}.$$

Proof. Let $g(x) \in F$. For each $n\in N$ $g(x)\in G_n$ and by the continuity of g there exists $D_m \ni x$ such that $g(D_m)\subset G_n$ and, a fortiori, $g(s_i) \in G_n$ for $s_i \in D_m$ $(m\in N)$. Conversely, if for fixed $n\in N$, $x\in D_m$ such that $g(s_i)\in G_n$ for $s_i\in D_m$ $(m\in N)$, then by the continuity of g, $g(x)\in \bar{G}_n$. Thus $g(x)\in F$ and Lemma 6.1 is proved.

(6.2) THEOREM (cf. 6.1 in Himmelberg (1975)). Let X be second countable, Y perfectly normal and let a function $f : T\times X \to Y$ be weakly a - measurable (weakly m - measurable) in t and continuous in x. Then f is amam - measurable (mam - measurable) with respect to the product of $\mathcal{T}$ and $\mathcal{G}$, i.e. $f^{-1}(F)\in \mathcal{T}\times\mathcal{G}$ amam $(\mathcal{T}\times\mathcal{G}_{mam})$, for every closed F in Y.

Proof. Applying Lemma 6.1 (and assuming its notations) for a closed set F we have

$$(6.3)\quad f^{-1}(F) = \bigcap_{n\in N} \bigcup_{m\in N} \bigcap_{s_i\in D_m} \{t \mid f(t,s_i)\in G_n\}\times D_m.$$

By our assumption for fixed $n,m \in N$, $s_i \in D_m$ the set $\{t | f(t,s_i) \in G_n\} \times D_m$ is a countable union (intersection) of elements of $\mathcal{T} \times \mathcal{G}$ and thus $f^{-1}(F) \in \mathcal{T} \times \mathcal{G}_{amam} (\mathcal{T} \times \mathcal{G}_{mam})$.

We recall that for a topological space T a multifunction $\varphi : T \to \mathcal{P}(X)$ is of class $\alpha_-(\alpha^-)$ $(\alpha < \Omega)$ if $\varphi^{-1}(Z)$ is of additive (multiplicative) class α whenever Z is open (closed), and, a function $f : T \to X$ is of Baire class α if $f^{-1}(Z)$ is of multiplicative class α for every closed Z.

From the formula (6.3) follows

(6.4) COROLLARY: Let T be a topological space, X second countable, Y perfectly normal and let a function $f : T \times X \to Y$ be of Baire class α in t and continuous in x. Then f is of Baire class $\alpha + 3$ in (t,x).

At the end of this section we formulate

(6.5) THEOREM (cf. Theorem 6.2 in Himmelberg (1975)). Let X,Y be topological spaces and X be separable, $f : T \times X \to Y$ a function weakly a - measurable in t and continuous in x and, U an open subset of Y. Then $\varphi(t) = \{x \in X | f(t,x) \in U\}$ defines an a - measurable multifunction from T to X (possibly empty valued).

A proof of Theorem 6.5 is similar to the proof of Theorem 6.2 in Himmelberg (1975) and we omit it.

IMPLICIT FUNCTION THEOREMS

Theorems which we shall prove in this section are of the type of Filippov's lemma (1959). Very general implicit function theorems were presented in the paper of Himmelberg (1975). Our first version concerns with implicit functions of Baire class α. The second one is similar to the version of Himmelberg (see Theorem 7.1 in Himmelberg (1975)) but it is not more general than the latter; it is of another kind. They coincide for separable metric space Y (cf. Theorem 7.6).

(7.1) DEFINITION. Let Y be an arbitrary set. We say that a family $\mathcal{B} = \{B_n\}_{n \in N} \subset P(Y)$ countably separates points if for every $y \in Y$ $\bigcap \{B_m | y \in B_m \in \mathcal{B},\ m \in N\} = \{y\}$.

(7.2) THEOREM. Let T be a topological space with the property that every open set is F_δ, X a separable metric space and, Y a topological space with a countable family of open subsets $\{B_n\}_{n\in N}$ such that the family $\{\overline{B}_n\}_{n\in N}$ countably separates points; $\overline{B}_n$ is the closure of B_n in Y $(n\in N)$. And let $f : T \times X \to Y$ be a function of Baire class α $(\alpha < \Omega)$ in t and continuous in x, $\varphi : T \to \mathcal{P}(X)$ is a multifunction of class $(\alpha+1)_-$ with compact values and $g : T \to X$ is a function of Baire class α such that $g(t) \in f(t,\varphi(t))$ for $t\in T$. Then there exists a selector $h:T \to X$ of class $\alpha + 3$ for φ such that $g(t) = f(t,h(t))$ for all $t\in T$.

To prove this theorem we need two lemmas.

(7.3) LEMMA. Let B be an open subset of Y. Under notations and assumptions of Theorem 7.2 the multifunction $\varphi_B : T \to \mathcal{P}(X)$ defined by

$$\varphi_B(t) = \begin{cases} X & \text{on } \{t \mid g(t) \notin B\} \\ \{x \mid f(t,x) \in B\} & \text{on } \{t \mid g(t) \in B\} \end{cases}$$

is of class $(\alpha+1)_-$.

Proof (of Lemma 7.3). For each $t\in T$ $g(t) \in f(t,\varphi(t))$ and if $g(t) \in B$, then $\{x\mid f(t,x) \in B\} \neq \emptyset$. Thus φ_B is well-defined. By Theorem 6.5 (substituting to $\mathcal{T}$ the field of sets which are simultaneously of additive and multiplicative class α) and Proposition 4.1 the set $\{t\mid \{x\mid f(t,x) \in B\} \cap C \neq \emptyset\}$ is of additive class α for every open set C in X. The set $\{t\mid g(t) \in B\}$ is of additive class α and the set $\{t\mid g(t) \notin B\}$ is of multiplicative class α. Finally, for a nonempty open set C, the set $\varphi_B^{-1}(C) = \{t\mid g(t)\notin B\}\cup\{t\mid \{x\mid f(t,x) \in B\}\cap C \neq \emptyset\} \cap\{t\mid g(t) \in B\}$ is of additive class $(\alpha+1)$.

(7.4) LEMMA. (see Corollary 2 in Idzik (1981)). Let T and X be as in Theorem 7.2 and let $\psi : T \to \mathcal{P}(X)$ be a multifunction of class β^- $(\beta < \Omega)$ with compact values. Then there exists a selector of Baire class $\beta+1$ for ψ.

Proof (of Theorem 7.2). For B_n the multifunction φ_{B_n} defined as in Lemma 7.3 is of class $(\alpha+1)_-$. By Proposition 4.4 the multifunction $\overline{\varphi}_{B_n}$ is also of class $(\alpha+1)_-$ $(n\in N)$. Let $\psi: T \to \mathcal{P}(X)$ be the multifunction defined by $\psi(t) = \varphi(t) \cap \bigcap_{n\in N} \overline{\varphi}_{B_n}(t)$ for $t\in T$. Applying Theorem 4.6 we see that the multifunction ψ is of class $(\alpha+2)^-$ and by Lemma 7.4 there exists a selector $h : T \to X$ of Baire class $\alpha+3$ for ψ. Of course h is also a selector for φ. Furthermore, for each $t \in T$ if $g(t) \in B_m \subset \bar{B}_m$ for some $m \in N$, then $h(t) \in \overline{\{x \mid f(t,x) \in B_m\}} \subset \{x \mid f(t,x) \subset \bar{B}_m\}$ and $f(t,h(t)) \in \bar{B}_m$. But the family $\{\bar{B}_n\}_{n\in N}$ countably separates points and thus $g(t) = f(t,h(t))$ for all $t \in T$.

To formulate the second version of Filippov's lemma we recall a theorem on the existence of m - measurable selectors, which was presented in Idzik (1981).

(7.5) LEMMA (see Theorem 6 in Idzik (1981)). Let $\varphi : T \to \mathcal{P}(X)$ be a measurable multifunction such that $\varphi(t)$ is countably compact for each $t \in T$. If there exists a family $\mathcal{B}_0 \subset \mathcal{F}$ which countably separates points, then there exists an m - measurable selector for φ.

In a similar way as Theorem 7.2, applying Lemma 7.5 instead of Lemma 7.4, we can prove

(7.6) THEOREM. Let X be a separable metric space and Y a topological space with a countable family of open subsets $\{B_n\}_{n\in N}$ such that the family $\{\bar{B}_n\}_{n\in N}$ countably separates points; $\bar{B}_n$ is the closure of B_n in Y $(n \in N)$. And let $f : T \times X \to Y$ be a function $\mathcal{S}$ - measurable in t and continuous in x, $\varphi : T \to \mathcal{P}(X)$ a weakly $\mathcal{S}$ - measurable multifunction with compact values, and $g : T \to X$ a $\mathcal{S}$ - measurable function such that $g(t) \in f(t,\varphi(t))$ for $t \in T$. Then there exists a $\mathcal{S}$ - measurable selector $h : T \to X$ for φ such that $g(t) = f(t,h(t))$ for all $t \in T$.

REFERENCES

Castaing C. (1967): Sur les multi-applications measurables, Rev. Francaise d'Informatique et de Recherche Operationnelle 1, 91-126.

Engelking R. (1977): General Topology, PWN - Polish Scientific Publishers, Warszawa.

Filippov A.F. (1959): On certain questions in the theory of optimal control, Vestnik Moskov. Univ. Ser. I Mat. Meh. 2, 25-32. English translation in J. Soc. Indust. Appl. Math. Ser. A Control 1 (1962), 76-84.

Himmelberg C.J. (1975): Measurable relations, Fund. Math. 87, 53-72.

Himmelberg C.J., T. Parthasarathy and F.S. Van Vleck (1981): On measurable relations, Fund. Math. 111, 161-167.
Idzik A. (1981): Theorems on selectors for measurable multifunctions. Bull. Acad. Polon. Sci. Sér. Sci. Math., 597-603.
(1983): Theorems on selectors in topological spaces I. In: Trans. of the Ninth Prague Conference on Information Theory,..., Academia, Prague, 287-292, Vol. A.
Kuratowski K. (1966): Topology, Vol. I, Academic Press, New York - London, PWN, Warszawa.

Institute of Computer Science,
Polish Academy of Sciences,
P.O. Box 22
00-901 Warsaw, PKiN
Poland

NON-DURABLE AND DURABLE ECONOMIC PROCESSES IN A DYNAMIC MODEL OF PRODUCTION AND CONSUMPTION

A. Idzik and P.B. Simonsen

Warsaw Copenhagen

Key words: Noncooperative game with constraints, Arrow-Debreu model

1. ABSTRACT

Generalizations of Arrow and Debreu's (1954) static model of production and consumption were made by many authors (e.g. Makarov and Rubinov (1977), Idzik and Simonsen (1983), Flåm (1981)).

A dynamic model of production and consumption, based on a game-theoretic approach, was presented at the 11th IFIP Conference on System Modelling and Optimization by Idzik and Simonsen (1983b).

In our paper we consider non-durable and durable economic processes in a dynamic model of production and consumption. We prove, under suitable assumptions, the existence of an equilibrium in this model for each of the economic processes.

2. A DYNAMIC ARROW-DEBREU MODEL

We say that a subset W of an Euclidean space is strongly convex if $m,n \in W$ and $m \neq n$ imply $\delta m+(1-\delta)n \in \mathrm{Int}(W)$ for $0<\delta<1$.

We apply a terminology of the paper of Idzik and Simonsen (1983a)*.

* There are two small errors in formulae (3.1.7) and (3.2.2) in this paper. The formulae read as follows:

(3.1.7) $\max\limits_{y_j \in \psi(\bar{z})} u_j(\bar{x}, y_j, \bar{y}_j', \bar{p}) = u_j(\bar{z})$ and $\bar{y}_j \in \psi_j(\bar{z})$, $(j \in N)$

(3.2.2) Y_j is a compact, strongly convex subset of E $(j \in N)$,

For simplicity we consider in our model only one producer and one consumer. A generalization to countably many of them does not present any difficulties. In the model there is also a pricing agency, a bank and a market.

Let the set of moments of time be indexed by the set $T=\{0,1,\ldots\}$. At each moment t the producer has a production set $X_t \subset E$ and the consumer has a consumption set $Y_t \subset E$, where E is a fixed 1-dimensional Euclidean space. Each of them has also a value-production function $v_t : Z \to R$ and a utility function $u_t : Z \to R$ at each moment t, respectively, which he maximizes. Here we denote $Z=X\times Y\times P$ and $X = \Pi\{X_t | t\in T\}$, $Y = \Pi\{Y_t | t\in T\}$, $P = \Pi\{P_1^t | t\in T\}$, $P_1^t = P_1 (t\in T)$. Furthermore, the producer has a production-constraining function $\varphi_t : Z \to 2^E$ ($\varphi_t(z)\subset X_t$ for $z\in Z$, $t\in T$) and there are fixed sequences of vectors: $\mu=\{\mu_t | t\in T\}$ and $\nu=\{\nu_t | t\in T\}$ which are obligations of, respectively, the producer and the consumer, concerning the market at each moment.

At each moment $t\in T$ the pricing agency establishes prices on all commodities and the bank pays to the consumer an amount $\alpha_t(z)$, where $\alpha_t : Z \to R$, if $\alpha_t(z) \geqslant 0$ (or receives from the consumer an amount $\alpha_t(z)$, if $\alpha_t(z) < 0$).

If planned production is x ($x\in X$), planned consumption is y ($y\in Y$) and planned prices are p ($p\in P$), then $\Sigma_{r=0}^{t}\mu_r \circ x_r$ of commodities will be supplied on the market and $\Sigma_{r=0}^{t}\nu_r \circ y_r$ of commodities will be consumed to the moment t.

From now on we assume that for each $t\in T$

(2.1) X_t is a compact, convex subset of E,

(2.2) Y_t is a compact, strongly convex subset of E,

(2.3) φ_t is a continuous function and $\varphi_t(z)$ is convex for $z\in Z$,

(2.4) α_t is a continuous real function,

(2.5) v_t and u_t are continuous functions concave with respect to the variables x_t and y_t respectively.

Now, we distinguish two kinds of economic processes: durable and non-durable. By non-durable processes we understand those that produce non-durable goods, i.e. production at the moment t can be sold exactly at this moment. And by durable economic processes we understand processes that produce durable goods, i.e. production at the moment t can be sold at an arbitrary moment in the future.

We assume that $p_{-1} = \nu_{-1} = y_{-1} = 0 \in E$.

3. NON-DURABLE ECONOMIC PROCESSES

We characterize these processes by the inequality

$$(3.1) \qquad \sum_{r=0}^{t} \alpha_r(z) - \sum_{r=-1}^{t-1} p_r(\nu_r \circ y_r) \leqslant p_t(\mu_t \circ x_t)$$

for $z \in Z$ $(t \in T)$.

The inequality (3.1) means that a stock of money of the consumer at the moment t is less than the value of the production supplied on the market at that moment.

An equilibrium in the model, when only the non-durable processes are assumed, is a point $\bar{z} = (\bar{x}, \bar{y}, \bar{p}) \in Z$ which satisfies the following conditions:

$$(3.2) \qquad \mu_t \circ \bar{x}_t \geqslant \nu_t \circ \bar{y}_z,$$

$$(3.3) \qquad \max_{x_t \in \varphi_t(\bar{z})} v_t(x_t, \bar{x}_t^{\hat{}}, \bar{y}, \bar{p}) = v_t(\bar{z}),$$

$$(3.4) \qquad \max_{y_t \in \psi_t(\bar{z})} u_t(\bar{x}, y_t, \bar{y}_t^{\hat{}}, \bar{p}) = u_t(\bar{z}),$$

where $\psi_t : Z \to 2^{Y_t}$ is defined by

$$(3.5) \qquad \psi_t(z) = \left\{ m \mid m \in Y_t, p_t(\nu_t \circ m) \leqslant \sum_{r=0}^{t} \alpha_r(z) - \sum_{r=-1}^{t-1} p_r(\nu_r \circ y_r) \right\}$$

An interpretation of (3.2) - (3.2) is similar to the interpretation of (3.1.5) - (3.1.7) in the static Arrow-Debreu model presented by Idzik and Simonsen (1983a).

The function ψ_t restricts the consumption at the moment t. When the prices are p_t the consumer cannot buy more than his stock of money $\sum_{r=0}^{t} \alpha_r(z) - \sum_{r=-1}^{t-1} p_r(\nu_r \circ y_r)$ allows him to.

However we assume that the consumer can always buy something at the moment t, i.e.

(3.6) $\Psi_t(z) \neq \emptyset$ for $z \in Z$ $(t \in T)$,

where Ψ_t is defined by (3.5).

(3.7) Theorem. If the conditions (2.1) - (2.5) and (3.6) are satisfied, then the dynamic Arrow-Debreu model has an equilibrium in the case of non-durable processes.

Proof. A noncooperative game can be associated with the model, the game having constraints as follows: the players are the producer, the consumer and the pricing agency at each moment $t \in T$. Sets of strategies of players are X_t, Y_t and P_1^t respectively. The constraints are defined by functions φ_t, Ψ_t and π_t, where $\pi_t(z)=P_1$ for $z \in Z$, and payoff functions by v_t, u_t and h_t, where $h_t(z)=p_t(\nu_t o y_t - \mu_t o x_t)$ for $z \in Z$, respectively.

By Lemma 3.2.8 of Idzik and Simonsen (1983a) Ψ_t $(t \in T)$ are the continuous functions and thus by our assumptions from Theorem 2.0.7 of Idzik and Simonsen (1983a) there follows the existence of an equilibrium point in the noncooperative game with constraints defined above.

Analogously as in the proof of Theorem 3.2.9 of Idzik and Simonsen (1983a), we show that this equilibrium point defines an equilibrium in our model.

Q.E.D.

4. DURABLE ECONOMIC PROCESSES

We characterize these processes by the inequality

(4.1) $$\sum_{r=0}^{t} \alpha_r(z) - \sum_{r=-1}^{t-1} p_r(\nu_r o y_r) \leq p_t\left(\sum_{r=0}^{t} \mu_r o x_r - \sum_{r=-1}^{t-1} \nu_r o y_r\right)$$

for $z \in Z$ $(t \in T)$.

The inequality (4.1) means that the stock of money of the consumer at the moment t is less than the value (in terms of prices

p_t) of the production on the market at the moment t and which was produced from the moment 0 to t.

An equilibrium in the model, when only the durable processes are assumed, is a point $\bar{z}=(\bar{x},\bar{y},\bar{p})\in Z$ which satisfies at each moment t the following conditions:

$$(4.2) \qquad \sum_{r=0}^{t} \mu_r \circ \bar{x}_r \geq \sum_{r=0}^{t} \nu_r \circ \bar{y}_r,$$

$$(4.3) \qquad \max_{x_t\in\varphi_t(\bar{z})} v_t(x_t,\bar{x}_t^-,\bar{y},\bar{p}) = v_t(\bar{z}),$$

$$(4.4) \qquad \max_{y_t\in\bar{\Psi}_t(\bar{z})} u_t(\bar{x},y_t,\bar{y}_t^-,\bar{p}) = u_t(\bar{z}),$$

where $\bar{\Psi}_t\colon Z \to 2^{Y_t}$ is defined by

$$(4.5) \qquad \bar{\Psi}_t(z) =\left\{m \mid m\in Y_t,\ p_t(\nu_t \circ m)\leq \sum_{r=0}^{t} \alpha_r(z) - \sum_{r=-1}^{t-1} p_r(\nu_r \circ y_r)\right\}.$$

An interpretation of (4.2) - (4.4) is similar to the interpretation of (3.1.5) - (3.1.7) in the static Arrow-Debreu model of Idzik and Simonsen (1983a).

The functions $\bar{\Psi}_t$ restrict the consumption. But we assume that the consumer can always buy something at each moment t, i.e.

$$(4.6) \qquad \bar{\Psi}_t(z) \neq \emptyset \quad \text{for} \quad z\in Z \quad (t\in T),$$

where $\bar{\Psi}_t$ is defined by (4.5).

(4.7) Theorem. If the conditions (2.1) - (2.5) and (4.6) are satisfied, then the dynamic Arrow-Debreu model has an equilibrium in the case of durable processes.

Since proof of Theorem (4.7) is similar to the proof of Theorem (3.7) we omit it here.

REFERENCES

Arrow, K.J. and Debreu, G. (1954): Existence of an equilibrium for a competitive economy. Econometrica, 22: 265-290.

Flåm, S.D. (1981): Equilibria in noncooperative games and competitive economies. Tamkang J. Math., 12: 47-57.

Idzik, A. and Simonsen, P.B. (1983a): A game-theoretic Arrow-Debreu model. In: Transactions of the Ninth Prague Conference on Information Theory, Academia, Prague, 293-299.

Idzik, A. and Simonsen, P.B. (1983b): A dynamic model of production and consumption. In: Abstracts of 11th IFIP Conference on System Modelling and Optimization. Copenhagen, July 25-29.

Makarov, V.L. and Rubinov, A.M. (1977): Mathematical theory of economic dynamics and equilibria. Springer-Verlag, New York.

Institute of Computer Science
Polish Academy of Sciences
PL-00-901 Warsaw PKiN, Poland

INFORMATION GEOMETRY OF THERMODYNAMICS

Roman S. Ingarden
Toruń

Key words: information theory, statistical thermodynamics, nonequilibrium thermodynamics, differential geometry, Finsler geometry

ABSTRACT

It is shown that a set of probability distributions (statistical states, states) of a nonequilibrium thermodynamical system can be equipped with the structure of a Finsler space. The Lagrange function L of the space is defined by means of the relative entropy between states. The time variable is specified as an additional position variable by requiring L be homogeneous in the directional variables. Such a Finsler model is a generalization of the Riemannian model of Fisher, well-known in mathematical statistics, recently applied in equilibrium thermodynamics by the author and his collaborators (A. Kossakowski, R. Mrugała, H. Janyszek and others).

1. FINSLER GEOMETRY

Finsler geometry, cf. Rund (1959), is a natural generalisation of both Riemannian and Minkowskian geometries, being in general inhomogeneous and anisotropic. It gives a geometrical interpretation of the general variational problem with first order derivatives in the Lagrangian. Therefore, such spaces have been called by their inventor, Paul Finsler (1918), "general spaces", although they are far from general from the present point of view of differential geometry, when non-metric and non-connected spaces are widely used. May be this explains why Finsler geometry is now not too fashionable among mathematicians and physicists.

In the present application we use the following definition of a Finsler space, a little bit more general as usually.

Definition 1. A Finsler space F_n is a triple (M^n, D, L), where M^n is an n-dimensional differential manifold, D is a closedset in TM^n such that for each $x \in M^n$ rays belonging to D in the tangential plane T_xM^n form a cone, and $L: D \to R^1_+$ is a function of class C^∞, positively homogeneous of the first order in the tangential (directional) variables $\dot{x}^i = y^i$ $(i=1,\dots,n)$ denoted also shortly by y :

$$L(x,ky) = k\,L(x,y) \qquad \text{for any } k>0. \tag{1}$$

Additionally, we assume that for each $x \in M^n$ L is convex (or concave) as a function of y. (T_xM^n is a linear space.)

2. GEOMETRY OF THERMODYNAMICS

The problem how to formulate a variational principle in a general, nonequilibrium thermodynamics was first discussed by a Polish physicist, Władysław Natanson (1896) who formulated his "thermokinetic principle". This question was then developed by L. Onsager, S.R. De Groot, I. Prigogine, P. Glansdorff, I. Gyarmati and others, cf. Gyarmati (1970), Gumiński (1983). Recently Yu. L. Klimontovich (1983) formulated and proved so-called "S-theorem" (in analogy to the "H-theorem" of Boltzmann) for selforganizing systems far from the usual equilibrium in the form

$$\frac{d}{dt}(F - F_0) = D\,\frac{d}{dt}\int_0^\infty f(E,t)\,\ln\frac{f(E,t)}{f_0(E)}\,dE \leqslant 0, \tag{2}$$

where F is the generalized free energy, E is energy, D is the generalized diffusion constant, and index o refers to the stationary state $f_0(E)$.

Generalizing (2) we would like to formulate a geodesic principle for an "information geometry" of thermodynamics. The principle consists in finding the shortest way between two states $f(x|\xi)$ and $f(x'|\xi)$ (x and x' are the "external" parameters of the states, i.e. their labels, while $\xi = (\xi_1,\dots,\xi_k)$ are the "internal" parameters of the distributions, e.g., coordinates of the phase space; we may write also $f_x(\xi), f_{x'}(\xi)$). Of course, the respective process is only virtually possible, but in this approximation we assume such an assumption. Roughly speaking this assumption corresponds to the assumption of local temperature in the usual gasodynamics and hydrodynamics. Mathematically, the "information distance" from state x or $f(x|\xi)$ to state x' or $f(x'|\xi)$ is defined by the relative entropy

$$(3) \quad S(x,x') := \int_\Omega d\mu(\xi) f(x|\xi) \ln \frac{f(x|\xi)}{f(x'|\xi)} = \left\langle \ln \frac{f(x)}{f(x')} \right\rangle \geq 0,$$

where $\mu(\xi)$ is a fixed measure in the internal space Ω. Putting $x' = x + dx$ we obtain under the well-known regularity conditions, cf. Kullback (1959) p. 26, Ingarden (1981) p.1630,

$$(4) \quad S(x,x+dx) = \frac{1}{2!} g_{ij}(x) dx^i dx^j + \frac{1}{3!} g_{ijk}(x) dx^i dx^j dx^k + \ldots =$$

$$= \sum_{r=2}^{\infty} \frac{1}{r!} ds_r^r < \infty,$$

$$(5) \quad g_{ij}(x) := \left\langle \frac{\partial \ln f}{\partial x^i} \frac{\partial \ln f}{\partial x^j} \right\rangle, \quad g_{ijk}(x) := \left\langle 3 \frac{\partial \ln f}{\partial x^i} \frac{\partial \ln f}{\partial x^j} \frac{\partial \ln f}{\partial x^k} + \sum_{cycl} \frac{\partial \ln f}{\partial x^i} \frac{\partial^2 \ln f}{\partial x^j \partial x^k} \right\rangle$$

where the first term in (5) is the well-known positive definite Fisher information matrix, and the second term is given in Ingarden (1981). Usually, cf. Čencov (1982), Amari (1985), people are satisfied in approximation "to within second-order terms", as Kullback writes (1959), p. 28. The problem is, however, if this approximation is always good enough. Principally, and in general we cannot look at it as an approximation at all. Each term in (4) is a scalar (the Einstein summation rule is assumed) and each corresponds to the r-th power of an element of lenght of a well-known geometry, for r=2 a Riemannian geometry R_n and for r=3,4,... of a Finsler geometry of the r-th power form, $F_n^{(r)}$. Since the corresponding differential forms are of different orders, they cannot be added as forms, but only as scalars after summation, and not before. Therefore, to say that the information or statistical problem (in our case a thermodynamical one) can be geometrically described by a Riemannian metric R_n is a strong exaggeration and in general is incorrect. Correctly, we have to say that it is described by a sequence of spaces

$$(6) \qquad R_n, F_n^{(3)}, F_n^{(4)}, F_n^{(5)}, \ldots$$

It seems that this sequence corresponds to a contact space which can be defined in phenomenological equilibrium thermodynamics, cf.

Mrugała (1984). Such a space is not metric nad has no connection. The reason is that the "distance" (3) or (4) has not property (1) and, therefore, cannot be used to definition of a curve integral. The Fisher idealization has thus no deeper meaning and changes the essential property of our object.

But it seems that the truth is nevertheless very near. The function S (3) can be "homogenized" by means of the following known procedure, cf. Rund (1966), Shibata (1978) as examples. We introduce a new dimension of the space $t = x^{n+1}$, interpreted as time, and write

$$(7)\quad ds = L(\bar{x}, d\bar{x}) = L(\bar{x}, \bar{y})du := S(x, x + \frac{dx/du}{dt/du})\frac{dt}{du}du = S(x, x + \frac{y}{v})v du,$$

where

$$(8)\quad v := \frac{dt}{du} = \frac{dx^{n+1}}{du} = y^{n+1}.$$

Now $\bar{x} = x^{\alpha} = (x,t)$, $\bar{y} = y^{\alpha} = (y,v)$, $\alpha = 1,\ldots,n+1$. Thus we formulate:

Definition 2. If $x = x^i$ $(i=1,\ldots,n)$ are thermodynamical parameters at time t of a state $f_t(x|\xi)$, we define its information distance to a state $f_{t+dt}(x+dx|\xi)$ (or, simpler, from a state (x,t) to a state $(x+dx, t+dt)$ using a "space-time" language) as ds in formula (7), i.e., as a metric of information geometry being a Finsler space F_{n+1} with Lagrangian function $L(\bar{x},\bar{y})$.

The point of this definition is that we go outside of the state space $\{x\}$ to the state space-time $\{(x,t)\}$ and then equip the latter with a Finsler structure.

3. AN EXAMPLE

As an example let us consider thermodynamical processes with an ideal gas consisting only in changing temperature with time. The processes are slow enough to have at any moment a constant temperature, but not infinitely slow. Thus we are almost at the limiting case with equilibrium thermodynamics. More instructive example will be with "generalized equilibria" and "higher order temperatures", but this is much more complicated mathematically.

For mathematical convenience we choose the temperature variable in the logarithmical scale (to have it changing from $-\infty$ to $+\infty$)

$$x = \ln(kT/kT_0) = \ln(T/T_0), \tag{9}$$

where T_0 is arbitrary (e.g., $T_0 = 1$ K). We obtain

$$S(x,x') = A(x'-x + e^{x-x'} - 1) =: AK(-a), A := \frac{3Nk}{2}, a := x'-x = \frac{y}{v}, \tag{10}$$

where T is absolute temperature, k is the Boltzmann constant, and N is the number of particles. Then

$$L(x,t,y,v) := AvK(-\frac{y}{v}) = 1 \tag{11}$$

is the equation of the indicatrix of our F_2 having the form of Fig.1.

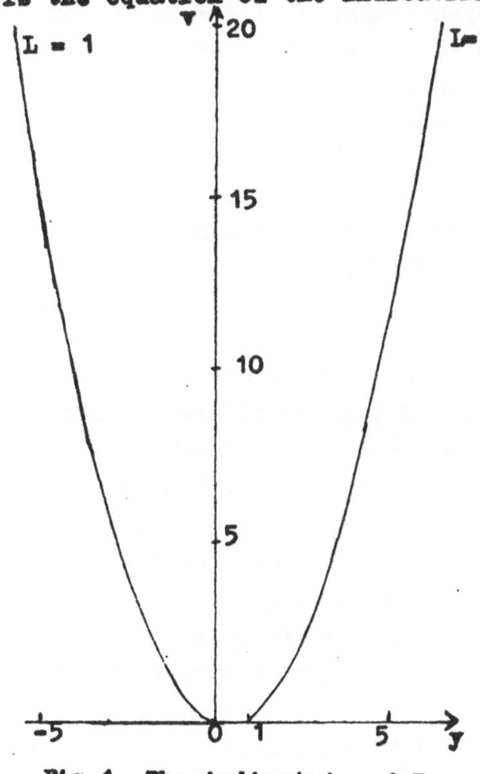

Fig.1. The indicatrix of F_2 ($A = 1$)

The indicatrix exists only for positive or zero v and in the first approximation (in a, when we develop K(-a) in a power series in a) has the form of parabola

$$v = Ay^2. \tag{12}$$

But near the point (0,0) the indicatrix has an interesting asymmetry connected with the asymmetry of S. In the interval (0,1) no value exists, as a function of y.

In such a way we obtain a "quasi-parabolic" in (y,v), homogeneous in (x,t) Finsler space F_2. All 3 Cartan curvature tensors of this space vanish, while the metric tensor g_{ij} and the torsion tensor C_{ijk} have in these coordinates the values

(13) $g_{xx} = A^2[2K^2+(1-3a)K+a^2], g_{xt} = A^2[(1+2a)K^2-3a^2K+a^3],$

(14) $g_{tt} = A^2[(1+2a+2a^2)K^2+a^2(1+3a)K+a^4], g = A^4(K-a+1)K^3,$

(15) $C_{xxx} = (1/2v)A^2a^2M = -C_{txx} = C_{ttx}, C_{ttt} = (1/2v)A^2a^3M,$

where

(16) $K := K(-a), M := 4K^2 + (4-7a)K + 3(1-a)a.$

From Fig.1 is seen that for $v = 0$ the "indicatrix" consists of two points, 0 and 1. So it is asymmetric and cannot represent any Finsler space. Although this example is not completely convincing since one-dimensional spaces cannot be rather considered as differential spaces, it seems to support our point of view that in equilibrium thermodynamics is no information geometry of Riemannian or Finsler type. We introduce, therefore, a special terminology:

Definition 3. A differential manifold $M^n = \{x\}$ with a function $S: M^n \times M^n \to \mathbb{R}^1_+$ of class C^∞ such that $S(x,x') = 0$ iff $x = x'$ and identity (4) holds where ds_r are elements of lenght of the Finsler spaces (6) for $r=2,3,\ldots$, respectively, is said to be a denumerably (countably) Finsler space Fd_n with a denumerable quasi-distance $S(x,x')$. We may write $Fd_n = (M^n, S, \{F_n^{(i)}\}_{i=2}^{\infty})$.

4. INEQUIVALENT REPRESENTATIONS

In our example we started from the temperature variable in the representation (9). If we were started from a different representation of this variable, e.g., T, we would obtain a different Finsler space, in general. Two such spaces are inequivalent in the sense that no transformation of variable can reduce one of them to the other. The understanding of this fact caused that the author delayed the publication of the present work, although it was already presented to two international conferences (Clausthat 1983, Debrecen 1984, both on methods of differential geometry). It seems that at present this obstacle is already surmounted.

The point is that in nonequilibrium thermodynamics we have a non-local dependence on boundary conditions which are not invariant under the change of scale of time. Even in classical mechanics time is a parameter which cannot be transformed arbitrarily (only in general relativity this is possible), but the macroscopic point of view of thermodynamics makes this point much stronger. In situations far from equilibrium (as such which lead to dissipative structures or selforganization) we have non-local conditions from outside which differentiate the systems otherwise similar. Therefore, it seems that the inequivalent representations which we have met in our theory are just needed for physical interpretation. We remind that inequivalent representations occur also in quantum field theory and in non-standard analysis of A. Robinson. May be, some roots of these situations have similar conceptual reasons. Further studies will show if this point of view is fruitful or not.

Anyhow, we hope that irreversible character of "thermodynamic time" which we got in our "parabolic geometry" is some indication of physical applicability of our model.

REFERENCES

Amari S. (1985): Differential-geometrical methods in statistics. Springer, Berlin.

Čencov N. N. (1982): Statistical decision rules and optimal inference, Transl.Math.Monographs 53, Am.Math.Soc. Providence (Russian original: Nauka, Moscow, 1972).

Finsler P. (1918): Über Kurven und Flächen in allgemeinen Räumen, Dissertation, Göttingen (also: Birkhäuser, Basel, 1951).

Gumiński K. (1983): Termodynamika procesów nieodwracalnych (Thermodynamics of irreversible processes), in Polish. PWN, Warszawa.

Gyarmati I. (1970): Non-equilibrium thermodynamics. Springer, Berlin.

Ingarden R.S. (1981): Information geometry in functional spaces of classical and quantum finite statistical systems. Int. J. Engng. Sci. 19, No.12, 1609-1633.

Kullback S. (1959): Information theory and statistics. Wiley, New York.

Mrugała R. (1984): On equivalence of two metrics in classical thermodynamics. Physica 125A, 631-639.

Natanson W. (1896): On the laws of irreversible phenomena. Phil. Mag. 41, 385-406.

Rund H. (1959): The differential geometry of Finsler spaces. Springer, Berlin. (Russian enlarged edition: Nauka, Moscow, 1981.)

(1966): The Hamilton-Jacobi theory in the calculus of variations. Van Nostrand, London.

Shibata C. (1978): On Finsler spaces with Kropina metric. Rep.Math. Phys. 13, 117-128.

Климонтович, Ю. Л. (1983): Уменшение энтропии в процессе самоорганизации. 5-Теоремы (на примере перехода через порог генерации). Письма в Ж. Техн. Физ. 9, вып. 23, 1412-1416.

Nicholas Copernicus University
Institute of Physics
Grudziadzka 5/7
PL-87100 Toruń
Poland

STATISTICAL ANALYSIS OF GIBBS RANDOM FIELDS

Martin Janžura

Prague

Key words: Gibbs random field, potential, parameter estimation, asymptotic properties of the estimate

ABSTRACT

Gibbs random fields are used to form the stochastic models for statistical analysis of dependent spatial data observed on a regular lattice. Thus, the problem of finding the unknown distribution corresponding to a given collection of data is transformed to a vector parameter estimation problem. A method for estimation is proposed and the asymptotic properties of the estimate are investigated.

1. INTRODUCTION

Suppose we are given a collection of observations obtained from the lattice points of a d-dimensional rectangular observation region. The observed variables assume values from some fixed finite set.

The collection of observations is considered to be generated by some random process indexed by d-tuples of integers, i. e. some random field. Our task is to find the distribution of the random field. Such problem, however, cannot be solved well in general. In order to make the problem reasonable we have to restrict our considerations to random fields with some kind of homogeneity and weak dependence.

Gibbs random fields, studied in frame of statistical mechanics, seem to satisfy the mentioned above conditions. The homogeneity is ensured by the translation invariance (i. e. spatial stationarity), and the dependence structure being described by finite range interactions,the requirement of rather weak dependence is satisfied as well.

On the other hand the class of Gibbs random fields is wide enough, including e. g. Markovian (of any order) random fields.

The Gibbsian description of random fields being applied, we are given the needed tool to view on our problem as on a parameter estimation problem. The way we follow consists in transforming the unknown parameter vector to another one which is easy to estimate. Transforming the estimate back, we obtain an estimate of the original parameter vector. The inverse transformation preserves the asymptotic properties of the estimate as consistency, asymptotic normality, and asymptotic efficiency.

The method is, of course its metodology aspect only, similar to the method of moments, widely used in mathematical statistics. Another analogy is represented by the time series approach when the parameters are derived from the sample covariances which are calculated first of all.

Special cases of the problem treated here in general were considered in the author's preceding papers (cf. Janžura (1986b) and (1986c)).

2. PRELIMINARIES

Considering a finite state space X with the σ-algebra of all its subsets $F = \exp X$, by *random field* (r.f.) we mean a probability measure on the product measurable space (X^T, F^T) where the parameter set T is given by the d-dimensional $(d \geq 1)$ lattice Z^d. Fixing the state space X, we denote by M the set of all r.f.'s.

Further we denote by M_{St} the subset of *stationary* r.f.'s, i. e. those which are invariant with respect to all shifts:

$$\mu \in M_{St} \quad \text{iff} \quad \mu\theta_t^{-1} = \mu \quad \text{for every } t \in T;$$

the corresponding *shift* θ_t is the transformation on X^T defined through: $[\theta_t(x)]_s = x_{t+s}$ for every $x \in X^T$, $s \in T$. (For any $V \subset T$ by

x_V we mean the projection of $x \in X^T$ into the space X^V and by the same symbol we denote also the corresponding measurable cylinder.)

A r.f. $\mu \in M_{St}$ is called *ergodic* (we write $\mu \in M_E$) if its restriction to the σ-algebra $S \subset F^T$ of invariant sets assumes only values zero or one, i. e.

$\mu \in M_E$ iff for every $F \in S$ it holds: if $\mu(F)>0$ then $\mu(F)=1$, where $S = \{F \in F^T;\ \theta_t^{-1}F=F$ for every $t \in T\}$.

Let us denote $k(T;t) = \{A \subset T;\ A \ni t,\ 0<|A|<\infty\}$, $k(T) = \bigcup_{t \in T} k(T;t)$. (For any set A by $|A|$ we mean its cardinality.)

Potential is a family $U = \{U_A\colon X^A \to R\}_{A \in k(T)}$ of real-valued maps called *interactions*.

In the sequel we shall consider only such potentials which are *stationary*: $U_A(x_A) = U_{A-t}(\theta_t(x)_{A-t})$ for every $A \in k(T)$, $t \in T$, $x \in X^T$; and *bounded*: $|U| = \sum_{A \in k(T;0)} |A| \cdot |U_A|_\infty < \infty$, where $|U_A|_\infty = \sup_{x_A \in X^A} |U_A(x_A)|$.

A potential U is of finite range if the interactions vanish for large index sets, i. e. there exists $r > 0$ so that $U_A \equiv 0$ for every A: $\mathrm{diam}(A) > r$. Obviously, any finite range potential is bounded.

The set U of stationary bounded potentials is a *Banach space* with the norm $\|.\|$. If we denote by $U_r \subset U$ the subset of potentials of range $r > 0$ then $\bigcup_{r>0} U_r$ is a dense subset of U.

For fixed $U \in U$ *specification* Π^U is the family of maps

$$\Pi^U = \{\Pi_A^U\colon X^A \times X^{T \setminus A} \to [0,1]\}_{A \in k(T)}$$

each Π_A^U being defined through

$$\Pi_A^U(x_A | x_{T \setminus A}) = [Z_A^U(x_{T \setminus A})]^{-1} . \exp\{-E_A^U(x_A | x_{T \setminus A})\},$$

where

$$E_A^U(x_A | x_{T \setminus A}) = \sum_{V \in k(T), V \cap A \neq \emptyset} U_V(x_V)$$

and

$$Z_A^U(x_{T \setminus A}) = \sum_{y_A \in X^A} \{\exp -E_A^U(y_A | x_{T \setminus A})\}$$

is the appropriate normalizing constant.

For every $t \in T$ let us define

$$\gamma_t(U) = \sup\{\tfrac{1}{2}\sum_{x_0 \in X} |\Pi_{\{0\}}^U(x_0 | y_{T \setminus \{0\}}) - \Pi_{\{0\}}^U(x_0 | z_{T \setminus \{0\}})|;\ y_s = z_s \text{ for } s \neq t\}.$$

If $\gamma(U) = \sum_{t \neq 0} \gamma_t(U) < 1$ the potential is said to satisfy the *Dobrushin's condition*.

Gross (1981) proved that the set $\mathcal{D} = \{U \in \mathcal{U};\ \gamma(U) < 1\}$ of potentials satisfying the Dobrushin's condition is an open subset of the space $\mathcal{U}$. The proof is based on the continuity of the function $\gamma: U \to \gamma(U)$.

A r. f. $\mu \in M$ is called *Gibbs r. f.* with respect to a potential $U \in \mathcal{U}$, we write $\mu \in G(U)$, if the specification Π^U forms the family of conditional finite-dimensional distributions of the r.f. μ, i. e.

$\mu \in G(U)$ iff $\mu(x_A | F^{T \setminus A})(y) = \Pi^U_A(x_A | y_{T \setminus A})$ holds for every $A \in k(T)$, $x_A \in X^A$, and a. e. $y \in X^T[\mu]$.

For every $U \in \mathcal{D}$ there is *exactly one* Gibbs r. f. μ_U (this is the famous Dobrushin's result - cf. e. g. Künsch (1982), Corollary 2.3) which is, moreover, stationary and ergodic (cf. Theorem 4.1 and Theorem 4.3 in Preston (1976)).

For fixed $a \in T$ we set $V(a) = \{t \in T;\ 0 \le t_i < a_i$ for $i=1,\dots,d\}$. Writing $a \to \infty$, we mean $\min_{i=1,\dots,d} a_i \to \infty$.

For every $U \in \mathcal{U}$ there exists (cf. e. g. Föllmer (1973), Formula 4.24)

$$\lim_{a \to \infty} |V(a)|^{-1} \log Z^U_{V(a)}(x_{T \setminus V(a)}) = p(U),$$

where $p(U)$, called *pressure* corresponding to the potential U, does not depend on what $x_{T \setminus V(a)}$'s are taken in.

The pressure, considered as the map $p: U \to p(U)$, is *continuous convex* real-valued function defined on $\mathcal{U}$ (cf. Lemma 8.6 and Lemma 8.7 in Preston (1976)), and on the Dobrushin's region $\mathcal{D}$ it is, moreover, *twice continuously differentiable* (cf. Theorem 5.1 in Künsch (1982)) in sense of Gâteau.

Namely, it holds $\frac{\partial}{\partial u} p(U^0 + uU^1)|_{u=0} = -\int g_{U^1} d\mu_{U^0}$

and

$$\frac{\partial}{\partial u \partial v} p(U^0 + uU^1 + vU^2)|_{u=v=0} = -\frac{\partial}{\partial u} \int g_{U^2} d\mu_{U^0 + uU^1}|_{u=0} =$$

$$= \sum_{t \in T} \mathrm{cov}_{\mu_{U^0}}(g_{U^2}, g_{U^1} \circ \theta_t)$$

for every $U^0 \in \mathcal{D}$; $U^1, U^2 \in \mathcal{U}$,

where *the function* g_U is for every $U \in \mathcal{U}$ defined through the formula

$$g_U(x) = \sum_{A \in k(T;0)} |A|^{-1} U_A(x_A) \quad \text{for every } x \in X^T.$$

3. FINITE-DIMENSIONAL SUBSPACE OF POTENTIALS AND THE TRANSFORMATION OF PARAMETERS

The procedure of constructing specifications generates an *equivalence relation* on the space $(\mathcal{U},\|.\|)$. We shall write $U^1 \approx U^2$ (saying the potentials are equivalent) if the corresponding specifications Π^{U^1}, Π^{U^2} are equal. Potentials $U^1,\dots,U^N \in \mathcal{U}$ are said to be mutually *non-equivalent* if the only linear combination of them equivalent to the zero potential $\mathcal{O}$ ($\mathcal{O}_V \equiv 0$ for every $V \in k(T)$) is the zero one, i. e. if $\sum_{i=1}^N c_i\, U^i \approx \mathcal{O}$ then $c_1 = \dots = c_N = 0$.

Let us denote by $L = \mathrm{Lin}\{U^1,\dots,U^N\}$ the finite-dimensional subspace of $\mathcal{U}$ spanned by a fixed collection $U^1,\dots,U^N \in \mathcal{U}$ of mutually non-equivalent potentials.

The pressure p *restricted to* L is *strictly convex* function (cf. Proposition 8.5 in Preston (1976)), and, moreover, according to Dobrushin and Nahapetian (1977), for every compact $K \subset L$ the pressure is *strongly convex*, i. e. for every $k > 0$ there exists a constant $c_k > 0$ such that for every U^0, $U^0+U \in K$, where $U \in L$, $\|U\| \le k$, and every $\lambda \in [0,1]$ it holds

$$\lambda p(U^0+U)+(1-\lambda)p(U^0)-p(U^0+\lambda U) \ge \lambda(1-\lambda)c_k \cdot \|U\|^2 .$$

As well-known, there is an isomorphism $\phi: L \to R^N$ between the N-dimensional Banach space L and the N-dimensional Euclidean space, hence $\phi(U^\alpha) = \alpha = (\alpha_1,\dots,\alpha_N) \in R^N$ iff $U^\alpha = \sum_{i=1}^N \alpha_i U^i \in L$.

Let us denote $\mathcal{E} = \phi(L \cap \mathcal{D})$. Clearly, $\mathcal{E}$ is an open subset of R^N. For every $\alpha = \phi(U) \in \mathcal{E}$ we shall write μ_α instead of μ_U.
Let us define transform $\beta: \mathcal{E} \to R^N$ through the following formula

$$\beta(\alpha) = (\beta_1(\alpha),\dots,\beta_N(\alpha)) \quad \text{for every } \alpha \in \mathcal{E},$$

where $\beta_j(\alpha) = \int g_{U^j}\, d\mu_\alpha$ for every $j = 1,\dots,N$.

We denote by $D(\alpha) = \left(\frac{\partial \beta_i}{\partial \alpha_j}(\alpha)\right)_{i,j=1}^N$ the Jacobian matrix of the transform β. According to the properties of the pressure p, as mentioned in Section 2, each

$$D(\alpha)_{ij} = \frac{\partial \beta_i}{\partial \alpha_j}(\alpha) = \sum_{t \in T} \mathrm{cov}_{\mu_\alpha}(g_{U^i}, g_{U^j} \circ \theta_t)$$

is continuous function of $\alpha \in \mathcal{E}$.

Proposition 1. The transform $\beta: E \to R^N$ is one-to-one, its Jacobian matrix $D(\alpha)$ being positive definite for every $\alpha \in E$.

Proof. Let $\beta(\alpha) = \beta(\bar{\alpha})$ hold. Then, according to Formula 4.25 in Föllmer (1973), it follows

$$0 \le H(\mu_\alpha | \mu_{\bar{\alpha}}) + H(\mu_{\bar{\alpha}} | \mu_\alpha) = \sum_{j=1}^N (\bar{\alpha}_j - \alpha_j)(\beta_j(\alpha) - \beta_j(\bar{\alpha})) = 0,$$

where $H(.|.)$ is the relative entropy rate (information gain) given by

$$H(\mu|\nu) = \lim_{a\to\infty} |V(a)|^{-1} \int \log \frac{\mu(x_{V(a)})}{\nu(x_{V(a)})} d\mu(x),$$

providing the expressions make sense and the limit exists. Therefore $U^\alpha \doteq U^{\bar{\alpha}}$ (cf. Theorem 4.27 in Föllmer (1973)), and from the non-equivalence of the basis potentials it follows $\alpha = \bar{\alpha}$.

Further, for every non-zero $c \in R^N$ it holds

$$c^T D(\alpha) c = \sum_{t\in T} \operatorname{cov}_{\mu_\alpha}(g_{U^c}, g_{U^c} \circ \theta_t) = \frac{\partial^2 p}{\partial u^2}(U^\alpha + uU^c)\Big|_{u=0} > 0.$$

due to the strong convexity of the pressure. ∎

Remark. Thus, we have proved that the transform β is so called regular mapping on the open set E. This especially yields that the image $\beta(O)$ of every open $O \subset E$ in an open subset of $\beta(E)$.

4. ESTIMATION PROBLEM

The problem to be solved is to find the unknown distribution on (X^T, F^T) corresponding to a given *collection of observations* $x_{V(a)} \in X^{V(a)}$ obtained from the rectangular *observation region* $V(a)$.

In the sequel we shall assume the unknown distribution to be given by a Gibbs r. f. with respect to some potential $U^0 \in L \cap D$, where $L = \mathrm{Lin}\{U^1,\ldots,U^N\}$ is again a linear subspace of U spanned by a finite sequence $U^1,\ldots,U^N \in U$ of mutually non-equivalent potentials. Moreover, we shall assume that all the basis potentials $U^1,\ldots,U^N$ are of some fixed finite range $r > 0$.

Hence, following the notation used in the preceding section, we may consider our problem as a *parameter estimation problem* with the *family* $\{\mu_\alpha\}_{\alpha\in E}$ of probability distributions.

The aim of introducing the transform β consists in the fact that the unknown parameter $\alpha^0 \in E$ will be estimated via estimating the transformed parameter $\beta^0 = \beta(\alpha^0) \in \beta(E) \subset R^N$.

Suppose the lattice $T = Z^d$ to be provided with some linear ordering "$\prec$", e. g. the lexicographical one.

Let us denote $K_r(T;0)^+ = \{A \in k(T;0);\ 0 \prec t \text{ for every } t \in A, \text{diam}(A) \le r\}$, and $V_A(a) = \{t \in V(a);\ A+t \subset V(a)\}$ for every $a > 0$, $A \in k(T)$. (Further we assume $\min_{i=1,\dots,d} a_i > r$.)

For every $j = 1,\dots,N$ let us define the estimate of β_j^0 through the formula

$$\tilde{\beta}_j(a) = \sum_{A \in K_r(T;0)^+} |V_A(a)|^{-1} \sum_{t \in V_A(a)} U^j_{A+t}(x_{A+t}).$$

In order to avoid the difficulties which arise whenever $\tilde{\beta}(a) \notin \beta(E)$ for some collection of observations we shall modify the estimate in the following way

$$\hat{\beta}(a) = \tilde{\beta}(a) \cdot I_{\{\tilde{\beta}(a) \in \beta(E)\}} + \beta(0) \cdot I_{\{\tilde{\beta}(a) \notin \beta(E)\}}$$

(By I we mean the indicator function. Obviously $\beta(0) \in \beta(E)$.)

Now, we may define the estimate $\hat{\alpha}(a) = \beta^{-1}(\hat{\beta}(a))$ of the original parameter $\alpha^0 \in E$.

We shall investigate the asymptotic properties of the estimate $\hat{\alpha}(a)$ as $a \to \infty$, namely the consistency, the asymptotic normality, and the asymptotic efficiency.

Let us recall the definitions. We say the estimate $\hat{\alpha}(a)$ is *consistent* if $\hat{\alpha}(a) \to \alpha$ a. s. $[\mu_{\alpha^0}]$, *asymptotically normal* if

$$L_{\mu_{\alpha^0}}(|V(a)|^{\frac{1}{2}}(\hat{\alpha}(a)-\alpha^0)) \Rightarrow N_N(0,V),$$

i. e. $|V(a)|^{\frac{1}{2}}(\hat{\alpha}(a)-\alpha^0)$ converges in distribution to the N-dimensional normal law with zero vector of mean values and covariance matrix V, and *asymptotically efficient* (in sense of Rao) if there exists a $N \times N$ matrix $B(\alpha^0)$ of constants possibly depending on α^0 so that

$$|V(a)|^{\frac{1}{2}}(\alpha(a)-\alpha^0-B(\alpha^0)\ell^0(a)) \to 0 \text{ in probability } [\mu_{\alpha^0}]$$

where $\ell_j^0(a) = |V(a)|^{-1} \frac{\partial}{\partial \alpha_j} \log \mu_\alpha(x_{V(a)})|_{\alpha=\alpha^0}$ for every $j=1,\dots,N$.

Theorem 1. The estimate $\hat{\alpha}(a)$ of the vector parameter $\alpha^0 \in E$ is consistent, asymptotically normal, and asymptotically efficient. Moreover, both the covariance matrix V of the limit normal distribution and the matrix $B(\alpha^0)$ in definition of the asymptotic efficiency are equal to $D(\alpha^0)^{-1}$.

Proof. First we shall derive properties of the estimate $\tilde{\beta}(a)$.

Since $\mu_{\alpha^0} \in M_E$, the consistency of $\tilde{\beta}(a)$ follows (after some easy rearrangements) from the multi-dimensional version of the ergodic theorem (cf. Theorem VIII.6.9 in Dunford and Schwartz (1958)).

The proof of asymptotic normality is based on the central limit theorem for functionals of Gibbs random fields (cf. e. g. Künsch (1982) or Janžura (1986a)), the assumptions of which are in our case satisfied. Hence $L_{\mu_{\alpha^0}}(|V(a)|^{\frac{1}{2}}(\tilde{\beta}(a)-\beta^0)) \Rightarrow N_N(0,D(\alpha^0))$.

Thus, it remains to prove that

$$|V(a)|^{\frac{1}{2}}(\tilde{\beta}(a)-\beta^0-\ell^0(a)) \to 0 \text{ in probability } [\mu_{\alpha^0}]$$

which, in fact, requires some more effort, but with the aid of the very ingenious estimates contained in Corollary 2.4 and Theorem 3.2 of Künsch (1982) it is a matter of direct calculation.

We must only realize, according to Proposition 5.3 in Künsch (1982), that we may write $\ell^0(a) = Q^1(a) + Q^2(a)$,

where $Q_j^1(a) = |V(a)|^{-1} \sum_{t\in V(a)} q_j(t)$, $Q_j^2(a) = |V(a)|^{-1} \sum_{t\in T\setminus V(a)} q_j(t)$ for every $j = 1,\dots,N$ and

$$q_j(t) = \int g_{Uj}\circ\theta_t \mu_{\alpha^0}(dy|F^{V(a)}) - \int g_{Uj}\,d\mu_{\alpha^0} \text{ for every } j=1,\dots,N \text{ and } t\in T.$$

Then we prove separately

$$|V(a)|^{\frac{1}{2}}(\tilde{\beta}(a)-\beta^0-Q^1(a)) \to 0 \text{ and } |V(a)|^{\frac{1}{2}}(-Q^2(a)) \to 0 \text{ in probability } [\mu_{\alpha^0}].$$

Since $\beta(E)$ is an open subset of R^N and $\mu_{\alpha^0}\{\tilde{\beta}(a)\in\beta(E)\} \to 1$ for $a \to \infty$, the same results remain valid for the estimate $\hat{\beta}(a)$ as well.

Finally, the properties of the transform β, derived in Proposition 1, yield the statement of the theorem. ¤

Remark 1. The concept of the Rao's asymptotic efficiency becomes meaningful in connection with a certain regularity condition on the parameter family $\{\mu_\alpha\}_{\alpha\in E}$. This condition, called local asymptotic normality, is satisfied (cf. Theorem 4.1 in

Janžura (1986a)), and therefore the lower bound for asymptotic local maximum risk is achieved. (For details see e. g. Hájek (1970).) In such sense the estimate introduced above is actually *asymptotically optimal.*

Remark 2. The crucial role within the method is played by the transform β. But no explicit formula for calculating the inverse transform has been offered. However, due to the variational principle for Gibbs r.f.'s (cf. Theorem 4.27 in Föllmer (1973)) the estimate $\hat{\alpha}(a)$ may be obtained by minimizing the function

$$F_a(\alpha) = p(\sum_{j=1}^{N} \alpha_j U^j) + \sum_{j=1}^{N} \alpha_j \hat{\beta}_j(a).$$

Here the problem is with calculating the values of the pressure. In one-dimensional case the method of so-called transfer-matrix (cf. Janžura (1986b)) is available. In multi-dimensional case some approximate method should be applied (cf. e. g. Janžura (1986c)).

Remark 3. From the preceding remark it follows that the proposed method may be viewed on as the "minimum distance method" (cf. e. g. Vajda (1983)). The given collection of observations generates some stationary "empirical r. f." represented by the vector statistic $\hat{\beta}(a)$. Minimizing the function F_a, we seek for the r. f. $\mu_{\hat{\alpha}(a)}$ which is nearest to the empirical r. f. in sense of the distance measured by the relative entropy rate $H(.|.)$ (for its definition see the proof of Proposition 1).

REFERENCES

Dobrushin R. L., Nahapetian B. S. (1974): Strong convexity of the pressure for lattice systems of classical statistical physics. Teor. Mat. Phys. 20, 223-234 (in Russian).

Dunford N., Schwartz J. T. (1958): Linear Operators I. New York, Interscience.

Föllmer H. (1973): On Entropy and Information Gain in Random Fields. Z. Wahrs. verw. Geb. 26, 207-217.

Gross L. (1981): Absence of second-order phase transition in the Dobrushin's uniqueness region. J. Stat. Phys. 27, 57-72.

Hájek J. (1970): Local asymptotic minimax and admissibility in estimation. Proc. 6th Berkeley Symposium, Vol. I, 175-194.

Janžura M. (1986a): Central limit theorem for random fields with application to locally asymptotic normality of Gibbs random fields. (Submitted to Z. Wahrscheinlichkeitstheorie verw. Gebiete.)

Janžura M. (1986b): Estimating interactions in binary data sequences. (To appear in Kybernetika.)

Janžura M. (1986c): Estimating interactions in binary lattice data with nearest neighbor property. (To appear in Kybernetika.)

Künsch H. (1982): Decay of correlations under Dobrushin's uniqueness condition and its applications. Commun. Math. Phys. 84, 207-222.

Preston C. (1976): Random fields. Springer-Verlag, Lecture Notes in Math. 534.

Vajda I. (1983): A new general approach to minimum distance estimation. Trans. of the Ninth Prague Conference,..., 1982, Academia, Prague, 103-112.

Czechoslovak Academy of Sciences
Institute of Information Theory and Automation
182 08 Prague 8
Pod vodárenskou věží 4
Czechoslovakia